Molecular Biology and Physiology of Water and Solute Transport

Molecular Biology and Physiology of Water and Solute Transport

Edited by

Stefan Hohmann
Göteborg University
Göteborg, Sweden

and

Søren Nielsen
Institute of Anatomy
University of Århus
Århus, Denmark

Kluwer Academic / Plenum Publishers
New York, Boston, Dordrecht, London, Moscow

Library of Congress Cataloging-in-Publication Data

Molecular biology and physiology of water and solute transport/edited by Stefan Hohmann and Søren Nielsen

p. cm.

Proceedings of the 3rd International Conference on Molecular Biology and Physiology of Water and Solute Transport, held July 1–4, 2000, in Gothenburg, Sweden.

Includes bilbiographical references (p.).

ISBN 0-306-46501-9

1. Biological transport. 2. Molecular biology. I. Hohmann, Stefan, 1956– II. Nielsen, Søren Nors. III. International Conference on Molecular Biology and Physiology of Water and Solute Transport (3rd: 2000 Gothenburg, Sweden).

QH509 .M66 2000
571.6′4—dc21

00-046614

Front cover: The three-dimensional structure of GlpF (*E. coli*) tetramer is shown as seen from the cytoplasmic side. The structure was reconstructed at a resolution of 6.5 Å (z:12 Å) from tilted images, recorded with a Jeol 3000 SFF (MPI Frankfurt) helium-cooled microscope at 1 Kelvin. The map in the background is the GlpF projection map at 3.7-Å resolution. The image was created using the program *DINO* (http://www.bioz.unibas.ch~xray/dino). (See also Braun, Philippsen, Wirtz, Borgnia, Agre, Kühlbrandt, Engel, and Stahlberg, Chapter 1.2, this volume).

Proceedings of the 3rd International Conference on Molecular Biology and Physiology of Water and Solute Transport: Fundamental and Applied Aspects, held July 1–14, 2000, in Göteborg, Sweden.

ISBN 0-306-46501-9

233 Spring Street, New York, N.Y. 10013

http://www.wkap.nl/

10 9 8 7 6 5 4 3 2 1

A C.I.P. record for this book is available from the Library of Congress.

Printed in the United States of America

PREFACE

Biophysical studies in the 1950's and 1960's led to the realization that the water permeability of certain biological membranes must be due to the presence of water transporting proteins. This hypothesis was confirmed in 1991 and 1992 with the pioneering discovery of the first molecular membrane water channel, CHIP28, by Agre and coworkers. This integral membrane protein, which is abundant in the erythrocyte membrane and in many epithelial cells, is now called aquaporin-1 or AQP1. Thus, the terms water channel or aquaporin are synonymous.

Already at that time several known proteins showed topological and sequence similarity to AQP1: MIP (now AQP0), the major intrinsic protein from mammalian lens fiber, Nod26 and TIP, two plant proteins, as well as the *Escherichia coli* glycerol facilitator GlpF. The existence of homologous proteins from completely different organisms suggested that those were members of an ancient family, the MIP family. Indeed, all eukaryotes and most bacterial or archeal organisms studied so far posses MIP channels.

Human and rat have at least 10 aquaporins that are expressed in different tissues, most prominently in the kidney. Kidney aquaporins play distinct and crucial roles in kidney function for tight regulation of body water balance. Analysis of knockout rat models is now providing further insights into the physiological roles of mammalian aquaporins. Present knowledge suggests that aquaporins are targets for novel drugs, which could be used in the treatment of imbalanced water homeostasis. Such imbalances are the consequence of various kidney, heart, liver and endocrine diseases.

Plants appear to have a large number of MIP channels; the present estimate for the model plant *Arabidospsis* some 30 different proteins expressed in different cell types, developmental stages and environmental conditions. Genetic analysis of the physiological roles is only starting but plant aquaporins are expected to play central roles in transcellular and intracellular water partitioning. Hence, plant aquaporins may be a key to understanding the delicate control of plant water homeostasis, which in turn is crucial for the engineering of numerous crop properties.

Unicellular organisms, i.e., bacteria and fungi, appear to have only few MIP channels. Glycerol facilitators are common in bacteria and mediate the uptake of glycerol and related compounds for catabolism. A yeast glycerol facilitator has been shown to mediate glycerol export in osmoregulation. Many microorganisms also posses aquaporins. Their exact physiological roles are still poorly understood although some evidence suggests that they may play roles in the control of cell expansion during growth.

Presently more than 200 MIP channels are known. Those have a common topology of six transmembrane domains. The two halves of the protein are similar suggesting an internal duplication during evolution. Sequences in loops 2 and 5, the NPA boxes, are well conserved and seem to be part of the aqueous pore through the membrane. Phylogenetic analysis and functional studies have separated MIP channels into two major subfamilies, water channels and glycerol facilitators. These subfamilies are characterized by specific conserved amino acid residues thought to determine substrate specificity. The structural analysis of MIP channels is advancing now to atomic resolution, which will be essential to understand and probe the interesting transport specificity: the fact that certain MIP channels transport glycerol but not water excludes size selection as a simple explanation.

In July 2000 more than 200 researchers came together in Gothenburg, Sweden, for the "3rd International Conference on the Molecular Biology and Physiology of Water and Solute Transport" to discuss progress in this emerging research field. 58 different presentations from this conference are the basis for this book. Cumulatively, these 58 short chapters provide a balanced overview complementing numerous recent reviews in this field. We have organized the chapters into different sections, i.e. structure function analysis, mammalian physiology, pathophysiology, plant biology and microbiology. The section titles illustrate the multidisciplinary character of the research area of water and solute transport and aquaporin function, where researchers studying very different biological systems communicate with each other. We hope that this volume will be a milestone in this exciting and rapidly emerging field with major breakthroughs in the understanding of MIP channel structure, function and physiological/pathophysiological roles just ahead of us.

Stefan Hohmann and Søren Nielsen, Göteborg and Århus

ACKNOWLEDGMENTS

The conference "Molecular Biology and Physiology of Water and Solute Transport" (MIP2000), which took place in July 2000 in Gothenburg, Sweden, was the basis of this book. We are grateful for support of this conference and hence of these proceedings from the European Commission (contract HPCFCT-1999-00196), the Swedish Natural Science Research Council NFR, the Swedish Council for Forestry and Agricultural Research, the Wenner-Gren Foundation, the Technology Link Foundation (certain parts of the conference), Pfizer, Baxter, Ferring and AstraZeneca.

CONTENTS

STRUCTURE FUNCTION ANALYSIS OF AQUAPORINS AND GLYCEROL FACILITATORS

FUNCTION, PHYSIOLOGICAL ROLES, AND REGULATION OF MAMMALIAN AQUAPORINS

WATER CHANNELS AND GLYCEROL FACILITATORS IN MICROORGANISMS

AQUAPORINS IN FISH

1

STRUCTURE FUNCTION ANALYSIS OF AQUAPORINS AND GLYCEROL FACILITATORS

1.1 Structure function analysis

FUNCTIONAL ANALYSIS OF THE UNUSUAL SIGNATURE MOTIFS OF THE YEAST MIP CHANNEL, Fps1p

[1]Roslyn M. Bill, [1,2]Kristina Hedfalk, [1]Sara Karlgren, [2]Jan Rydström and [1]Stefan Hohmann

Departments of [1]Cell and Molecular Biology/Microbiology and [2]Biochemistry and Biophysics, Göteborg University, Box 462, S-405 30 Göteborg, Sweden. E-mail: roslyn.bill@gmm.gu.se

1. INTRODUCTION

Major Intrinsic Protein (MIP) channels – aquaporins and glycerol facilitators – occur in all types of organisms ranging from bacteria to humans and facilitate the diffusion of water, glycerol and other uncharged compounds across biological membranes. To date, more than 200 MIPs have been identified comprising an amazing number of isoforms expressed in different subcellular compartments and tissues, under different environmental conditions or during different developmental stages. For example, ten genes encoding MIP channels are currently known in humans (Borgnia *et al.*, 1999), 30 have been described in the plant *Arabidopsis thaliana* (Kjellbom *et al.*, 1999) and the nematode *Caenorhabditis elegans* has nine (Kuwahara *et al.*, 1998). These channel proteins are involved in biological processes as diverse as urine concentration in the mammalian kidney, (Deen and Knoers 1998) root development in *Arabidopsis thaliana* (Kaldenhoff *et al.*, 1998; Kjellbom *et al.*, 1999) and osmoregulation in *Saccharomyces cerevisiae* (Luyten *et al.*, 1995; Tamás *et al.*, 1999).

The central importance of Fps1p in yeast osmoregulation is well known. Although it is closely related to bacterial glycerol facilitators such as *E. coli* GlpF, it is – so far – unique in the MIP family for a number of reasons (Hohmann *et al.*, 2000). For example, it has long amino- and carboxy-terminal hydrophilic extensions, resulting in a protein of 669 amino acids, compared with 281 amino acids for GlpF. In addition, neither of the family's signature Asn-Pro- Ala (NPA) motifs in the presumed channel-forming loops B and E are fully preserved, being Asn-Pro-Ser (NPS) and Asn-Leu-Ala (NLA), respectively.

The sequences of only two other microbial MIPs contain motifs other than NPA, and in both cases NPA is preserved in loop B but not in loop E. In *Enterococcus faecalis,* the presumed glycerol facilitator (gef 6176) contains an NQA motif in loop E, while in *Chloribium tepidum*, the putative aquaporin (gct 5) has an NPV motif (Hohmann *et al.*, 2000). Unfortunately, these MIPs are incompletely characterized and thus their transport characteristics cannot be used to aid our understanding of the role of these atypical features. In this study we have therefore chosen to examine the consequences of Fps1p's NPS and NLA sequences on channel function. This is of particular relevance to a general understanding of MIP channels since their generic NPA motifs are believed to be integral to the formation of a continuous solute channel, the so-called 'hourglass' (Jung *et al.*, 1994).

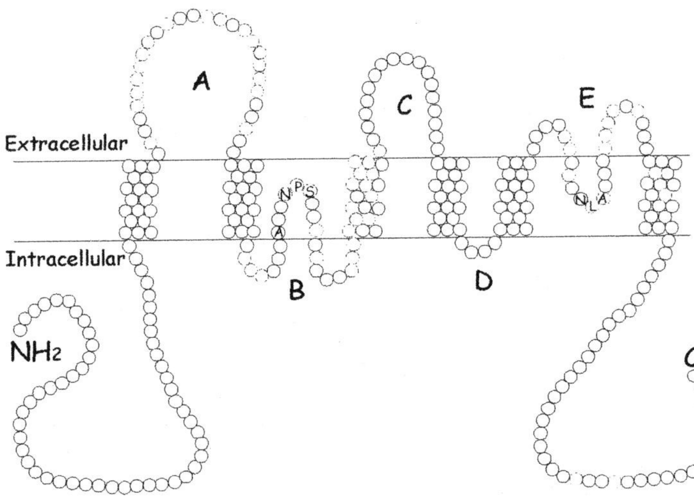

Figure 1. Schematic diagram of Fps1p. Fps1p differs from more typical members of the MIP family, such as GlpF, in a number of ways. For example the family's signature NPA motifs are replaced by NPS and NLA in Fps1p, as indicated. The letters A to E denote Fps1p's loops of which B and E are believed to be involved in the formation of the glycerol channel by dipping into the membrane to form an 'hourglass.'

In order to understand the functional consequences of their atypical sequence composition, we have analyzed loops B and E of Fps1p by site-directed mutagenesis. Rather than employing the commonly-used *Xenopus laevis* oocyte system, our mutants have been assayed in whole yeast cells. Even though it has not been possible to functionally express Fps1p in oocytes (as is the case for many mutant aquaporins, which are often mislocalized in this system), we believe that yeast is the obvious choice for the analysis of the yeast protein, Fps1p. We have focused on two types of mutations. First, we have made mutations in the B and E loops of Fps1p in order to directly observe their effect on function. Second, we have taken advantage of the fact that Fps1p's closest homolog, *E. coli* GlpF (which contains two NPA motifs), can be functionally expressed in yeast (Luyten *et al.*, 1995), thus enabling mutations to be made in this typical MIP so that it resembles Fps1p. We have independently analyzed both glycerol efflux and glycerol influx for our mutants, as well as observing growth phenotypes on plates. Our data indicate that the Ser residue of the NPS motif in loop B of Fps1p is important for both glycerol efflux *and* glycerol influx. However, replacing NPA with NPS in loop B of GlpF affects glycerol influx but not glycerol efflux. Our data for the mutant in which NPS is changed to QPS also suggest that the presence of the Asn residue is important for proper efflux. We therefore propose that the unique deviation from the family's signature NPA motif in loop B of Fps1p may be involved in determining the relative initial rates of glycerol efflux and glycerol influx in this channel, and that this is reflected in Fps1p's physiological role as a mediator of glycerol export.

2. RESULTS AND DISCUSSION

2.1 Yeast is a robust functional analysis system for MIP channels

Numerous studies have been published describing analyses of the transport function of MIP channels. In the majority of cases, *Xenopus laevis* oocytes have been used as the expression and assay system. However, this permits only a limited insight into the behavior of MIPs and their mutants presumably because many of them fail to traffic to the oocyte plasma membrane (Mathai and Agre 1999). Agre and colleagues (Lagrée *et al.*, 1998) have already proposed that *Saccharomyces cerevisiae* is a more appropriate system for the analysis of MIP channels. Indeed, mutants of the insect aquaporin, AQP_{cic}, could not be expressed in an active form in *Xenopus laevis*, but were successfully functionally expressed in yeast (Lagrée *et al.*, 1998). Furthermore, the same group has indicated that most mutations in residues near the NPA motifs in loops B and E of AQP1 (and therefore probably in other MIPs) are incorrectly localized in oocyte plasma membranes, and have instead assayed their mutants in yeast (Mathai and Agre 1999). We have chosen to analyze the transport of glycerol by Fps1p in the most relevant expression system: the homologous host, *S. cerevisiae*. We find that all our mutants, made either in loop B or loop E, are correctly localized to the yeast plasma membrane and that mutants of the *E. coli* glycerol facilitator, GlpF, are similarly correctly localized. Efflux and influx of glycerol can be quantitated independently in different assays for both Fps1p and GlpF, thereby providing us with a robust system for the analysis of these channels. Furthermore, we have been able to perform qualitative plate growth assays that support our quantitative data.

2.2 Comparison of Fps1p and GlpF suggests a role for Fps1p's NPS motif in glycerol transport

Our data indicate that Fps1p is fully functional if both NPA motifs are 'restored' (Fig. 2). This provides direct biochemical evidence for an interaction between loops B and E, since the introduction of the mutation, L481P, to form a double mutant can compensate for the reduced transport observed in the single mutant, S354A. However, these data do not resolve the question of why Fps1p's sequence should deviate from the highly-conserved family signature sequences in loops B and E.

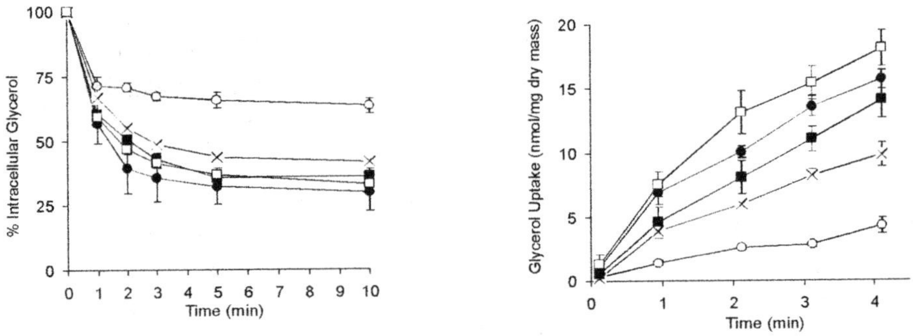

Figure 2. Functional analysis of point mutants of Fps1p. Glycerol transport measurements on whole yeast cells were performed for both efflux (left) and influx (right) in order to determine the transport characteristics of cells with either Fps1p (●), no glycerol channel (○), Fps1p S354A (×), Fps1p L481P (■) or Fps1p S354A L481P (□).

As shown in Fig. 3, this cannot be simply explained by mirroring the mutations in GlpF. Here we find that NPA to NPS (A70S) has a complex effect on the function of GlpF since glycerol efflux is unaffected within experimental error, but the amount of glycerol taken up is reduced to approximately 50% of the wild-type activity 4 min after exposing the cells to radio-labeled glycerol. In contrast, function is completely abolished by mutating NPA to NLA in loop E (P204L), and unlike Fps1p, the function of GlpF is completely abolished when NPS *and* NLA (A70S P204L) are present. It is therefore plausible that there might be sequences in the B and E loops of Fps1p that are required if the interaction of these NPS- and NLA-containing loops are to form a fully functional channel in GlpF. However, GlpF is less highly expressed compared to wild-type when it contains either loop B or loop E of Fps1p (data not shown). In fact, only the mutant in which loop B has been swapped from Fps1p onto GlpF has any glycerol transport activity (data not shown). This would imply that loops B and E alone do not hold the key to functional interaction of NPS and NLA. Of course, it is possible that certain sequences in the transmembrane domains of Fps1p rather than its loops could restore function to GlpF-A70S P204L, and this is currently under study.

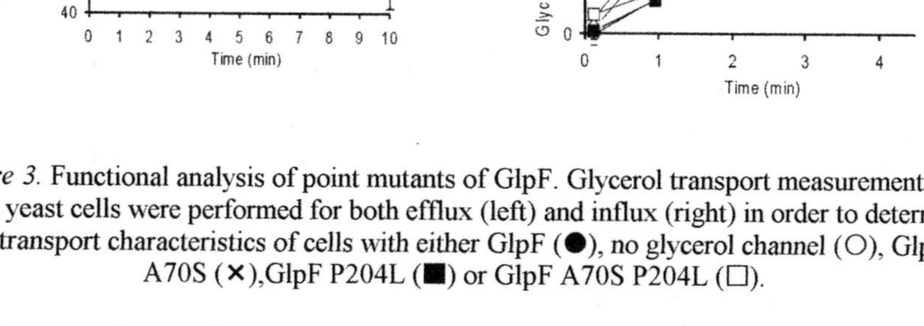

Figure 3. Functional analysis of point mutants of GlpF. Glycerol transport measurements on whole yeast cells were performed for both efflux (left) and influx (right) in order to determine the transport characteristics of cells with either GlpF (●), no glycerol channel (○), GlpF A70S (×),GlpF P204L (■) or GlpF A70S P204L (□).

Overall, then, our data are suggestive of a specific role for the serine residue in the NPS motif of Fps1p. Its replacement in Fps1p reduces both glycerol efflux and glycerol influx compared to wild-type, which cannot be explained by a reduced expression level for the protein. Two possible explanations for this observation are (1) that serine is required as a hydrogen-bonding partner for glycerol as it passes through the solute channel or (2) that there is some steric effect that influences the shape of the channel. Interestingly, when a serine residue is introduced into GlpF, only glycerol efflux is reduced. The asymmetric effect of such a mutation has already been hinted at, not only in measurements of initial rates for Fps1p (Luyten *et al.*, 1995; Tamás *et al.*, 1999), but also with the acknowledgement that loops B and E are not symmetrical with respect to function in AQP2 (Kuwahara *et al.*, 1999). It is conceivable that serine could have a different influence on the initial rates of efflux and influx, compared to wild-type, in the two independent assays without affecting the thermodynamics of the protein as a passive channel. In fact, markedly different initial rates are not an uncommon phenomenon in enzymatic reactions despite a $\Delta G°$ around zero. This only means that the rates of the two directions must become similar when equilibrium is approached.

2.3 The use of yeast as an assay system allows further quantitative assessment of glycerol transport by Fps1p

The role of loop B in influencing glycerol transport is supported by the data shown in Fig. 4. Von Heijne (Monné *et al.*, 1999) has demonstrated that certain amino acids, particularly Asn and Pro, can cause helix turning in the membrane. This is relevant to both loops B and E, which are believed to dip into the membrane to form the solute channel. In the case of loop B, both Asn and Pro are present in Fps1p's pseudo-signature motif: amino acids that are theoretically able to cause the loop to turn. In loop E, only Asn is present in the NLA motif. Consequently, the N to Q mutation is likely to have a more dramatic effect on channel formation in loop E than loop B, and this is indeed reflected in the influx data for N352Q compared with N480Q.

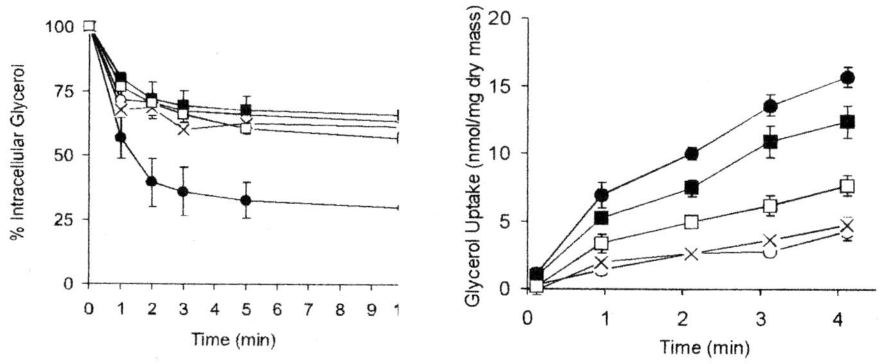

Figure 4. Functional analysis of further mutants of Fps1p. Glycerol transport measurements on whole yeast cells were performed for both efflux (left) and influx (right) in order to determine the transport characteristics of cells with either Fps1p (●), no channel (○), Fps1p A349M (×), Fps1p N352Q (■) or Fps1p N480Q (□).

Surprisingly, the efflux data indicates that the effect of these mutations is more complex than this. In both cases, the selective abolition of glycerol efflux activity suggests that the presence of Asn is required for Fps1p to be an effective channel. This effect is more striking in loop B, as discussed above, and may contribute to the determination of initial rates of glycerol transport.

3. PERSPECTIVES

The challenge now is to interpret these observations with the limited amount of structural data available for MIP family members. To date, low-resolution structural data (at 4.5 Å) are only available for AQP1 (Mitsuoka *et al.*, 1999). However, even though the most divergent members of the MIP family are less than 20% identical (Park and Saier Jr 1996), it is expected that the gross structural features apparent in AQP1 will also be present in other water and glycerol channels. For example, it is believed that loop B is a narrow, kinked structure, whereas loop E is a broader funnel (Mathai and Agre 1999). It is therefore conceivable that in Fps1p, the A to S substitution in loop B has had a more dramatic effect than the P to L substitution in loop E, and caused the channel to realign to become an efflux channel. Detailed structural explanations for our mutations await confirmation from crystallographic studies, and also a more detailed kinetic analysis following the purification and reconstitution of Fps1p, which is currently in progress. Nonetheless, our data provide a rationale in particular for the presence of

NPS in Fps1p, and suggest that loop B is involved in dictating glycerol efflux in this unusual member of the MIP family.

ACKNOWLEDGEMENTS

This work was supported by grants from the European Commission: BIO4-CT98-0024 and FMRX-CT96-0128.

REFERENCES

Borgnia M Nielsen S Engel A and Agre P 1999 Cellular and molecular biology of the aquaporin water channels *Ann Rev Biochem* **68**: 425-458

Deen PM and Knoers NV 1998 Physiology and pathophysiology of the aquaporin-2 water channel *Curr Opin Nephrol Hypertens* **7**: 37-42

Hohmann S Bill RM Kayingo G and Prior BA 2000 Microbial MIP channels *Trends Microbiol* **8**: 32-36

Jung JS Preston GM Smith BL Guggino WB and Agre P 1994 Molecular structure of the water channel through aquaporin CHIP The hourglass model *J Biol Chem* **269**: 14648-14654

Kaldenhoff R Grote K Zhu JJ and Zimmermann U 1998 Significance of plasmalemma aquaporins for water-transport in Arabidopsis thaliana *Plant J* **14**: 121-8

Kjellbom P Larsson C Johansson I Karlsson M and Johanson U 1999 Aquaporins and water homeostasis in plants *Trends Plant Sci* **4**: 308-314

Kuwahara M Ishibashi K Gu Y Terada Y Kohara Y Marumo F and Sasaki S 1998 A water channel of the nematode C elegans and its implications for channel selectivity of MIP proteins *Am J Physiol* **275**: C1459-64

Kuwahara M Shinbo I Sato K Terada Y Marumo F and Sasaki S 1999 Transmembrane helix 5 is critical for the high water permeability of aquaporin *Biochemistry* **38**: 16340-6

Lagrée V Pellerin I Hubert JF Tacnet F Le Caherec F Roudier N Thomas D Gouranton J and Deschamps S 1998 A yeast recombinant aquaporin mutant that is not expressed or mistargeted in Xenopus oocyte can be functionally analyzed in reconstituted proteoliposomes *J Biol Chem* **273**: 12422-6

Luyten K Albertyn J Skibbe WF Prior BA Ramos J Thevelein JM and Hohmann S 1995 Fps1 a yeast member of the MIP family of channel proteins is a facilitator for glycerol uptake and efflux and is inactive under osmotic stress *EMBO J* **14**: 1360-1371

Mathai JC and Agre P 1999 Hourglass pore-forming domains restrict aquaporin-1 tetramer assembly *Biochemistry* **38**: 923-8

Mitsuoka K Murata K Walz T Hirai T Agre P Heymann JB Engel A and Fujiyoshi Y 1999 The structure of aquaporin-1 at 45-A resolution reveals short alpha- helices in the center of the monomer *J Struct Biol* **128**: 34-43

Monné M Nilsson I Elofsson A and von Heijne G 1999 Turns in transmembrane helices: determination of the minimal length of a "helical hairpin" and derivation of a fine-grained turn propensity scale *J Mol Biol* **293**: 807-14

Park JH and Saier Jr MH 1996 Phylogenetic characterisation of the MIP family of transmembrane channel proteins *J Membrane Biol* **53**: 171-180

Tamás MJ Luyten K Sutherland FCW Hernandez A Albertyn J Valadi H Li H Prior BA Kilian SG Ramos J Gustafsson L Thevelein JM and Hohmann S 1999 Fps1p controls the accumulation and release of the compatible solute glycerol in yeast osmoregulation *Mol Microbiol* **31**: 1087-1104

1.2 Structure function analysis

GLPF: A STRUCTURAL VARIANT OF THE AQUAPORIN TETRAMER

The GlpF projection at 3.5 Å

[1]Thomas Braun, [2]Ansgar Philippsen, [1]Sabine Wirtz, [3]Mario J. Borgnia, [3]Peter Agre, [4]Werner Kühlbrandt, [1]Andreas Engel and [1]Henning Stahlberg

[1]M. E. Müller Institute for Microscopy, Biozentrum, University of Basel, Klingelbergstrasse 70, CH-4056 Basel, Switzerland. [2]Department of Structural Biology, Biozentrum, University of Basel, Klingelbergstrasse 70, CH–4056 Basel, Switzerland. [3]Department of Biological Chemistry, Johns Hopkins University School of Medicine, Baltimore, MD 21205, USA. [4]Max-Planck-Institute for Biophysics, Kennedyallee 70, D-60596 Frankfurt, Germany. E-mail: Henning.Stahlberg@unibas.ch

1. INTRODUCTION

The glycerol uptake facilitator of *E. coli* (GlpF) (Boos *et al.*, 1990) is one of the few known diffusion facilitators in the inner membrane of this bacterium. Glycerol diffuses into the cell through GlpF and is phosphorylated by the glycerol kinase (GlpK), which prevents back-diffusion. Besides glycerol transport, the diffusion of polyols and urea derivatives through GlpF have been reported (Maurel *et al.*, 1994), but none of these substrates are transported in a phosphorylated state. In contrast, it remains unclear whether GlpF allows the passage of water.

GlpF is a member of the aquaporin protein super-family (formerly known as the MIP family) which includes water channels (AQPs) and the facilitators of small solutes like glycerol (GLPs) (Heymann and Engel, 2000; Park and Saier, 1996; Reizer *et al.*, 1993). All of these channels are strictly selective for non-ionic compounds. Sequence comparisons have revealed an internal homology between the C- and N-terminal halves of all aquaporins, with highly conserved regions in the cytoplasmic loop B and the extracellular loop E (Heymann and Engel, 2000). According to the model of

Jung *et al.* (1994), these two loops fold back into the membrane to form a channel akin to an hourglass. This region is believed to be the site of solute selectivity. A comparison between AQP0 (AQP subcluster) and GlpF (GLP subcluster) is given in Figure 1. Loop E and C of GlpF contain additional amino acids compared to AQP1 and MIP0. Other conserved residues have been found in the aquaporin super-family by (Froger *et al.*, 1998) and demonstrated by (Lagree *et al.*, 1999) to discriminate AQPs from GLPs. In addition (Heymann and Engel, 2000) have identified a pair of residues in helix VI that may determine the size of the pore. Note that all important differences are located at the extracellular surface.

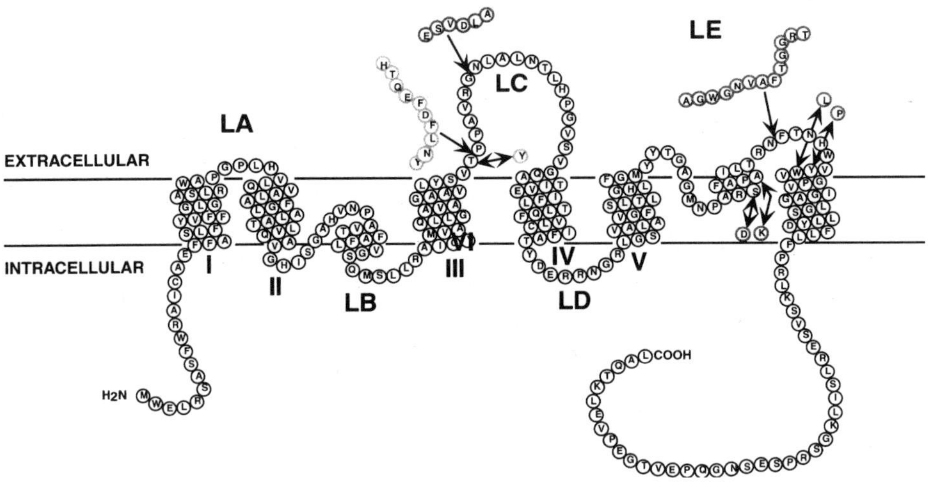

Figure 1. Sequence comparison between the AQP subcluster and the GLP subcluster. Most important differences of GLPs (in this case GlpF from E. Coli) are indicated with bold outlines. The reference sequence is AQP0 from the AQP subcluster. Single arrowheaded lines are indicating inserts, double headed lines are indicating aminoacids substitutions. Taken from Froger *et al.* (1998), Heymann and Engel (2000) and Mulders *et al.* (1995).

2. RESULTS

2.1 Protein expression and single particle analysis

GlpF carrying ten C-terminal histidine residues was overexpressed in *E. coli,* solubilized and purified in octyl-β-D-glucopyranoside (OG) as described in (Borgnia *et al.*, 1999). Analysis of purified and solubilized GlpF in SDS-PAGE revealed a prominent band at an apparent molecular weight of 30 kDa. Negatively stained preparations of the solubilized, purified GlpF were imaged with a transmission electron microscope (TEM). *Figure s*hows the homogeneous size distribution of the purified protein that often exhibited a square shape. Occasionally larger complexes can be seen (arrows). Single particles were windowed and subjected to reference-free alignment, resulting in a square-shaped average projection that had a side-length of about 80 Å (Figure 2, inset). Four weak densities can be distinguished at the periphery of the square.

Figure 2. Overview of negatively stained solubilized GlpF in the transmission electron microscope. Square shaped particles are predominant, but complexes of approx. twice the size are also seen (arrows). Scale bar corresponds to 450 Å. Insets: Four selected square-shaped particles and a 4 fold symmetrized average after reference free single particle analysis. Image side length corresponds to 230 Å.

2.2 Two-dimensional crystallization

Solubilized GlpF was reproducibly crystallized using a continuous flow dialyzing device (Jap *et al.*, 1992). A short description of this setup is given in Figure 3. Analysis in the TEM showed polycrystalline vesicles with diameters up to 40 μm and mostly mono-crystalline vesicles with rectangular shapes and diameters up to 8 μm. The crystals diffracted the electron beam beyond 3.3 Å, but weak structure factors between 7 and 5 Å were observed (Braun *et al.*, submitted).

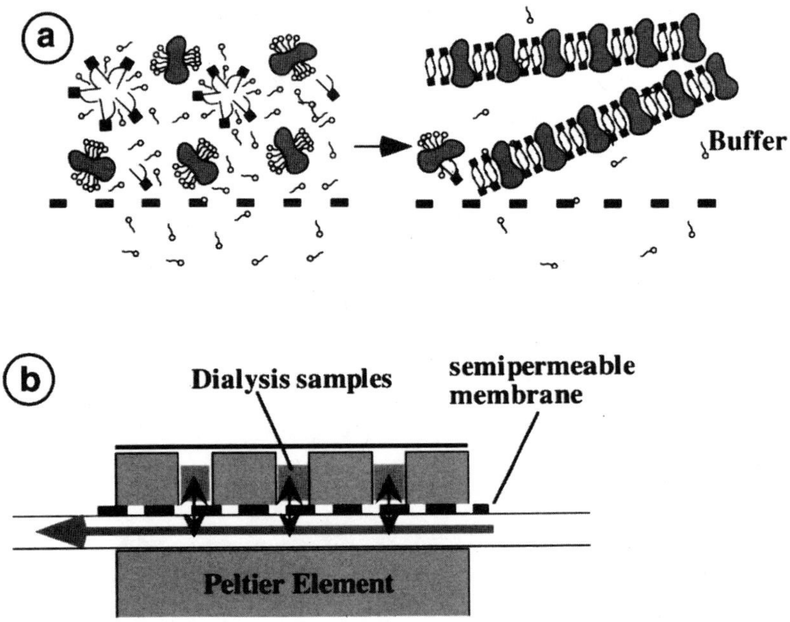

Figure 3 Principle of the used crystallization approach. **a:** Mixed lipid/detergent micelles and solubilized protein is present. Only free detergent can pass the dialysis membrane (dashed line). By dialysing against a buffer without any detergent, the latter is subsequently replaced by lipid in the protein/detergent micelles. Membranes are favoured by the lipids. With the right buffer conditions 2 dimensional crystals are formed. **b:** Principle of a continuous flow dialyzing device as described in Jap *et al.*, (1992). The dialyzing buffer is continuously exchanged. The temperature is controlled by a peltier-element, which allows to run temperature profiles.

2.3 Electron Microscopy and Image treatment

Images of frozen hydrated crystals recorded at low dose revealed sharp spots when examined by optical diffraction out to a resolution of 7 Å. The 9 best images were digitized and processed by the MRC program package (Henderson *et al.*, 1990; Henderson *et al.*, 1986). A general overview over the image processing is given in Figure 4. The phase residuals of the merged data obtained after lattice unbending and transfer function correction indicated significant information up to a resolution of 3.5 Å. The resulting projection map is shown in Figure 5.

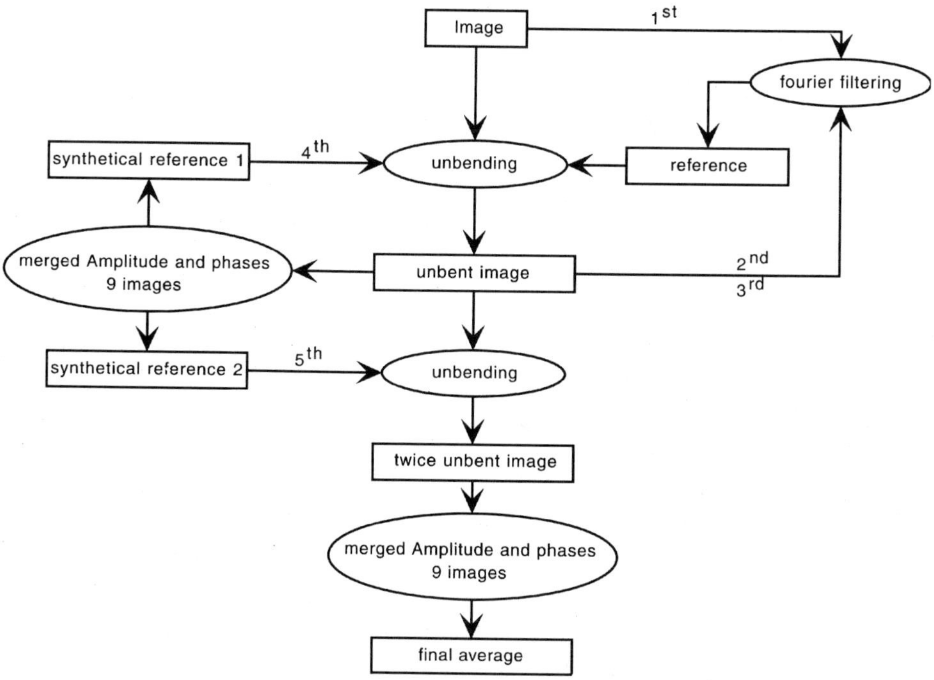

Figure 4. Principle of the applied image processing. In a first run, images were unbent three times, using the Fourier-filtered images themselves as a reference (Henderson *et al.*, 1990; Henderson *et al.*, 1986). The merged amplitudes and phases from 9 images were then used to create a synthetic reference (program MAKETRAN) , applying two different negative temperature factors to correct the envelope function. All images were unbent in a fourth run using the reference 1 with the smaller negative temperature factor. The images unbent in this way were then fine-corrected by unbending them again, using the reference 2 with the stronger negative temperature factor, this time allowing only very small (5 pixel/unit–cell) displacements of the unit cells.

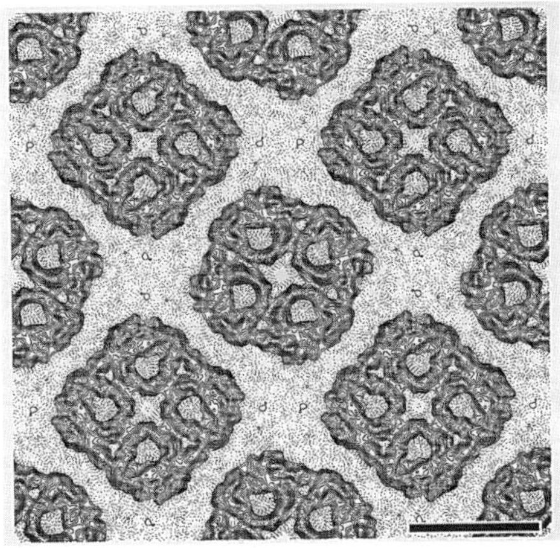

Figure 5. The p4-symmetrized 3.5 Å projection structure of GlpF. The map was calculated by merging 9 electron micrographs after unbending crystal distortions and correcting the transfer function. The phase residuals indicated significant information up to 3.5 Å resolution. Neighboring tetramers are up down oriented as visible in their mirrored handedness. A P4 symmetry was found and not a $p42_12$ as in AQP1 (Walz *et al.*, 1994). Unit cell length is 104 Å. A negative temperature factor of –20 Å^2 was applied. Scale bar corresponds to 50 Å.

3. DISCUSSION

GlpF showed a tetrameric organization throughout the purification. This was monitored by electron microscopy of negatively stained samples taken at various steps and was corroborated by mass-measurements of freeze dried unstained preparations in the STEM (Braun *et al.*, submitted). Thus, GlpF exhibits the same tetrameric structure as other aquaporins previously studied: AQP1 (Smith and Agre, 1991; Walz *et al.*, 1994), AqpZ (Ringler *et al.*, 1999), MIP (Hasler *et al.*, 1998). The tetrameric nature of GlpF supports the model proposed by (Voegele *et al.*, 1993) that GlpF interacts with the tetrameric glycerol kinase GlpK to stimulate glycerol phosphorylation.

It is interesting to compare the first high-resolution projection structure of the archetypal member of the GLP sub-family with that of AQP1, the first aquaporin structurally analyzed to high resolution (Cheng *et al.*, 1997; Li *et al.*, 1997; Mitsuoka *et al.*, 1999; Walz *et al.*, 1997; Walz *et al.*, 1994) (Figure 6). The density maxima marked in the AQP1 map appear to be slightly shifted in the GlpF map. They are related to the highly tilted helices that surround a central density produced by loops B and E, marked by an X in Figure 6 (Mitsuoka *et al.*, 1999). According to the hourglass model (Jung *et al.*, 1994) this is the site of the channel and differences between GlpF and AQP1 in this region are of particular interest. While the projection map of

AQP1 shows a weak density at this position, GlpF seems to have a much larger hole with no inner structure discernable in the projection map. We may speculate that the weak density seen in the AQP1 channel is related to the phenylalanine in helix VI (Figure 1, Tyrosine in the case of AQP0) found in many water channels. Most GLPs have a leucine at this position (Heymann and Engel, 2000), and indeed, no corresponding density is seen for GlpF.

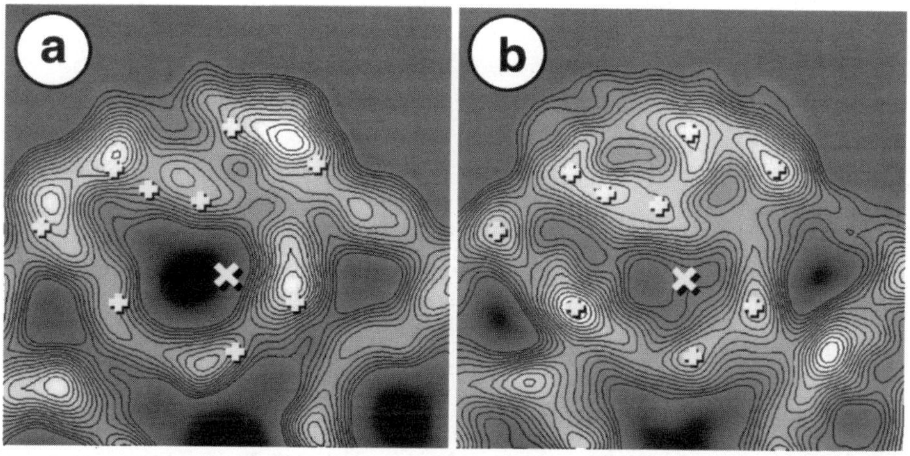

Figure 6. Comparison of GlpF and AQP1 at 4 Å resolution. **a**, GlpF monomer. **b,** AQP1 monomer. Overlaid crosses mark the position of density maxima found in AQP1. The depression in the AQP1 monomer, thought to represent the pore, is marked by an "X". The most prominent difference revealed by the GlpF monomer is the larger central depression of approx. 10 Å diameter. The one in AQP1 has a diameter of roughly 4 Å and a complex shape. Minor differences are seen in the surrounding density maxima. They correspond to the projection of overlapping highly tilted helices. The minima around the four–fold axis and between the monomers are rather similar. The image side length corresponds to 41 Å.

In summary, the projections of GlpF and AQP1 are comparable in their overall architecture but present interesting differences, which may be related to their different biological functions. It is compelling to speculate that the larger depression seen in the GlpF monomer represents the pore where glycerol and similar solutes pass (Maurel *et al.*, 1994). This would in turn support the hourglass model. However, to understand the molecular mechanisms and specificity of water and solute transport in aquaporins, the atomic structures of both AQP1 and GlpF are required. The highly ordered 2D crystals presented here are a first essential step towards this goal.

ACKNOWLEDGMENTS

We thank Deryck Mills for the introduction and expert support with the Jeol 3000SFF microscope and Vinzenz Unger for advice on the use of the MAKETRAN routine. We are indebted to Yoshinori Fujiyoshi, Kaoru Mitsuoka and Thomas Walz for providing the AQP1 data, Lorenz Hasler and Shirley Müller for their expert help. The work was supported by Swiss National Foundation for Scientific Research (NF grant No. 4036–44062 to AE), the M.E. Müller Foundation of Switzerland, and the National Institutes of Health (to PA).

REFERENCES

Boos W Ehmann U Forkl H Klein W Rimmele M and Postma P 1990 Trehalose transport and metabolism in Escherichia coli *J Bacteriol* **172**: 3450-61

Borgnia MJ Kozono D Calamita G Maloney PC and Agre P 1999 Functional reconstitution and characterization of AqpZ the E coli water channel protein *J Mol Biol* **291**: 1169-1179

Cheng A van Hoek AN Yeager M Verkman AS and Mitra AK 1997 Three-dimensional organization of a human water channel *Nature* **387**: 627-630

Froger A Tallur B Thomas D and Delamarche C 1998 Prediction of functional residues in water channels and related proteins *Protein Sci* **7**: 1458-1468

Hasler L Walz T Tittmann P Gross H Kistler J and Engel A 1998 Purified lens major intrinsic protein MIP forms highly ordered tetragonal two-dimensional arrays by reconstitution *J Mol Biol* **279**: 855-864

Henderson R Baldwin JM Ceska TA Zemlin F Beckmann E and Downing KH 1990 Model for the structure of *Bacteriorhodopsin* based on high-resolution electron cryo-microscopy *J Mol Biol* **213**: 899- 929

Henderson R Baldwin JM Downing KH Lepault J and Zemlin F 1986 Structure of purple membrane from *Halobacterium halobium* : recording measurement and evaluation of electron micrographs at 35 Å resolution *Ultramicroscopy* **19**: 147-178

Heymann B and Engel A 2000 Structural Clues in the Sequences of the Aquaporins *J Mol Biol* **295**: 1039-1053

Jap BK Zulauf M Scheybani T Hefti A Baumeister W Aebi U and Engel A 1992 2D crystallization: from art to science *Ultramicroscopy* **46**: 45-84

Jung JS Preston GM Smith BL Guggino WB and Agre P 1994 Molecular structure of the water channel through aquaporin CHIP The hourglass model *J Biol Chem* **269**: 14648-14654

Lagree V Froger A Deschamps S Hubert JF Delamarche C Bonnec G Thomas D Gouranton J and Pellerin I 1999 Switch from an aquaporin to a glycerol channel by two amino acids substitution *J Biol Chem* **274**: 6817-6819

Li H Lee S and Jap BK 1997 Molecular design of aquaporin-1 water channel as revealed by electron crystallography [letter] [see comments] *Nature Struct Biol* **4** 263-265

Maurel C Reizer J Schroeder JI Chrispeels MJ and Saier MH Jr 1994 Functional characterization of the Escherichia coli glycerol facilitator GlpF in Xenopus oocytes *J Biol Chem* **269**: 11869-11872

Mitsuoka K Murata K Walz T Hirai T Agre P Heymann JB Engel A and Fujiyoshi Y 1999 The Structure of Aquaporin-1 at 45-# Resolution Reveals Short alpha- Helices in the Center of the Monomer *J Struct Biol* **128**: 34-43

Mulders SM Preston GM Deen PM Guggino WB van Os CH and Agre P 1995 Water channel properties of major intrinsic protein of lens *J Biol Chem* **270**: 9010-9016

Park JH and Saier MH Jr 1996 Phylogenetic characterization of the MIP family of transmembrane channel proteins *J Membr Biol* **153**: 171-180

Reizer J Reizer A and Saier MH Jr 1993 The MIP family of integral membrane channel proteins: sequence comparisons evolutionary relationships reconstructed pathway of evolution and proposed functional differentiation of the two repeated halves of the proteins *Crit Rev Biochem Mol Biol* **28**: 235-257

Ringler P Borgnia MJ Stahlberg H Maloney PC Agre P and Engel A 1999 Structure of the water channel AqpZ from Escherichia coli revealed by electron crystallography *J Mol Biol* **291**: 1181-1190

Smith BL and Agre P 1991 Erythrocyte Mr 28000 transmembrane protein exists as a multisubunit oligomer similar to channel proteins *J Biol Chem* **266**: 6407-6415

Voegele RT Sweet GD and Boos W 1993 Glycerol kinase of Escherichia coli is activated by interaction with the glycerol facilitator *J Bacteriol* **175**: 1087-1094

Walz T Hirai T Murata K Heymann JB Mitsuoka K Fujiyoshi Y Smith BL Agre P and Engel A 1997 The three-dimensional structure of Aquaporin-1 *Nature* **387**: 624-627

Walz T Smith BL Agre P and Engel A 1994 The three-dimensional structure of human erythrocyte aquaporin CHIP *EMBO J* **13**: 2985-2993

1.3 Structure function analysis

DIFFERENT BEHAVIOURS OF MIP PROTEINS IN N-LAUROYLSARCOSINE

[1]Laurence Duchesne, [1]Isabelle Pellerin, [1]Valérie Lagrée, [1]Alexandrine Froger, [1]Christian Delamarche, [2]Jean-Pierre Le Caer, [1]Patrick Bron, [1]Daniel Thomas, [1]Jean-François Hubert and [1]Stéphane Deschamps

[1]"Canaux et récepteurs membranaires", UMR-CNRS 6026, Université de Rennes I, 35042 Rennes cedex, Bretagne, France and [2]UMR-CNRS 7637, ESPCI, 10, rue Vauquelin, 75231 Paris cedex 05, France. E-mail: laurence.duchesne@univ-rennes1.fr

1. INTRODUCTION

Purification of MIP proteins is of fundamental interest for further reconstitution in artificial lipid bilayers and determination of functional properties or structural analysis by cryoelectron microscopy of 2D-crystals. It has been shown that native or recombinant AQP1 can be purified by a few steps procedure using N-lauroylsarcosine (NLS) as detergent (Smith & Agre, 1991). A first extraction with NLS solubilizes almost all the membranous proteins, except AQP1 which remains in 100,000g pellet. The aquaporin can then be obtained at a good degree of purity by resuspending the NLS insoluble extract in n-octyl glucoside (OG). We successfully used this procedure to purify the insect aquaporin expressed in yeast. This protocol allows the protein to exhibit water channel properties in proteoliposomes and to retain its homotetrameric native state (Lagrée *et al.*, 1998, 1999). In the MIP family, we observed that the *E. coli* glycerol facilitator, GlpF, is monomeric in the non-denaturing detergents OG (2%, w/v) and Triton X-100 (1%, v/v), while aquaporins are tetrameric. We thus hypothesised the existence of a relationship between the functional specificity and the quaternary structural organisation of MIPs. As mentioned earlier, NLS resistance is a well established physicochemical characteristic of the aquaporins AQP1 and AQPcic. In the present study, we analysed the behaviour in 2% NLS (w/v) of the solute facilitator GlpF and several mutants of AQPcic.

2. GLPF, AQPCIC, MUTANTS AND CHIMERAS EXPRESSED IN S. CEREVISIAE.

Mutants and truncated forms of the aquaporin AQPcic used in this study are presented on figure 1. Amino acid mutations in loop E correspond to the substitution of the AQP residue by the residue present in the corresponding position in GlpF (Froger *et al.*, 1998). The mutant AQP-C134S is described in Lagrée *et al.*, (1998). Deletion of the extramembraneous C-terminal part was achieved in the protein AQP-245Z and deletion of the end of the loop E and sixth transmembrane domain in AQP-216Z. In addition, in the chimera protein AAG, the sixth transmembrane domain and the cytoplasmic C-terminal part of the aquaporin were replaced by those from GlpF.

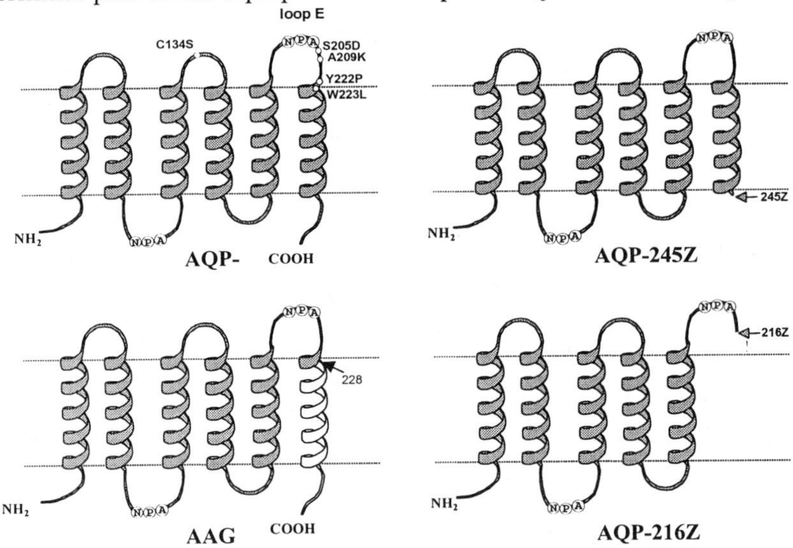

Figure 1. Mutants and chimeras of AQPcic and GlpF proteins. The AAG protein contains the N-terminal of AQPcic and the sixth transmembrane domain of *E. coli* GlpF (AQPcic 1-228/GlpF 242-281).

All proteins were overexpressed in *S. cerevisiae* according to Lagrée *et al.*, (1998). Quantitative western blots using anti-AQPcic antiserum demonstrated that wild type AQPcic and all mutants are expressed in yeast with similar level. The use of an antiserum directed against the C-terminal part of GlpF evidenced identical expression levels for GlpF and AAG chimera proteins (data not shown).

2.1 NLS comportment and oligomeric state of recombinant MIPs

The content of insoluble material following incubation of yeast membranes in NLS is shown in figure 2.

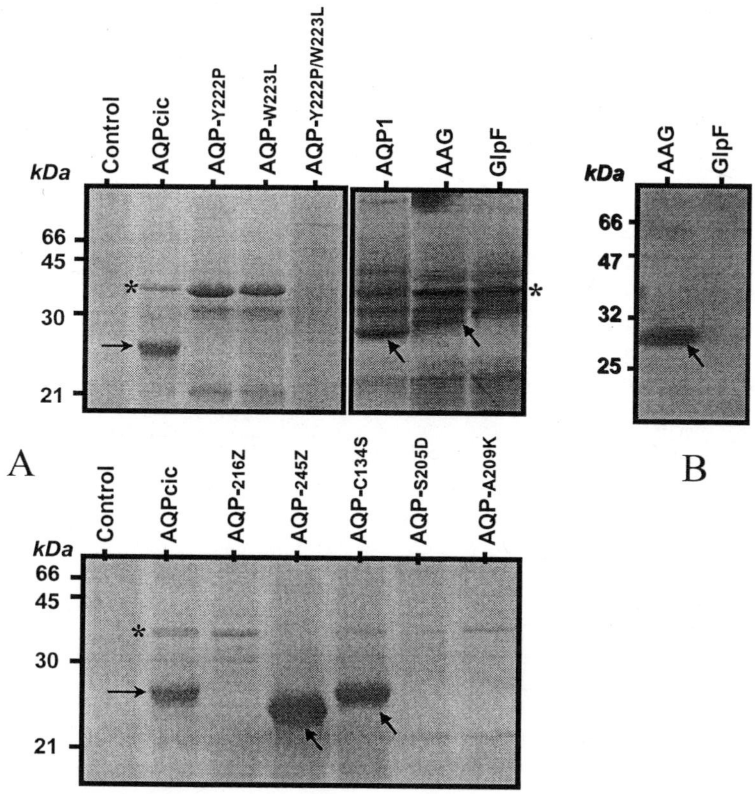

Figure 2 : NLS insoluble recombinant MIP proteins. Yeast membranes were incubated in 2% N-lauroyl-sarcosine for 4 hours and centrifuged for 45 min at 100000g. Pellets were analysed by SDS-PAGE followed by Coomassie blue staining (A) or Western blot using anti-C-terminus GlpF antiserum (**B**). MIP resistant to NLS are indicated by arrows. The GlpF protein is NLS soluble but not AAG. Another abundant 35 kDa NLS resistant protein (*) was often detected in yeast extracts. Mass Finger printing by MALDI TOF and peptide sequencing by ESI-MS/MS allowed us to identify this protein as heat shock protein Hsp30 described by Regnacq and Boucherie (1993).

Results are summarised in table I. AQP1, AQPcic and AQP-C134S are functional water channels, exhibit a tetrameric oligomerisation state in OG and are NLS resistant. GlpF, AQP-S205D, AQP-Y222P and AQP-Y222P/W223L are monomeric in OG, are unable to transport water and are soluble in NLS detergent. The AQPcic mutations A209K and W223L do not modify the tetrameric organisation observed in wild type aquaporin but abolish the water transport function and modify the particular comportment of the protein in NLS. Additionally, we show that the truncated AQP-245Z is functional and NLS resistant. Inversely, deletion of the sixth transmembrane domain (AQP-216Z) leads to a protein expressed in yeast, monomeric in OG and NLS soluble. AQP-216Z expression was not detected in *Xenopus* oocyte. However, an aquaporin in which the C-terminal part is replaced by that of GlpF (AAG chimera) remains tetrameric and NLS resistant.

Table 1. Comparison of water channel function, oligomerisation state in OG and NLS solubilisation resistance of MIP proteins and AQPcic mutants or chimera. Water permeability measurements were performed in the *Xenopus* oocyte system 48h following cRNA microinjection (Le Cahérec *et al.,* 1996) or by stopped flow experiments on liposomes containing the purified recombinant proteins (Lagrée *et al.,* 1998). Oligomeric state in OG were deduced from sedimentation velocity on 2-20% sucrose gradients (Beuron *et al.,* 1995). NLS solubility was deduced from results observed in figure 2. ip : in progress.

MIP and mutants	Water permeability * : in oocytes □ : in proteoliposomes	Oligomeric state in 2% OG	NLS resistance
AQP1	+* □	Tetramer	+
AQPcic	+* □	Tetramer	+
AQP-C134S	+□	Tetramer	+
AQP-245Z	+□	ip	+
AAG	ip	Tetramer	+
GlpF	–*	Monomer	–
AQP-S205D	–*	Monomer	–
AQP-A209K	–*	Tetramer	–
AQP-Y222P	–*	Monomer	–
AQP-W223L	–*	Tetramer	–
AQP-Y222P/W223L	–*	Monomer	–
AQP-216Z	Unexpressed *	Monomer	–

3. CONCLUSION

S. cerevisiae had been previously used for the production of aquaporins. In this study, we have successfully expressed in yeast the solute facilitator

GlpF. Our data shows that the differential solubilisation protocol based on NLS partial resistance of AQP1 or AQPcic for further protein purification is not applicable to GlpF or some nonfunctional mutants of AQPcic.

The molecular basis of the partial resistance to NLS of AQP1, AQPcic or other proteins such as connexins (Cascio *et al.*,1990) remains not understood. According to our results, the NLS resistance of aquaporins could not be explained only by their native tetrameric state. In AQP-A209K and AQP-W223L mutants, three-dimensional modifications inside the pore itself likely abolish channel function and NLS particular comportment but do not change the tetrameric organisation. However, GlpF and the other AQPcic mutants (AQP-S205D, AQP-Y222P or AQP-Y222P/W223L) exhibit monomeric forms with structural features accessible to the detergent.

The cytoplasmic C-terminal part of the water channel seems not essential neither for the function of the protein nor for the NLS resistance. Inversely, the presence of the sixth transmembrane domain appears necessary for the oligomeric state and for the particular comportment of AQPcic in NLS.

However, the replacement of this sixth transmembrane domain of AQPcic by the analogous domain from GlpF in the AAG chimera leads to a tetrameric and NLS resistant protein.

All together, these data bring new evidences for the importance of the loop E and sixth transmembrane domain in structure and function of MIP proteins.

We observe a close correlation between the water transport function and the resistance of MIP proteins to NLS solubilisation. Additionally, all water permeable proteins are found tetrameric in OG. We show that GlpF or MIP proteins bearing characteristics from GlpF are NLS soluble and consequently not suitable for rapid purification. However, our data demonstrate that expression and purification of some mutated and truncated AQP or AQP/GlpF chimeras remain possible through N-lauroylsarcosine differential extraction.

We have thus engaged the production of AQPcic, AQP-245Z and the AAG chimera for 2D crystallisation and structural studies. Comparison with established AQP1 structural models will provide new data for molecular organisation of aquaporins.

ACKNOWLEDGMENTS

This work was supported by the Fondation Langlois (Rennes, France). We would like to thank G. Bonnec and C. Heichette for technical assistance.

REFERENCES

Beuron F Le Cahérec F Guillam M-T Cavalier A Garret A Tassan J-P Delamarche C Schulz P Mallouh V Rolland J-P Hubert J-F Gouranton J Thomas D 1995 Structural analysis of a MIP family protein from the digestive tract of *Cicadella viridis JBiol Chem* **270**: 17414-17422

Cascio M Gogol E and Wallace BA 1990 The secondary structure of Gap junctions : influence of isolation methods and proteolysis *J Biol Chem* **265**: 2358-2364

Froger A Tallur B Thomas D and Delamarche C 1998 Prediction of functional residues in water channels and related proteins *Protein Science* **7**: 1458-1468

Lagrée V Pellerin I Hubert J-F Tacnet F Le Cahérec F Roudier N Thomas D Gouranton J and Deschamps S 1998 A yeast recombinant aquaporin mutant that is not expressed or mistargeted in *Xenopus* oocyte can be functionally analyzed in reconstituted proteoliposomes *J Biol Chem* **273**: 12422-12426

Lagrée V Froger A Deschamps S Pellerin I Delamarche C Bonnec G Gouranton J Thomas D and Hubert J-F 1998 Oligomerization state of water channels and glycerol facilitators Involvement of loop E *J Biol Chem* **257**: *33949-33953*

Lagrée V Froger A Deschamps S Hubert J-F Delamarche C Bonnec G Thomas D Gouranton J and Pellerin I 1999 Switch from an aquaporin to a glycerol channel by two amino acids substitution *J Biol Chem* **274**: 6817-6819

Le Cahérec F Deschamps S Delamarche C Pellerin I Bonnec Guillam M-T Gouranton J Thomas D and Hubert J-F 1996 Molecular cloning and characterization of an insect aquaporin Functional comparison with aquaporin 1 *Eur J Biochem* **241**: 707-715

Regnacq M and Boucherie H 1993 Isolation and sequence of HSP30 a yeast heat-shock gene coding for a hydrophobic membrane protein *Curr Genet* **23**: 435-442

Smith BL and Agre P 1991 Erythrocyte Mr 28000 transmembrane protein exists as a multisubunit oligomer similar to channel proteins *J Biol Chem* **266**: 6407-6415

1.4 Structure function analysis

OVEREXPRESSION AND PURIFICATION OF THE GLYCEROL TRANSPORT FACILITATORS, Fps1p AND GlpF, IN *SACCHAROMYCES CEREVISIAE* AND *ESCHERICHIA COLI*

[1]Kristina Hedfalk, [2]Roslyn M Bill, [2]Stefan Hohmann and [3]Jan Rydström
[1]Department of Molecular Biotechnology, Chalmers University of Technology, [2]Department of Cell and Molecular Biology - Microbiology, Göteborg University, [3]Department of Biochemistry and Biophysics, Göteborg University, Box 462, SE-405 30 Göteborg, Sweden. E-mail: kristina.hedfalk@molbiotech.chalmers.se

1. INTRODUCTION

Fps1p is an osmoregulated glycerol transport facilitator, located in the *Saccharomyces cerevisiae* plasma membrane, where it is mainly used for efflux of glycerol in the adaptation of yeast cells to lower external osmolarity (Luyten *et al.*, 1995). Fps1p is an unusual member of the MIP (Major Intrinsic Protein)-family. The protein is much larger than most of the other proteins in the family, due to long hydrophilic extensions at both termini. The N-terminus (~250 aa) is involved in the regulation of the channel (Tamas *et al.*, 1999), but the function of the C-terminus (~150 aa) is not yet known. Fps1p is also one of the few members that differs in the well-conserved NPA-motifs in the two putative channel forming loops B and E; serine instead of alanine in Loop B and leucine instead of proline in Loop E.

Because of the special features of Fps1p and the fact that the channel is regulated, we are interested in studying the structure-function relationship of this MIP-protein in more detail. Our goal is to obtain pure Fps1p for two purposes; to obtain 2D crystals for structural studies and to reconstitute protein into liposomes and measure function with stopped-flow techniques. We are also interested in exploring the structure of the hydrophilic N-terminus by NMR. In addition we are interested in comparing Fps1p with its

Molecular Biology and Physiology of Water and Solute Transport.
Edited by Hohmann and Nielsen, Kluwer Academic/Plenum Publishers, New York, 2000.

closest homologue, GlpF, the glycerol transport facilitator in *Escherichia coli*. GlpF can replace the transport function of Fps1p in yeast, but the transport is not regulated.

2. OVEREXPRESSION OF Fps1p

Fps1p has been overexpressed in *S.cerevisiae*, *Pichia pastoris* and *E.coli* (lac- and T7/lac promoters). Fps1p is expressed and localised to the membrane when expressed in all three systems.

2.1 Overexpression in Yeast

The homologous host *S.cerevisiae* was the first system we used for overexpression of Fps1p. Fps1p is cloned into pYX212 (TPI promoter) and the pYX213 (GAL promoter) with a C-terminal HA_3-tag in both cases. The protein was expressed at the same level from the TPI promoter and the inducible GAL promoter, respectively. A band of the correct molecular weight was detected on Western Blot as well as some bands of higher molecular weight, which are probably glycosylated forms.

After fractionation of the yeast extract, Fps1p was localised to the membrane, as detected with the HA-antibody on a Western Blot. Although expressed and localised, no overexpressed band could be seen in the membrane fraction on a Coomassie stained gel. This indicated that the expression level was rather low and therefore we decided to explore an alternative yeast system for overexpression, namely *P.pastoris*.

P.pastoris has been successfully used for overexpression of many proteins; as much as 12 g/L has been achieved for the soluble protein Tetanus Toxin, Fragment C (Clare *et al.*, 1991). More relevant is that expression levels of >1 mg/L culture have been reported for membrane bound proteins, i.e. the mouse 5-HT_{5A} serotonin receptor (Weiss *et al.*, 1995). In *P.pastoris* the protein of interest is expressed from the very strong and inducible AOX (Alcohol Oxidase)1 promoter, which is induced by growth on methanol as the sole carbon source. In theory, a high expression level can be achieved when cells are grown under controlled conditions with sufficient aeration and agitation. An additional advantage is the possibility of expressing the gene from multiple copies integrated into the yeast genome, which could improve the expression level.

Fps1p was cloned into the pPICZA vector and the protein expressed with a C-terminal myc- and 6xHis-tag in the *P.pastoris* strains X-33, GS115, and KM71, respectively, where the different strains have different methanol utilisation rates. Seven or eight clones were tested from each strain and the highest expressing clone was found from the wildtype strain, X-33. This clone was grown in a fermentor and Fps1p was localised to the plasma membrane (the 53%/43.5% interface of a discontinuous sucrose gradient). Only one band was detected on a Western Blot and the product was stable during further solubilisation and purification procedures. Fps1p was also expressed together with the α-factor signal sequence from *S.cerevisiae*, since this has been shown to positively influence the expression level of membrane proteins (Yellen *et al.*, 1990). It was noted that the expression level in the strain X-33 was improved in this construct. However, the recombinant Fps1p could not be detected on a Coomassie gel. Since the closest homologue to Fps1p, GlpF, as well as the N-terminus of the protein had already been successfully expressed in *E.coli* (see below) we decided to try that system.

2.2 Overexpression in E.coli

Two *E.coli* systems have been tried for the overexpression of Fps1p; the lac promoter and the strain JM109 and the T7/lac promoter in the strain BL21(DE3). The two constructs have an N-terminal 6xHIS tag and a C-terminal HA-tag. Fps1p is expressed and membrane localised in both systems, even though the expression level seems to be higher in the latter system. This is estimated from Western Blot signals and also from the fact that a band can be seen on a Coomassie stained gel after a small-scale purification. In both *E.coli* systems, it was observed that a minor portion of the overexpressed Fps1p is localised to the membrane when grown at 37 °C, which indicates the presence of inclusion bodies. However, Fps1p expressed in BL21(DE3) seems to be stable compared to the lac system, which gives a more degraded product. The yield from the T7/lac system has been estimated to 25-50 μg/L culture after the Ni-column.

2.3 Overexpression of the Fps1p N-terminus in E.coli

The Fps1p N-terminus is expressed from the T7/lac promoter in the strain BL21(DE3) and the yield of this soluble peptide is about 4-5 mg/L after the Ni-column.

3. OVEREXPRESSION AND PURIFICATION OF GlpF

GlpF is expressed in the *E.coli* strain BL21(DE3) from the T7/lac promoter and the protein is tagged with a 6xHis residue at the N-terminus. The protein is solubilised in 1 % Triton X-100 and purified on a Ni-column. The estimated yield is about 1 mg/L culture.

4. SOLUBILISATION AND INITIAL PURIFICATION OF Fps1p

The solubilisation and initial purification work was performed on the membrane pellet from the *P.pastoris* strain X-33 overexpressing Fps1p and the procedure was followed on Western Blots. Three different detergents in different concentrations were used to solubilise Fps1p from the total membrane; 1 % Triton X-100, 2 % Octylglucopyranoside (OG), and 3 % Decylmaltoside (DeM). OG was the most efficient detergent with respect to Fps1p, but the protein aggregates and degrades, while DeM is gentle but not very efficient for the protein of interest. Therefore Triton X-100 was chosen for solubilisation, but was exchanged to DeM (0.3%) later on in the purification procedure, since this detergent is a better choice for structural work. Fps1p was then purified on the Ni-column, with 100% binding efficiency. The protein was eluted with an imidazole gradient: Fps1p eluted at about 80 mM imidazole together with a high proportion of contaminating peptides.

5. SUMMARY OF EXPRESSION OF Fps1p AND GlpF

Table 1. Summary of Expression of Fps1p and GlpF

Target	Expression host	Strain	P[a]	S.P[b]	E[c]	L[d]	Yield[e]
Fps1p	*S.cerevisiae*	W303-1A	TPI	-	Yes	tm	-
Fps1p	*S.cerevisiae*	W303-1A	GAL	-	Yes	tm	-
Fps1p	*P.pastoris*	X-33	AOX1	-	Yes	pm	-
Fps1p	*P.pastoris*	X-33	AOX1	Yes	Yes	nd	-

Target	Expression host	Strain	P[a]	S.P[b]	E[c]	L[d]	Yield[e]
Fps1p	*P.pastoris*	GS115	AOX1	-	Yes	nd	-
Fps1p	*P.pastoris*	KM71	AOX1	-	Yes	nd	-
Fps1p	*E.coli*	JM109	lac	-	Yes	tm	-
Fps1p	*E.coli*	BL21(DE3)	T7/lac	-	Yes	tm	25-50 μg/L
Fps1p N-terminus	*E.coli*	BL21(DE3)	T7/lac	-	Yes	sf	4-5 mg/L
GlpF	*E.coli*	BL21(DE3)	T7/lac	-	Yes	tm	1 mg/L

[a]Promoter.

[b]Signal Peptide.

[c]Expression verified with Western Blot.

[d]Localisation of expressed target; total membrane (tm), plasma membrane (pm), not determined (nd), soluble fraction (sf).

[e]Yield estimated from Coomassie stained gel.

6. DISCUSSION

Although the expression levels were rather low, a band corresponding to Fps1p was detected on a Western Blot in all cases. Of the two yeast systems, *P.pastoris* was the most successful, since this system could be optimised at many levels. The methanol induction could be varied and the expression continued as long as the aeration and agitation were sufficient so that the carbon source itself was the limiting factor. *P.pastoris* also grows at very high cell density (200-400 g/L), which can partly compensate for a low expression level. Another advantage of *P.pastoris* is that there is one stable product, where there are several products of higher molecular weight in addition to the expected product in *S.cerevisiae*.

When comparing the expression systems tested the highest expression level would appear to be achieved in *E.coli* when expressed from the T7/lac promoter. This is supported by a very strong Western Blot signal, comparable to the one from the highly expressed N-terminus, from the crude extract and the fact that a band could be detected on a Coomassie stained gel after a small-scale purification. The estimated yield after a Ni-column is 25-50 μg/L, which is comparable to reported data for membrane proteins expressed in similar systems (Purdy *et al.*, 1999). The next step is to verify the band on the gel with mass spectrometry. The expression will be improved by trying different growth temperatures during the IPTG induction, in an attempt to allow a higher proportion of Fps1p to be localised to the membrane. The expression will then be scaled up and the purification will be optimised. Protein will then be produced for functional and structural studies.

7. CONCLUSION

It is widely acknowledged that successful overexpression of membrane proteins is a difficult task. However, when expressed from the T7/lac promoter in *E.coli*, Fps1p is expressed at relatively high levels. When optimised and scaled up it should be possible to produce enough protein to enable structural information. This will be of great value, since this glycerol transport facilitator is the only gated channel in the MIP-family of proteins.

ACKNOWLEDGMENTS

The work from our laboratory was supported by grants from the European Commission; BIO4-CT98-0024.

REFERENCES

Clare JJ Rayment FB Ballantine SP Sreekrishna K Romanos MA 1991 High-level expression of tetanus toxin fragment C in Pichia pastoris strains containing multiple tandem integrations of the gene *Biotechnology* **9**: 455-460

Luyten K Albertyn J Skibbe WF Prior BA Ramos J Thevelein JM Hohmann S 1995 Fps1 a yeast member of the MIP family of channel proteins is a facilitator for glycerol uptake and efflux and is inactive under osmotic stress *EMBO J* **14**: 1360-1371

Purdy M D Wiener M C 2000 Expression purification and initial structural characterization of YadQ a bacterial homolog of mammalian ClC chloride channel proteins *FEBS Lett* **466**: 26-28

Tamas MJ Luyten K Sutherland FC Hernandez A Albertyn J Valadi H Li H Prior BA Kilian SG Ramos J Gustafsson L Thevelein JM Hohmann S 1999 Fps1p controls the accumulation and release of the compatible solute glycerol in yeast osmoregulation *Mol Microbiol* **31**: 1087-1104

Weiss HM Haase W Michel H Reiländer H 1995 Expression of functional mouse 5HT5A seretonin receptor in the methylotrophic yeast Pichia pastoris: pharmacological characterization and localization *FEBS Lett* **377**: 451-456

Yellen G Migeon JC 1990 Expression of Torpedo nicotinic acetylcholine receptor subunits in yeast is enhanced by use of yeast signal sequences *Gene* **86**: 145-152

1.5 Structure function analysis

THREE-DIMENSIONAL FOLD OF HUMAN AQP1 WATER CHANNEL DETERMINED BY ELECTRON CRYO-CRYSTALLOGRAPHY OF 2-DIMENSIONAL CRYSTALS EMBEDDED IN ICE

Structural, Functional Correlates of AQP1

[1]Alok K. Mitra, [1]Gang Ren, [1]Anchi Cheng, [2]Vijay Reddy and [1]Peter Melnyk
[1]Department of Cell Biology and [2]Department of Molecular Biology, The Scripps Research Institute, La Jolla, CA 92037, USA. E-mail: mitra@scripps.edu

1. INTRODUCTION

The entry and exit of water across the cell membrane is a fundamental physiological process. The observed high permeability of water in the case of mammalian erythrocytes and renal proximal tubules led to the discovery of water channels (aquaporins, AQP) (Denker *et al.*, 1988; Smith and Agre, 1991). Aquaporins are members of the MIP (major intrinsic protein) superfamily (Gorin *et al.*, 1984) and are found in both eukaryotes and prokaryotes where they serve as channels for rapid dissipation of osmotic gradients across the lipid bilayer (Agre *et al.*, 1995; Verkman *et al.*, 1996). In order to understand the structural elements that constitute and control the water-selective permeability of AQP1, a detailed knowledge of the high-resolution 3-D structure in the bilayer is essential. To this end, 2-D crystals of AQP1 have been used to examine the structure using electron crystallography. Projection maps of AQP1 were determined by electron cryo-crystallography by three groups (Jap and Li, 1995; Mitra *et al.*, 1995; Walz *et al.*, 1995) who used different conditions for 2-D crystallization and specimen preservation for microscopy. Subsequently, a 7Å resolution, 3-D density map of AQP1 was determined by us (Cheng *et al.*, 1997) from an analysis of minimal dose images and electron-diffraction patterns recorded

Molecular Biology and Physiology of Water and Solute Transport.
Edited by Hohmann and Nielsen, Kluwer Academic/Plenum Publishers, New York, 2000.

from ice-embedded 2-D crystals. 3-D density maps, at a similar resolution were reported by the other 2 groups (Li *et al.*, 1997; Walz *et al.*, 1997) which were determined using specimens preserved in trehalose and the tilted projections comprised of images and electron diffraction patterns (Walz *et al.*, 1997) and only images (Li *et al.*, 1997). All three maps reveal a barrel formed by six α-helices surrounding additional density near the middle of the bilayer; the helices pack with a right-handed twist in the models of Cheng *et al.* (1997) and Walz *et al.* (1997) but with a left-handed twist in the model of Li *et al.* (1997).

2. RESULTS AND DISCUSSION

We present an improved 3-D density map of AQP1 which was generated by including data up to 4Å resolution recorded from electron diffraction patterns and images of 2-D crystals tilted up to 60°. The present work and the recent 4.6Å resolution analysis of acetyl-choline receptor reported by Unwin and colleagues (Miyazawa *et al.*, 1999) represent the highest resolution 3-D structural analyses of ice-embedded specimens determined by electron cryo-crystallography. This is noteworthy in view of the fact that, to date, structural studies to derive atomic or near-atomic resolution structure by electron crystallography have been exclusively carried out on specimens preserved in sugars such as tannic acid (Kühlbrandt *et al.*, 1994), trehalose (Kimura *et al.*, 1997; Nogales *et al.*, 1998) or glucose (Grigorieff *et al.*, 1996).

Consistent with the improved resolution of the map, densities for the α-helical segments, some more than the others, display discernible protrusions along their lengths. The region near the 4-fold axis at the interface of the 4 monomers appears to lack significant protein density (Fig. 1); however, its diameter is smaller than seen before (Cheng *et al.*, 1997) because of the appearance of protrusions on density for helices C and D. In order to demonstrate the helical periodicity of the observed protrusions, C^{α} - backbone (in white) in a helical configuration (Fig. 1) was visually placed in each of the helical segments.

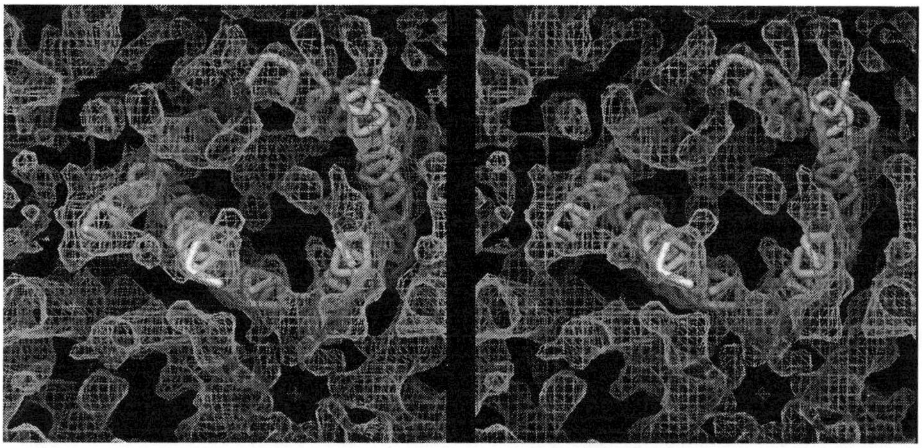

Figure 1. Stereo pair of 3-D density map of AQP1 viewed normal to the bilayer. One monomer with a polyalanine model for the helices (white, A-F clockwise) and portions of adjacent monomers are shown.

We carried out a careful analysis of the current, higher-resolution density map to scrutinize the presence of an in-plane pseudo 2-fold axis of symmetry within the hydrophobic core of the molecule previously discovered by us (Cheng *et al.*, 1997) and also suggested by Li *et al.* (1997). Our analysis showed that the presence of in-plane pseudo 2-fold axis of symmetry is strongly revealed only when data upto 6Å resolution was used. This result indicates that the local symmetry is significant only when features such as the spatial locations and presumably the backbones of the α-helical segments of two halves of the polypeptide chain are compared (Mitsuoka *et al.*, 1999).

In the current map, strong bands of density, some of, which are continuous and comparable in strength to that for the transmembrane helices, can be identified at the extremities of some of the helices. These segments of density are located either at the extrabilayer surfaces or are disposed within the helix barrel defining the putative NPA box region and suggest the locations of some of the ordered inter-helical loops. Synthesis of these observations, led to two possible arrangements for the six major, transmembrane segments in the sequence to the observed 6 helices in the 3-D density map. One of these 2 models is identical to that deduced from our earlier 7Å-resolution map and reported in Verkman and Mitra (2000). The earlier derivation was based on a "strict" imposition of the 2-fold pseudo symmetry relating the tandem repeats in the sequence.

All published lines of evidence (Preston *et al.*, 1993; Shi *et al.*, 1994; van Hoek *et al.*, 1991; Zhang *et al.*, 1993) indicate that the AQP1 monomer contains the functional channel. The volume of low density (negative density) representing the solvent-accessible region enclosed by the helix barrel has a vestibular shape which is relatively wide on both the cytoplasmic and the extracellular sides but narrows down to a diameter of ~6Å near the center of the molecule. This narrow region located in the hydrophobic core of the molecule may represent the physical constriction defining the selectivity filter. The vestibule region is bounded by the tightly-packed helices C and D near the 4-fold axis and by the wall of density within the barrel attributed to the NPA boxes thereby ensuring entry and exit to the narrow constriction only from the top and bottom of the monomer. Consistent with the biophysical measurements which suggest that the passage of water through the channel is similar to bulk flow, a hydrated pore (Welling *et al.*, 1996), within the hydrophobic core of the molecule, that is lined with polar or charged residues and sufficiently narrow for only water molecules to enter and exit by diffusive movements could describe the solute barrier. A model for the azimuthal orientations of the helices was generated based on consensus secondary structure assignment, alignment of the sequences of and the criterion that conserved and polar residues are likely to be involved in protein-protein contact. Such an analysis has allowed us to generate a list of residues that are likely to be functionally important. Structure/function studies using mutagenesis experiments pending an atomic resolution structure could provide further insight into how the archetype into MIP family of proteins plays a fundamental role in cell homeostasis.

ACKNOWLEDGMENTS

This research was supported by a grant from the NIH, in part by a grant from the NSF and a Grant-in-Aid from the American Heart Association, and a special fellowship from the Cystic Fibrosis Research Foundation to AKM and a post-doctoral fellowship from the American Heart Association, Western Affiliate to GR. AKM is an Established Investigator of the American Heart Association.

REFERENCES

Agre P Brown D and Nielsen S 1995 Aquaporin water channels: unanswered questions and unresolved controversies *Curr Opin Cell Biol* **7**: 472-483

Cheng A van Hoek AN Yeager M Verkman S and Mitra AK 1997 Three-dimensional organization of a human water channel *Nature* **387**: 627-30

Denker B M Smith BL Kuhajda FP and Agre P 1988 Identification purification and partial characterization of a novel Mr 28000 integral membrane protein from erythrocytes and renal tubules *J Biol Chem* **263**: 15634-15642

Gorin MB Yancey SB Cline J Revel JP and Horwitz J 1984 The major intrinsic protein MIP of the bovine lens fiber membrane: characterization and structure based on cDNA cloning *Cell* **391**: 49-59

Grigorieff N Ceska TA Downing KH Baldwin JM and Henderson R 1996 Electron-crystallographic refinement of the structure of bacteriorhodopsin *J Mol Biol* **259**3: 393-421

Jap BK and Li H 1995 Structure of the osmo-regulated H2O-channel AQP-CHIP in projection at 35 A resolution *J Mol Biol* **251**3: 413-420

Kimura Y Vassylyev DG Miyazawa A Kidera A Matsushima M Mitsuoka K Murata K Hirai T and Fujiyoshi Y 1997 Surface of bacteriorhodopsin revealed by high-resolution electron crystallography *Nature* **389**: 206-11

Kühlbrandt W Wang DN and Fujiyoshi Y 1994 Atomic model of plant light-harvesting complex by electron crystallography *Nature* **367**: 614-21

Li H Lee S and Jap BK 1997 Molecular design of aquaporin-1 water channel as revealed by electron crystallography *Nat Struct Biol* 44: 263-265

Mitra AK van Hoek AN Wiener MC Verkman AS and Yeager M 1995 The CHIP28 water channel visualized in ice by electron crystallography *Nat Struct Biol* **29**: 726-729

Mitsuoka K Murata K Walz T Hirai T Agre P Heymann JB Engel A and Fujiyoshi Y 1999 The structure of aquaporin-1 at 45-Å resolution reveals short α-helices in the center of the monomer *J Struct Biol* **128**: 34-43

Miyazawa A Fujiyoshi Y Stowell M and Unwin N 1999 Nicotinic acetylcholine receptor at 46Å resolution: transverse tunnels in the channel wall *J Mol Biol* **288**: 765-786

Nogales E Wolf SG and Downing KH 1998 Structure of the alpha beta tubulin dimer by electron crystallography *Nature* **391**: 1999-2003

Preston GM Jung JS Guggino WB and Agre P 1993 The mercury-sensitive residue at cysteine 189 in the CHIP28 water channel *Journal of Biological Chemistry* **268**: 17-20

Shi LB Skach WR and Verkman AS 1994 Functional independence of monomeric CHIP28 water channels revealed by expression of wild-type mutant heterodimers *J Biol Chem* **269**: 10417-22

Smith BL and Agre P 1991 Erythrocyte Mr 28000 transmembrane protein exists as a multisubunit oligomer similar to channel proteins *J Biol Chem* **266**: 6407-6415

van Hoek AN Hom ML Luthjens LH de Jong MD Dempster JA and van Os CH 1991 Functional unit of 30 kDa for proximal tubule water channels as revealed by radiation inactivation *J Biol Chem* **266**: 16633-16635

Verkman AS and Mitra AK 2000 Structure and function of aquaporin water channels *Am J Physiol* **278**: F13-F28

Verkman AS van Hoek AN Ma T Frigeri A Skach WR Mitra A Tamarappoo BK and Farinas J 1996 Water transport across mammalian cell membranes *Am J Physiol* **270**: C12-30

Walz T Hirai T Murata K Heymann J B Mitsuoka K Fujiyoshi Y Smith B L Agre P and Engel A 1997 The three-dimensional structure of aquaporin-1 *Nature* **387**6633 624-7

Walz T Typke D Smith BL Agre P and Engel A 1995 Projection map of aquaporin-1 determined by electron crystallography *Nat Struct Biol* **2**: 730-732

Welling DJ Welling PA and Welling LW 1996 Filled pore approximation: a theoretical framework for solute-solvent coupling in narrow water channels *Am J Physiol* **270**: C1246-1254

Zhang R van Hoek AN Biwersi J and Verkman AS 1993 A point mutation at cysteine 189 blocks the water permeability of rat kidney water channel CHIP28k *Biochemistry* **32**: 2938-2941

1.6 Structure function analysis

VOLUME FLUX ACROSS RED CELL AQP1 AND *E. COLI* AQPZ WATER CHANNEL PROTEINS RECONSTITUTED INTO PLANAR LIPID BILAYERS

[1]Sapar M. Saparov, [2]Ulrich Rothe, [3]Mario J. Borgnia, [3]Peter Agre, and [1]Peter Pohl

[1]Institut für Medizinische Physik und Biophysik, Martin-Luther-Universität, 06097 Halle, Germany; [2]Institut für Physiologische Chemie, Martin-Luther-Universität, 06097 Halle, Germany; [3]Department of Biological Chemistry, The Johns Hopkins University School of Medicine, Baltimore, Maryland 21205-2185, USA. E-mail: peter.pohl@medizin.uni-halle.de

1. INTRODUCTION

Several members of the aquaporin familiy of protein water channels have been reported to carry ionic currents. The major intrinsic protein (MIP, AQP0) of the lens functioned as an ion channel when reconstituted into planar lipid bilayers. In 0.1 M KCl, it had two preferred conductance states with amplitudes of 380 and 160 pS (Ehring *et al.,* 1990). In contrast, *Xenopus* oocytes expressing AQP0 failed to exhibit ion channel activity but facilitated water transport (Mulders *et al.,* 1995). Nodulin 26 (an integral symbiosome membrane protein of nitrogen- fixing soybean nodules) was the second member of the MIP channel protein family that was reported to form ion conducting channels with a large single channel conductance and weak anion selectivity (Weaver *et al.,* 1994). In planar bilayers reconstituted nodulin 26 channels showed sensitivity to high applied voltages, including more active gating and the tendency to occupy discrete lower subconductance states (Lee *et al.,* 1995). However, experiments involving expression of nodulin 26 in *Xenopus* oocytes have led to a reevaluation of the channel properties and it is now clear that nodulin 26 possesses aquaporin activity in this system (Rivers *et al.,* 1997).

Reports of ion conductance across aquaporins expressed in *Xenopus* oocytes are extremely rare. By voltage-clamp studies, the intracellular AQP6 was shown to be an anion selective channel when treated with the known water channel inhibitor, Hg^{2+} (Yasui *et al.*, 1999). When treated with forskolin, AQP1 transformed into a cation conducting channel (Yool *et al.*, 1996). However, other research-groups could not confirm these results for AQP1 or AQP2 (Agre *et al.*, 1997; Deen *et al.*, 1997; Sasaki *et al.*, 1997; Verkman and Yang, 1997).

It is difficult to establish whether aquaporin reconstitution into planar lipid bilayers alters their properties or whether expression in oocytes leads to regulatory events that also may alter their transport properties. To resolve these inconsistencies in function, the conductance to water of two different aquaporins (AQP 1 and AQPZ) that were reconstituted into planar bilayers was investigated simultaneously to their ability to mediate ion fluxes.

2. AQUAPORIN RECONSTITUTION AND ELECTROPHYSIOLOGICAL METHODS

The water flux across planar membranes was assessed from the accompanying solute concentration changes in the immediate membrane vicinity (Pohl *et al.*, 1997). Water passing through the membrane dilutes the solution it enters and concentrates the solution it leaves. Water flow induced concentration changes of sodium ions were monitored with the help of microelectrodes. The solute polarisation adjacent to lipid bilayers increases with an increasing osmotic gradient. The magnitude of the near-membrane concentration shift is also sensitive to the hydraulic conductivity of the lipid bilayer (Pohl *et al.*, 1997).

In the present study, measurements of ion and water fluxes across purified AQP1 and AQPZ were undertaken. AqpZ protein was expressed in *E. coli*. To facilitate purification, a poly(His) containing recombinant was constructed (Borgnia *et al.*, 1999). AQP1 was purified from human red blood cells (Agre *et al.*, 1999). Proteoliposomes were prepared by protein reconstitution into preformed vesicles from a phosphatidyl-choline phosphatidyl-ethanolamine mixture or *E.coli* lipid by a dialysis method. After dialysis, proteoliposomes were sedimented at 120,000 g and resuspended in buffer solutions.

2.1 Aquaporin incorporation by proteoliposome fusion with planar bilayers

To insure the likelihood of channel incorporation, a protein to lipid mole ratio of 1:500 was used in these experiments. The vesicles were added to the cis side of Muller-Rudin planar lipid bilayers. Their reliable and nonselective fusion to the planar membrane was achieved due to an increase of the salt concentration (from 100 to 450 - 650 mM) at one side of the bilayer. To visualise the fusion events, proteoliposomes containing additionally ergosterol and nystatin were prepared in some of the experiments. Under these conditions, the fusion of every single vesicle with the BLM was detected as a sudden jump in the transmembrane current induced by the simultaneous insertion of many nystatin channels from the ergosterol containing liposome (Woodbury and Miller, 1990). The increase in conductance was transient, i.e. it decayed back to baseline before the next fusion event because ergosterol, required for the integrity of the nystatin pore, diffuses away into the sterol-free planar bilayer after liposome fusion. During the experiments, the stirring rate was varied to modulate the fusion rate. In the absence of bulk movements, the vesicles did not fuse with the bilayer. In some of the experiments rare channel activity was observed that cannot be contributed to nystatin (Fig. 1). However, it was not correlated to the amount of aquaporin (i.e. to the number of vesicles fused with the planar bilayer) inserted into the membrane. Rather, membrane defects masquerade as channel-like current fluctuations because it is known that step conductance changes can be observed in planar lipid bilayers following brief exposure to large concentrations of lipid vesicles devoid of protein (Woodbury, 1989).

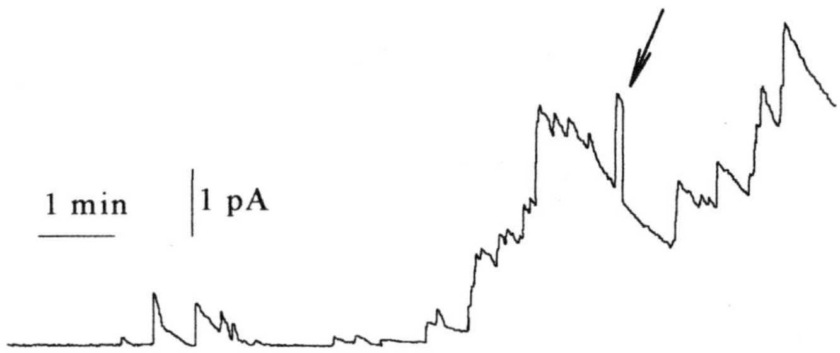

Figure 1. Fusion of ergosterol and nystatin containing proteoliposomes with ergosterol free planar bilayers. Each fusion spike represents the fusion of a single liposome carrying a package of preformed nystatin channels. After more than 20 vesicles have fused with the bilayer, one channel-like current fluctuation (arrow) of about 50 pS was detected. It can not be contributed neither to nystatin nor to AQP1 but rather, it represents a membrane defect that can be observed in planar bilayers following brief exposure to pure lipid vesicles. The accompanying insertion of AQP1 water channels was electrically silent. A NaCl-concentration gradient of 0.1 : 0.5 M was established. The membrane was clamped at 30 mV.

Fusion of proteoliposomes did not lead to an increase in membrane water permeability also. This result was expected because the surface area of the planar bilayer is 10^7 times the vesicle surface, so that fusion of about 1000 liposomes results in a molar protein : lipid ratio less than 1 : 10^6. Obviously, the small protein concentration cannot significantly contribute to membrane water permeability. To enhance the protein : lipid ratio, a method proposed by Schindler (1989) was used.

2.2 Functional AQP1 and AQPZ in planar membranes

Planar bilayer membranes containing functional aquaporin water channel protein were formed from proteoliposomes. The vesicles were transformed into monolayers which subsequently were apposed into planar bilayers. For both AQP1 And AQPZ containing bilayers, the membrane conductance was similar to that of protein free membranes (about 2 $\mu S/cm^2$). Current fluctuations did not occur. At the same time, reconstitution of AQP1 doubled the hydraulic conductivity (Fig. 2).

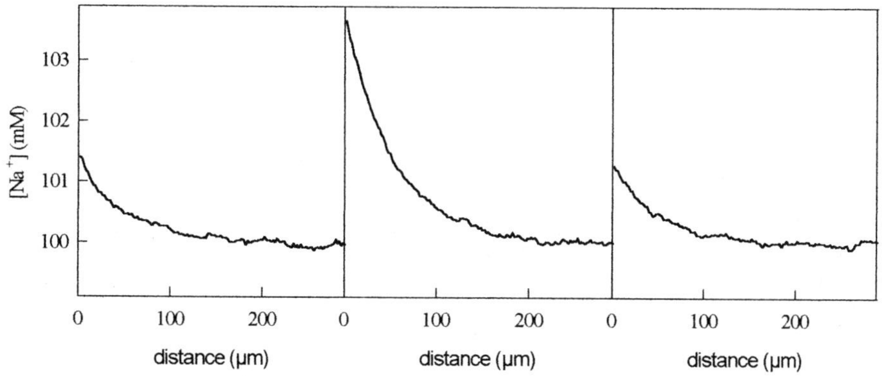

Figure 2. Sodium concentration as a function of the distance to the membrane. The water permeability increased from 23 μm/s in the absence of the protein (left panel) to 50 μm/s after the reconstitution of AQP1 (panel in the middle). After the addition of 1 mM of the inhibitor pCMBS (right panel) the hydraulic conductivity decreased to the level of a pure lipid bilayer. The lipid to protein ratio used to prepare the proteoliposomes was 70 : 1. The osmotic flux was driven by 1 M urea. pH was 7.5.

Because AqpZ is not sensitive to mercury compounds (Calamita *et al.*, 1995), the presence of a porous water transport pathway was established by measurements of the activation energy. Steady state water flow was monitored both across protein free Schindler membranes and aquaporin containing bilayers (Fig. 3). The corresponding Arrhenius plot for AqpZ induced water movement revealed an activation energy of 4 kcal/mol in comparison to 14 kcal/mol for pure lipid bilayers.

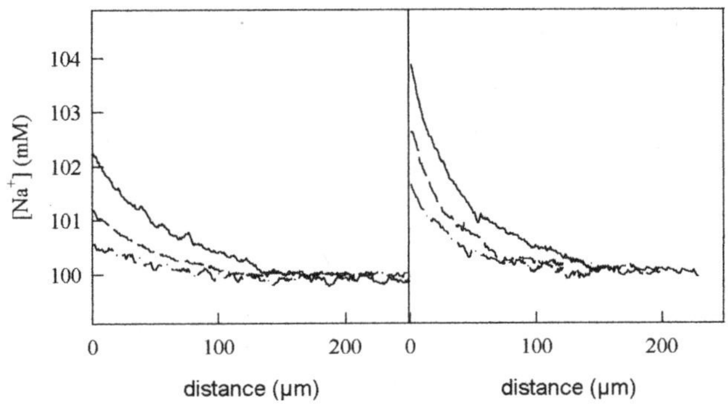

Figure 3. Representative sodium concentration profiles at the hypotonic side of a planar membrane. With an increase in temperature from 16 to 20 and finally to 26 °C, a water permeability of 12, 18 and 35 μm/s, respectively, was measured for pure lipid bilayers (left panel). Under the same conditions, the permeability of the bilayer containing reconstituted AQPZ increased from 53 to 67 up to 88 μm/s. The lipid to protein ratio used to prepare the proteoliposomes was 50 : 1. The osmotic flux was driven by 1 M urea. pH was 7.5.

3. DISCUSSION

When reconstituted into solvent free planar bilayers, AQP1 and AQPZ water channel proteins purified from human red blood cells and *E. coli*, respectively, are shown to facilitate water transport. The proteins remain functionally active including the sensitivity to inhibitors, and the Arrhenius activation energy. A porous pathway requires an energy of about 4 kcal/mol that was found in the present study for AQPZ, whereas activation energies of 10-15 kcal/mol were reported for diffusion of water across unmodified planar phospholipid bilayers (see review by Fettiplace and Haydon, 1980).

Simultaneous measurements of electrical conductivity and water permeability demonstrate an extremely high selectivity for water and exclusion of ions. This result is in agreement with studies of AQPZ and AQP1 in oocyte membranes (Calamita *et al.*, 1995; Preston *et al.*, 1992). Water transport across planar lipid bilayers containing both AQP1 and AqpZ is shown to be electrically silent. Along with the development of expression systems for the purification of large quantities of modified aquaporins (Borgnia *et al.*, 1999; Coury *et al.*, 1998), water channel reconstitution into planar bilayers should allow investigation of the molecular basis for the extremely high water selectivity, since both water and ion movement can be measured simultaneously. Importantly, this method should permit resolution of contradictory studies reporting putative ion permeation of aquaporins studied in other systems.

ACKNOWLEDGMENTS

Financial support of the Deutsche Forschungsgemeinschaft (Germany) (Po 533/2-2) and the National Institutes of Health is grateful acknowledged.

REFERENCES

Agre P Lee MD Devidas S and Guggino WB 1997 Aquaporins and ion conductance *Science* **275:** 1490-1490

Agre P Mathai JC Smith BL and Preston GM 1999 Functional analyses of aquaporin water channel proteins *Methods Enzymol* **294:** 550-572

Borgnia MJ Kozono D Calamita G Maloney PC and Agre P 1999 Functional reconstitution and characterization of AqpZ, the E. coli water channel protein *J Mol Biol* **291:** 1169-1179

Calamita G Bishai WR Preston GM Guggino WB and Agre P 1995 Molecular cloning and characterization of AqpZ, a water channel from Escherichia coli *J Biol Chem* **270:** 29063-29066

Coury LA Mathai JC Prasad GR Brodsky JL Agre P and Zeidel ML 1998 Reconstitution of water channel function of aquaporins 1 and 2 by expression in yeast secretory vesicles *Am J Physiol* **43:** F34-F42

Deen PM Mulders SM Kansen SM and van Os CH 1997 Aquaporins and ion conductance *Science* **275:** 1491

Ehring GR Zampighi G Horwitz J Bok D and Hall JE 1990 Properties of channels reconstituted from the major intrinsic protein of lens fiber membranes *J Gen Physiol* **96:** 631-664

Fettiplace R and Haydon DA 1980 Water permeability of lipid membranes *Physiol Rev* **60:** 510-550

Lee JW Zhang Y Weaver CD Shomer NH Louis CF and Roberts DM 1995 Phosphorylation of nodulin 26 on serine 262 affects its voltage-sensitive channel activity in planar lipid bilayers *J Biol Chem* **270:** 27051-27057

Mulders SM Preston GM Deen PMT Guggino WB van Os CH and Agre P 1995 Water channel properties of major intrinsic protein of lens *JBiolChem* 9010-9016

Pohl P Saparov SM and Antonenko YN 1997 The effect of a transmembrane osmotic flux on the ion concentration distribution in the immediate membrane vicinity measured by microelectrodes *Biophys J* **72:** 1711-1718

Preston GM Carroll TP Guggino WB and Agre P 1992 Appearance of water channels in Xenopus oocytes expressing red cell CHIP 28 protein *Science* **256:** 385-387

Rivers RL Dean RM Chandy G Hall JE Roberts DM and Zeidel ML 1997 Functional Analysis of Nodulin 26, an Aquaporin in Soybean Root Nodule Symbiosomes *J Biol Chem* **272:** 16256-16261

Sasaki S Uchida S Kuwahara M Fushimi K and Marumo F 1997 Aquaporins and ion conductance *Science* **275:** 1490-1491

Schindler H 1989 Planar lipid-protein membranes: strategies of formation and of detecting dependencies of ion transport functions on membrane conditions *MethodsEnzymol* **171:** 225-253

Verkman AS and Yang B 1997 Aquaporins and ion conductance *Science* **275:** 1491

Weaver CD Shomer NH Louis CF and Roberts DM 1994 Nodulin 26, a nodule-specific symbiosome membrane protein from soybean, is an ion channel *J Biol Chem* **269:** 17858-17862

Woodbury DJ 1989 Pure lipid vesicles can induce channel-like conductances in planar bilayers *J Membrane Biol* **109:**145-150

WoodburyDJ and Miller C 1990 Nystatin-induced liposome fusion. A versatile approach to ion channel reconstitution into planar bilayers *Biophys J* **58:** 833-839

Yasui M Hazama A Kwon TH Nielsen S Guggino WB and Agre P 1999 Rapid gating and anion permeability of an intracellular aquaporin *Nature* **402:** 184-187

Yool AJ Stamer WD and Regan JW 1996 Forskolin stimulation of water and cation permeability in aquaporin 1 water channels *Science* **273:** 1216-1218

1.7 Structure function analysis

BIOGENESIS AND FOLDING OF AQUAPORIN WATER CHANNELS IN THE ENDOPLASMIC RETICULUM

[1]Isaiah Turnbull, [1,3]Yun Lu, [3]Alan S. Verkman and [1]William R. Skach
[1]Division of Molecular Medicine, Oregon Health Sciences University, Portland, OR:
[2]Cardiovascular Research Institute, University of California, San Francisco, CA:
[3]Current Address: Genetics and Biochemistry Branch, National Institutes of Health.
E-mail: skachw@ohsu.edu

1. INTRODUCTION

The transmembrane topology of most eukaryotic polytopic membrane proteins is established in the endoplasmic reticulum (ER). This process involves multiple translocation and membrane integration events that are facilitaed coincident with protein synthesis by the Sec61 translocation machinery. Little is known, however, regarding the molecular mechanisms by which multiple transmembrane (TM) helices and peptide loops are precisely oriented into the ER membrane, ER lumen, and cytosol. Aquaporins have provided a unique opportunity to study this phenomenon because related family members acquire their topology through distinctly different mechanisms. The six TM segments of AQP4 are established sequentially and cotranslationally as the nascent polypeptide emerges from the ribosome (Shi *et al.,* 1995). In contrast, the initial topology of AQP1 at the ER membrane contains only four TM segments (Skach *et al.,* 1994). To better understand aquaporin folding and maturation, we have examined: 1) the process by which AQP1 acquires its mature, six-spanning orientation, and 2) the structural basis for different AQP1 and AQP4 folding pathways.

Molecular Biology and Physiology of Water and Solute Transport
Edited by Hohmann and Nielsen, Kluwer Academic/Plenum Publishers, New York, 2000.

2. AQP1 TOPOLOGICAL MATURATION IN THE ER MEMBRANE

Previous studies have demonstrated that topogenic determinants (e.g. signal anchor and stop transfer sequences) encoded within AQP1 cotranslationally establish a four-spanning topology at the ER membrane (Figure 1A). Yet at the plasma membrane, mature AQP1 contains six TM segments (Preston *et al.,* 1994, Cheng *et al.,* 1997, Walz, *et al.,* 1997) (Figure 1C). This suggested that AQP1 folding might involve an unusual topological reorientation in which TM3 is converted from a type I to a type II topology by a 180° rotation about the plane of the membrane (Figure 1B). To determine if and when such a reorientation occurred, we examined the dynamic topology of TM3 as a function of nascent chain length. A 10 residue c-myc epitope was engineered into peptide loops flanking the N- or C-terminus of TM3 at residues Pro77 or Thr120, respectively (Figure 1). Epitope tagged constructs were then truncated after TM segments 3, 4, 5 and 6 (residues Leu139, Pro169, Val215, and Val264, respectively) and fused to a C-terminus translocation reporter derived from bovine prolactin. These truncated constructs thus mimic topologic intermediates of AQP1, and the myc epitope provides a means to follow changes in TM3 topology at sequential stages of synthesis. Topology the myc epitope and C-terminus reporter was determined in each construct by protease protection and/or antibody accessibility in right-side-out ER membranes derived from cell-free and Xenopus oocyte expression systems.

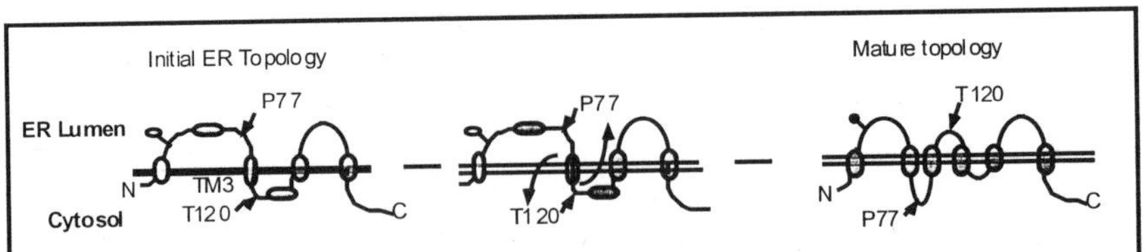

Figure 1. Predicted conversion of AQP1 from its initial four-spanning topology to its mature six-spanning topology. Myc epitope insertions sites(P77 and T120) are indicated. Shaded ovals represent TM segments. Open circles indicate glycosylation site at residue N42.

In microinjected Xenopus oocytes we observed a progressive shift of TM3 from a type I to a type II topology following the synthesis of each additional TM segment (Figure 2A). When nascent were chains truncated after TM3, <10% of the myc epitope at residue T120 was protected from protease, confirming that TM3 C-terminus flanking residues were initially directed into a cytosolic orientation. Following synthesis of TM6, however, these residues were translocated into the ER lumen in nearly 50% of nascent chains, consistent with TM3 acquiring its mature type II topology. Moreover, ~80% of newly synthesized full length (functional) myc-tagged AQP1 protein achieved a six-spanning topology within five hours after RNA injection. Thus, TM3 reorientation requires the synthesis of all six TM segments and continues even after protein synthesis has been completed.

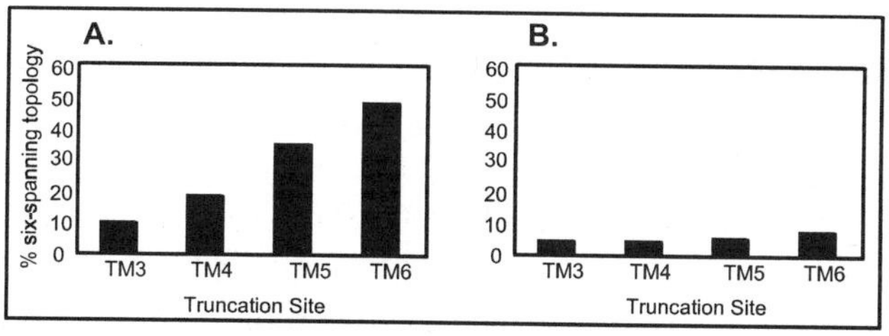

Figure 2. AQP1 topological maturation in **(A)** Xenopus oocytes and **(B)** rabbit reticulocyte lysate supplemented with canine pancreas microsomes. Shown is the percent of nascent chains in which TM3 has been converted to a type II topology immediately following synthesis of indicated TM segments.

Interestingly, the efficiency of AQP1 topological maturation was dependent on the source of ER membranes. AQP1 failed to acquire its six-spanning topology *in vitro* when expressed in RRL supplemented with translocation competent canine pancreatic microsomes. Under these conditions 8% of full length AQP1 exhibited a mature six-spanning topology within one hour of protein synthesis (Figure 2B), and this fraction increased to only 28% during a twelve hour chase period. However, when RRL was supplemented with oocyte-derived ER membranes, 75% of nascent AQP1 chains acquired a six-spanning topology within 5 hours, similar to that observed for intact oocytes. Using this *in vitro* system, we showed that reorientation of TM3 C-terminus flanking residues into the ER lumen was associated with a reciprocal movement of TM3 N-terminus residues (e.g. residue P77) from the ER lumen to the cytosol consistent with a true conversion of TM3 topology. Topologic reorientation was followed by compaction of the hydrophobic core into a protease resistant conformation.

These experiments thus begin to define a complex folding pathway for AQP1 that involves at least three distinct structural (i.e. topological) intermediates: i) a loosely folded four spanning topology, ii) a loosely folded six spanning topology, and iii) a compact, protease resistant six-spanning topology. In contrast to simple models of polytopic protein biogenesis, AQP1 maturation requires the posttranslational movement of large peptide regions about the plane of the membrane. Remarkably, AQP1 maturation also appears to exhibit more complex requirements within the ER compartment than are needed for simple protein translocation.

3. SEQUENCE DETERMINANTS RESPONSIBLE FOR DIFFERENT AQP BIOGENESIS PATHWAYS

Unlike AQP1, AQP4 does not undergo a topological reorientation to achieve its six-spanning topology. Rather, each TM segment is oriented sequentially and cotranslationally during protein synthesis (Shi *et al.,* 1995). Therefore, the major difference between AQP1 and AQP4 folding pathways is that the correct topology of AQP4-TM2 and TM3 is established prior to the synthesis of additional, downstream TM segments. To identify primary structural features responsible these different folding pathways, we generated a series of AQP1/AQP4 chimeric proteins by systematically exchanging TM segments and their connecting peptide loops. Chimeras were sequentially truncated, fused to a C-terminus translocation reporter, and analysed by protease accessibility in RRL. Topological examination of these translocation intermediates identified two peptide regions that were responsible for different translocation events, one at the N-terminus of TM2 and a second at the C-terminus of TM3.

During AQP4 synthesis, TM2 cotranslationally terminated translocation and spanned the membrane in its predicted orientation, whereas AQP1-TM2 failed to terminate translocation and passed into the ER lumen. Exchanging two hydrophilic residues at the N-terminus of AQP1-TM2 (N49/K51) with corresponding hydrophobic residues from AQP4 (M48/L50) converted AQP1-TM2 into an efficient stop transfer sequence and enabled it to cotranslationally span the membrane. We propose that these hydrophilic residues in AQP1 (N49/K51) decrease the length and/or hydrophobicity of TM2 below a critical threshold needed for recognition by ER translocation machinery. As a result, AQP1-TM2 transiently translocates into the ER lumen and must be repositioned into its proper membrane-spanning orientation during later events of protein (Figure 3, panels A, B).

A second difference in AQP biogenesis was due to the topogenic behavior of TM3. AQP4-TM3 efficiently reinitiated translocation and cotranslationally spanned the membrane in a type II topology. AQP1-TM3, however, failed to reinitiate translocation because it lacked signal sequence activity. In contrast to TM2, this difference in topogenic behavior was not caused by residues within the hydrophobic core of the TM segment, but rather was due to sequence differences within the hydrophilic TM3-4 peptide loop. Exchanging AQP1-TM3 C-terminus flanking residues (T118-I141) with corresponding residues from AQP4 (T115-V140) converted TM3 into a functional signal sequence and enabled it to reinitiate translocation and independently direct its proper type II topology.

Results of these experiments are illustrated in Figure 3 and show the initial topological orientation of AQP1 as the first three TM segments emerge from the ribosome at the ER membrane. TM2 from wild type (WT) AQP1 is initially located in the ER lumen, and TM3 exhibits a type I topology (panel A). When the mutations N49M/K51L were introduced, TM2 spanned the membrane but TM3 failed to reinitiate translocation (Panel B). Introducing N49M/K51L mutations together with exchanging the TM3-4 peptide loop resulted in all three TM segments independently and cotranslationally acquiring their correct topology (Panel C). Thus it was possible to convert AQP1 into a cotranslational biogenesis pathway similar to that of AQP4 by appropriate substitution of just two peptide regions. However, all chimeric substitutions that altered AQP1 biogenesis events, also markedly decreased water channel activity in microinjected *Xenopus laevis* oocytes. Taken together, these studies indicate that the complex folding pathway utilized by AQP1 is directly influenced by primary structural features that also play a role in later events of protein maturation and/or function.

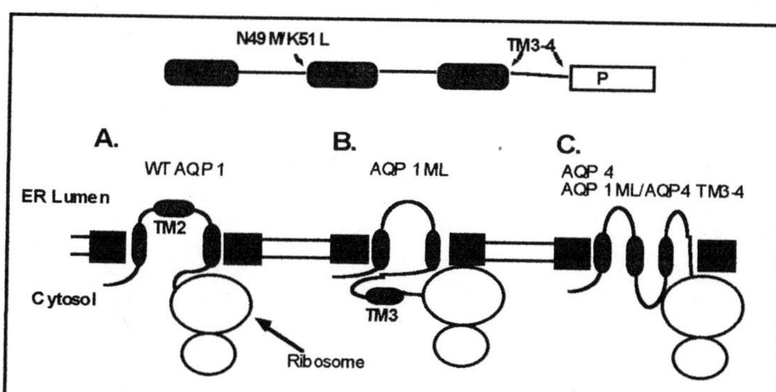

Figure 3. Topologic intermediates of WT and mutant AQP1 following synthesis of TM3. (top) N49M/K51L and TM3-4 refer to AQP1 residues changed to corresponding residues of AQP4. P refers to the prolactin-derived C-terminus translocation reporter. (bottom) Experimentally derived topologies of nascent chains for **(A)** WT AQP1, and **(B)** N49M/K51L (ML) AQP1. **(C)** TM2 N-terminus and TM3 C-terminus substitutions convert AQP1 topology to that observed for AQP4. Nascent polypeptides are shown attached to the ribosome. The precise relationship of TM helices to the Sec61 translocation channel (hatched boxes) is unknown.

4. CONCLUSIONS

Different aquaporin water channels acquire their final topology through markedly different translocation events. For AQP1 this involves a multistep assembly pathway in which initial protein topology, established cotranslationally, is dynamically modified by later steps of maturation. Importantly, the final topology of TMs 2-4 requires folding information encoded within C-terminus TM segments (e.g. TMs 5-6). It is possible that AQP1 topological maturation may involve helical packing of fully synthesized protein within the lipid bilayer. Alternatively, maturation could take place within the Sec61 translocon or at an intermediate stage while the protein is being transferred from the translocation channel into the bilayer. It is interesting that a similar topological reorientation within the translocon has recently been suggested for other native and engineered polytopic membrane proteins (Wilkinson *et al.,* 1996; Goder *et al.,* 1999), suggesting that AQP1 folding may represent a relatively common biogenesis variation. Understanding the role of ER translocation machinery in directing these different AQP folding pathways is now the major challenge.

Our studies also establish the structural basis for different AQP folding pathways and demonstrate that relatively minor sequence variations can profoundly influence protein folding events without altering topological outcome. We propose that the presence of alternate biogenesis pathways such as those observed for AQP1 and AQP4, provides a means by which proteins can acquire critical structural residues that would disrupt protein folding if biogenesis were restricted to a single rigid cotranslational pathway (Blobel 1980). Understanding the molecular mechanisms that underlie these folding pathways is of significant importance both in developing more accurate models to describe polytopic protein biogenesis as well as in predicting the effects of inherited mutations on aquaporin structure and function in disorders such as nephrogenic diabetes insipidus.

ACKNOWLEDGMENTS

This work was supported by grants from National Institutes of Health, GM53457 and DK51818.

REFERENCES

Blobel G 1980 Intracellular protein topogenesis *Proc Natl Acad Sci USA* **77**: 1496-1500

Cheng A van Hoek AN Yeager M Verkman AS and Mitra AK 1997 Three-dimensional organization of a human water channel *Nature* **387**: 627-630

Goder V Bieri C and Spiess M 1999 Glycosylation can influence topogenesis of membrane proteins and reveals dynamic reorientation of nascent polypeptides within the translocon *J Cell Biol* **147**: 257-265

Preston G Jung J Guggino W and Agre P 1994 Membrane topology of aquaporin CHIP - analysis of functional epitope-scanning mutants by vectorial proteolysis *J Biol Chem* **269**: 1668-1673

Shi L-B Skach W Ma T and Verkman AS 1995 Distinct biogenesis mechanisms for water channels MIWC and CHIP28 at the endoplasmic reticulum *Biochemistry* **34**: 8250-8256

Skach W Shi L-B Calayag MC Frigeri A Lingappa V and Verkman AS 1994 Biogenesis and transmembrane topology of the CHIP28 water channel in the endoplasmic reticulum *J Cell Biol* **125**: 803-815

Walz T Hiraim T Murata K Heymann JB Mitsuoka K Fujiyoshi Y Smith BL Agre P and Engel A 1997 The three-dimensional structure of aquaporin-1 *Nature* **387**: 624-627

Wilkinson B Critchley A and Stirling C 1996 Determination of the transmembrane topology of yeast Sec61p an essential component of the endoplasmic reticulum translocation complex *J Biol Chem* **271**: 25590-25597

2

FUNCTION, PHYSIOLOGICAL ROLES AND REGULATION OF MAMALIAN AQUAPORINS

2.1 Mammalian physiology

THE KIDNEY IN THE INNER EAR

[1,2]Eric Beitz, [3]Hidetaka Kumagami, [3]J. Peter Ruppersberg, [2]Joachim E. Schultz and [1]Peter Agre

[1]Johns Hopkins University, School of Medicine, Department of Biological Chemistry, 725 N. Wolfe Street, Baltimore, MD 21205, USA: [2]University of Tübingen, School of Pharmacy, Morgenstelle 8, D-72076 Tübingen, FRG: [3]University of Tübingen, Department of Physiology II, Gmelinstr. 4, D-72076 Tübingen, FRG. E-mail: ebeitz@jhmi.edu

1. INTRODUCTION

The mediation of fast osmotically driven water transport is an important function of aquaporins in many tissues, such as lung, brain and kidney (Beitz and Schultz 1999). In addition to these more obvious examples a closed fluid compartment of only 40µl in volume located in the human inner ear needs a tight ion and volume regulation in order to provide optimal conditions for the highly sensitive perception of sound and gravity/acceleration. It bears the so-called endolymph which is an extracellular fluid with high potassium and low sodium content (120-150mM and 5-15mM, respectively). This fluid bathes the apical portion of the sensory hair cells of the cochlea and the vestibulum, whereas the basolateral parts are exposed to the low potassium perilymph. This arrangement establishes a potassium gradient, which is necessary for a rapid signal transmission in the µsec range. Therefore mechanically gated potassium channels are in place in the stereocilia of the hair cells. A dysregulation of the endolymph volume leading to hair cell damage due to an overpressure in the endolymph compartment (endolymphatic hydrops) is the most favored hypothesis underlying the pathology of Ménière's disease. This syndrome is characterized by hearing loss, imbalance, tinnitus and a sensation of fullness in the inner ear.

2. MOLECULAR BASE OF WATER REGULATION IN THE INNER EAR

We prepared poly(A)$^+$ RNA from functionally and regionally distinct rat inner ear tissues; endolymphatic sac, stria vascularis and Reissner's membrane (these are involved in the endolymph regulation), plus vestibulum and organ of Corti (the sensory tissues). This mRNA was analyzed qualitatively and quantitatively by RT-PCR using specific primers for several genes involved in fluid regulation (Beitz *et al.*, 1999).

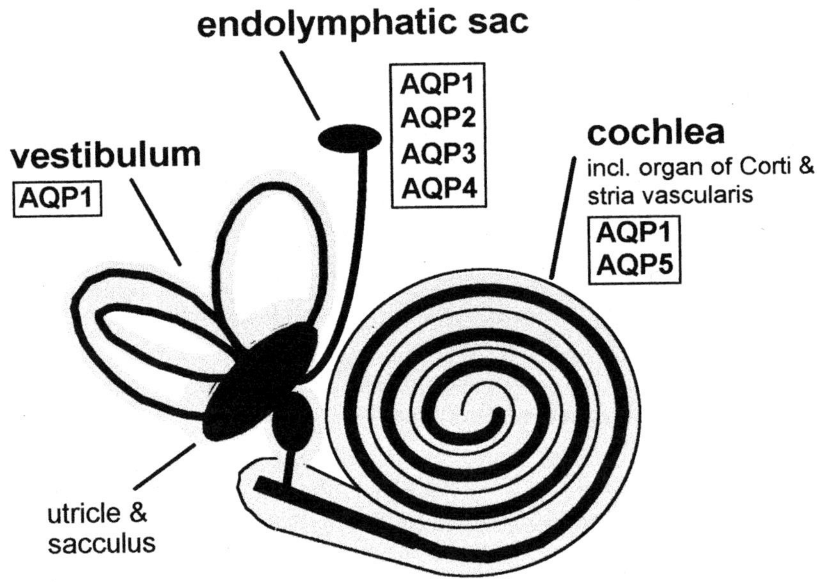

Figure 1. Expression pattern of aquaporins in the inner ear of the rat. The endolymph compartment is drawn in solid black, the perilymph is shown in gray.

2.1 Aquaporin Expression Pattern in the Inner Ear

AQP1 is ubiquitous. AQP2, AQP3, and AQP4 are present in the endolymphatic sac. AQP5 appears in the organ of Corti and Reissner's membrane. The expression pattern is depicted in Figure 1. Most striking was the identification of transcripts of AQP2 in the endolymphatic sac, which is thought to regulate the endolymph volume by re-absorption. AQP2 has so far been considered to be locally restricted to the principal cells of the renal collecting duct where its transcription as well as its trafficking to the apical membrane is regulated by vasopressin V_2 receptors via cAMP signaling. AQP3 and AQP4 are also present in principle cells yet are localized to the basolateral membrane. These similarities on the molecular level are further stressed by ultrastructural morphological similarities between the epithelia of the kidney collecting duct and the inner ear endolymphatic sac.

2.2 Expression of the V_2 Receptor and Vesicle Transport Components in the Endolymphatic Sac

Once stimulated by vasopressin the trafficking of AQP2 bearing vesicles requires an apparatus of guidance and regulation proteins, such as the vesicle associated membrane protein type 2 (VAMP-2), syntaxin-4 and the small GTPase Rab3a. All components were easily amplified from endolymphatic sac cDNA suggesting that the regulation of the endolymph water homeostasis may occur by similar mechanisms as in the collecting duct principal cells. In addition, our results reveal another player in the endolymph regulation, the calcium-sensing receptor (CaSR) which is also present in kidney AQP2-vesicles. This receptor might detect changes in the lumenal Ca^{2+} concentration and subsequent modulate membrane trafficking.

3. THE EFFECT OF VASOPRESSIN ON THE ENDOLYMPHATIC SAC

Ultrastructural analysis of the endolymphatic sac epithelium shows two types of cells, named mitochondrial- and ribosomal-rich cells. These are reminiscent of the principal cells and intercalated cells of the collecting duct. Using organotypical cultures of rat postnatal day 4 endolymphatic sacs, the effect of vasopressin on the membrane turnover was studied (Kumagami *et al.*, 1998). The endocytosis process was imaged by fluorescent dye containing fluid-phase. In contrast to the kidney principal cells, vasopressin and cAMP stimulating agents (cholera toxin, forskolin) inhibited membrane turnover in the mitochondria rich cells of the endolymphatic sac.

The inverted effect of vasopressin on the inner ear may result in decreased fluid reabsorption facilitating endolymphatic hydrops. Indeed, 7 out of 10 guinea pigs treated chronically with a systemic dose of vasopressin (0.2 units/g) showed moderate to severe endolymphatic hydrops accompanied by massive balance problems. None of the control animals was affected. As a measure of hydrops the distention of Reissner's membrane was taken. This membrane consists of only two cell layers and separates the endolymph from the perilymph throughout the cochlea.

4. CONCLUSION AND OUTLOOK

Aquaporins are most likely involved in the regulation of inner ear fluids. Especially the vasopressin regulated AQP2 system of the endolymphatic sac might be pivotal for a proper function of hearing and balance. A physiological function of an inverted mechanism of water reabsorption in the inner ear may be in the protection of the sensory system by maintenance of the inner ear pressure during hypovolemia. On the other hand, abnormally high plasma levels of vasopressin, which are present in some patients suffering from Ménière's disease, may cause endolymphatic hydrops.

The biochemical and physiological study of inner ear tissues is hampered by the difficult preparation of samples from a tight bone capsule. Furthermore, due to the tiny size and complexity of the inner ear the yield of pure tissue samples is very scarce. To overcome those problems we established a conditionally immortalized cell line from the endolymphatic sac of postnatal day 12 mice. These cells are going to be used for further studies of the AQP2 vesicle transport regulation in the endolymphatic sac.

ACKNOWLEDGEMENTS

This work was supported by the Deutsche Forschungsgemeinschaft (Klinische Forschergruppe Hörforschung and Schwerpunkt Molekulare Sinnesphysiologie) and by the Fonds der chemischen Industrie.

REFERENCES

Beitz E and Schultz JE 1999 The mammalian aquaporin water channel family: a promising new drug target *Curr Med Chem* **6**: 465-475

Beitz E Kumagami H Krippeit-Drews P Ruppersberg JP Schultz JE 1999 Expression pattern of aquaporin water channels in the inner ear of the rat *Hearing Res* **132**: 76-84

Kumagami H Loewenheim H Beitz E Schwartz H Yamashita K Schultz JE Paysan J Zenner H-P Ruppersberg JP 1998 The effect of anti-diuretic hormone on the endolymphatic sac of the inner ear *Pflügers Arch – Eur J Physiol* **436**: 970-975

2.2 Mammalian physiology

RENAL AQUAPORIN EXPRESSION IN AGING RAT

Sophie Combet, Nancy Geffroy, Laurent Teillet, Véronique Berthonaud, Renée Gobin, Alexis Beckmann, Bruno Corman and Jean-Marc Verbavatz
DBCM/SBCe, CEA/Saclay, F-91191 Gif-sur-Yvette CEDEX, France. E-mail: scombet@cea.fr

1. INTRODUCTION

Aging is frequently associated with impaired ability to retain water. This is accounted for by changes in thirst and altered kidney concentrating ability. The mechanisms involved in this process hasve been studied in many different models. In humans, the thirst threshold appears to increase with age which results in impaired water homeostasis (see Vellas *et al.*, 1998 for a review). In addition, a defect in renal water reabsorption has been described. In rodents, in particular, aging is characterized by polyuria and decreased urinary osmolality (Corman & Michel, 1987). Because water reabsorption in kidney is regulated by vasopressin (AVP), these changes can be due to impaired AVP secretion or to inappropriate kidney response to AVP.

A defect in AVP release in the course of aging has indeed been reported in some strains of rats (Zbuzek *et al.*, 1983, Swenson *et al.*, 1997). However, polyuria was also observed in strains of rats showing no change in plasma AVP concentration, suggesting that kidney structure or function was involved (Geelen & Corman, 1992). Aging is frequently associated with progressive nephrosis and loss of nephrons and results in hyperfiltration of remaining glomeruli which could also account for impaired water reabsorption. Nevertheless, a defect in water reabsorption was still observed in 30 mo-old female WAG/Rij rats which remained free of glomerulosclerosis. In kidney, the bulk of water reabsorption is mediated by aquaporin-1 (AQP1) in proximal tubules and thin descending limbs of Henle's loop. The final concentration of urine occurs in the collecting duct,

Molecular Biology and Physiology of Water and Solute Transport
Edited by Hohmann and Nielsen, Kluwer Academic/Plenum Publishers, New York, 2000.

where AVP is involved in the regulation of the water permeability of principal cells. Whereas aquaporin-3 and –4 (AQP3, AQP4) are constitutively localized to principal cell basolateral plasma membranes, aquaporin-2 (AQP2) is inserted into the apical plasma membrane of these cells, following AVP-mediated cAMP accumulation. Thus, a defect in aging kidney water reabsorption might result from impaired AQP expression or localization or from ineffective AVP-regulated insertion of AQP2 into principal cell plasma membranes. The purpose of this work was to investigate this hypothesis in WAG/Rij female rats.

2. RESULTS

Urinary concentration, AVP-induced cAMP accumulation and AQP expression and localization were studied in 10, 30 mo-old rats fed ad libitum and in 30 mo-old rats that were food restricted by 30% for 20 months. The survival rate of food restricted animals was similar to that of rats fed ad libitum.

2.1 Urinary concentration

Urinary volume and osmolality were 3.7 ± 0.4 ml/24h and 2389 ± 112 mosm/kg H_2O in adult rats and 8.3 ± 1.7 ml/24h and 1339 ± 157 mosm/kg H_2O in 30 mo-old rats, demonstrating age-related polyuria. Interestingly, chronic food restriction largely prevented polyuria (3.0 ± 0.4 ml/24h and 1960 ± 154 mosm/kg in 30 mo-old restricted rats).

2.2 AVP-mediated cAMP accumulation

Vasopressin V2 receptor mRNA, quantified by RT-PCR in the whole papilla or in dissected collecting tubules, did not significantly differ with age but ^{3}H-vasopressin binding sites were reduced in membrane papilla of 30 mo-old (51.7 ± 49 fmol/mg protein) as compared to the 10 mo-old rats (81.9 ± 9.4 fmol/mg proteins). However, this decrease in V2 receptor density did not affect cAMP content of the papilla (1.51 ± 0.10 pmol/mg in 10 mo-old animals and 1.47 ± 0.21 pmol/mg in 30 mo-old animals). Thus, these results suggest that the concentrating defect in 30 mo-old animals fed ad libitum is independent from changes in AVP-mediated cAMP accumulation.

On the other hand, papillary osmolality (1084 ± 103 mosm/kg H_2O in 10 mo-old animals), was significantly reduced in senescent animals (645 ± 49 and 678 ± 35 mosm/kg H_2O in rats fed ad libitum and food restricted respectively). This decrease in interstitial osmolality, observed in both senescent groups may not account per se for the polyuria observed only in animals fed ad libitum, however.

2.3 AQP expression

AQP1,2,3,4 expression was quantified by Western blotting analysis in the different regions of the kidney. The expression of AQP1 and AQP4 in cortex and medulla was mostly unchanged in all groups (not shown). In contrast, AQP2 and AQP3 expression was down regulated by 80% and 50%, respectively, in inner medulla of senescent animals fed ad libitum (Fig. 1), and to a small extent in outer medulla, but not in renal cortex (not shown). Similarly, AQP2 expression was markedly decreased in the inner medulla of food restricted animals, although to a slightly lesser extent, whereas AQP3 expression remained stable in food restricted 30 mo-old rats (Fig. 2).

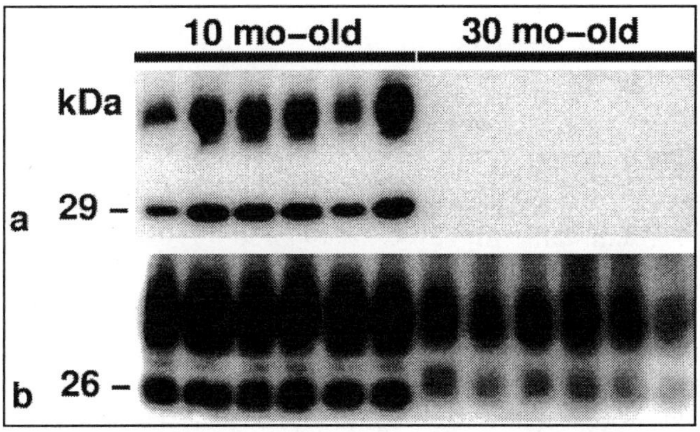

Figure 1. Semi-quantitative Western blotting of AQP2 (a) and AQP3 (b) in kidney inner medulla of 10 mo-old and 30 mo-old rats fed ad libitum shows a dramatic decrease in AQP2 expression and a small but significant decrease in staining for AQP3 in senescent animals.

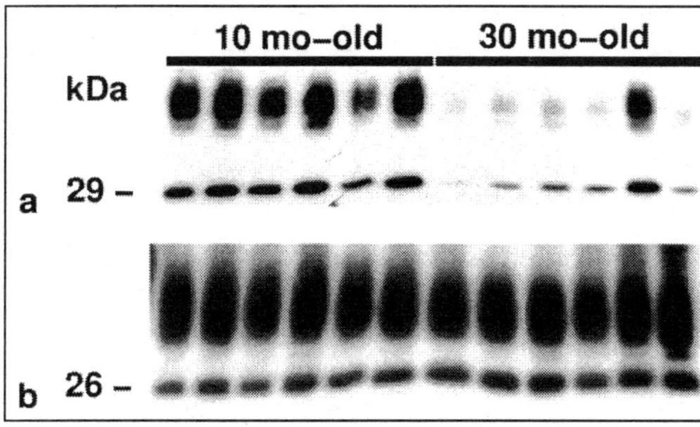

Figure 2. Semi-quantitative Western blotting of AQP2 (a) and AQP3 (b) demonstrates a marked decrease in AQP2 expression in food-restricted 30 mo-old animals compared to 10 mo-old adults in kidney inner medulla. No decrease in AQP3 expression was detected.

2.4 AQP localization

Indirect immunofluorescence of kidney sections showed a similar distribution of AQP1,3,4 in all groups and kidney regions (not shown). AQP2 staining was markedly reduced in the inner medullary collecting duct of all 30 mo-old animals (Fig. 3b,c), compared to 10 mo-old rats (Fig. 3a). In addition, AQP2 was improperly distributed throughout principal cell cytoplasm in 30 mo-old animals fed ad libitum (Fig. 3b). The absence of AQP2 in inner medullary principal cell plasma membranes of 30 mo-old rats fed ad libitum was confirmed by immunogold labeling at the EM level (not shown). In contrast, AQP2 was properly localized to the apical region of principal cells in food restricted 30 mo-old animals like in 10 mo-old rats (Fig. 3c).

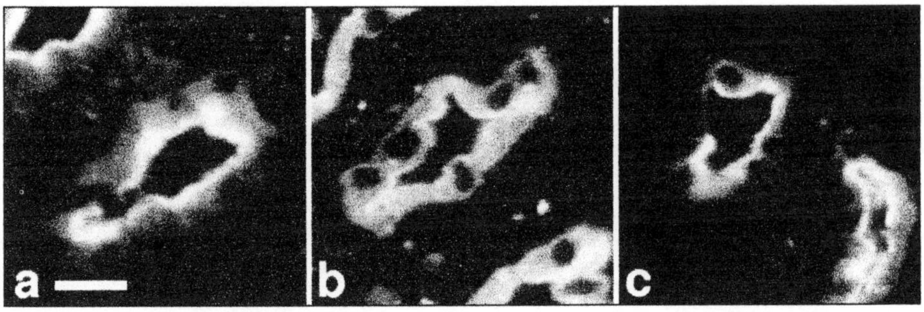

Figure 3. Immunofluorescence staining of AQP2 in (**a**) 10 mo-old, (**b**) 30 mo-old rats fed ad libitum and (**c**) food-restricted senescent animals shows a marked intracellular redistribution of AQP2 in 30 mo-old animals fed ad libitum only. Bar=20μm.

3. DISCUSSION

Our results confirm the AVP-independent polyuria previously observed in aging WAG/Rij female rats (Geelen and Corman, 1992). Interestingly, food-restriction in this strain of lean rats can prevent this defect.

In addition, our results demonstrate that aging is associated to a marked decrease in inner medullary AQP2 expression, which is independent from vasopressin-mediated intracellular cAMP accumulation but correlated to a significant decrease in papillary osmolality. The expression of other aquaporins is mostly unchanged in the course of aging. However, the decrease in inner medullary osmolality and AQP2 expression are not sufficient to account for the polyuria observed in 30 mo-old animals fed ad libitum. As observed in food-restricted rats, it might only account for a very small decrease in urinary concentration.

On the other hand, the polyuria observed only in 30 mo-old rats fed ad libitum, is associated to a marked intracellular redistribution of AQP2, which was not observed in food restricted animals.

Altogether, these observations suggest that AQP2 expression and AQP2 localization in the apical region of principal cells are regulated by two independent mechanisms in this experimental model.

ACKNOWLEDGMENTS

This work was supported in part by TMR FMRX-CT97-0128 from the European Commission.

REFERENCES

Corman B Michel JB 1987 Glomerular filtration renal blood flow and solute excretion in conscious aging rats *Am J Physiol* **253**: R555-R560

Geelen G and Corman B 1992 Relationship between vasopressin and renal concentrating ability in aging rats *Am J Physiol* 262:R826-R833

Swenson K Sands J Jacobs J and Sladeck C 1997 Effect of aging on vasopressin and aquaporin responses to dehydration in Fisher 344-Brown-Norway F1 rats *Am J Physiol* **273**: R35-40

Vellas B Albarede JL and Garry PJ 1998 Hydration and aging Paris & New York: Serdi & Springer

Zbuzek VK Zbuzek V and Wu W 1983 The effect of aging on vasopressin system in Fisher 344 rats *Exp Gerontol* **18**: 305-311

2.3 Mammalian physiology

EXPRESSION AND REGULATION OF AQUAPORIN-1 AND ENDOTHELIAL NITRIC OXIDE SYNTHASE IN RELATIONSHIP WITH WATER PERMEABILITY ACROSS THE PERITONEUM

[1]Olivier Devuyst, [2]Sophie Combet, [3]Jean-Luc Balligand, [1]Eric Goffin and [2]Jean-Marc Verbavatz
Div. of [1]Nephrology and [3]Pharmacology and Therapeutics, Université catholique de Louvain Medical School, Brussels, Belgium; [2]Dept. of Cell Biology, CEA Saclay, Gif-sur-Yvette, France. E-mail: devuyst@nefr.ucl.ac.be

1. PERITONEAL DIALYSIS AND THE PERITONEAL MEMBRANE

Over the past 25 years, peritoneal dialysis (PD) has become an established treatment for end-stage renal disease (ESRD), accounting for about 15% of the total number of patients on dialysis worldwide (Gokal and Mallick 1999). To perform PD, a catheter is inserted into the peritoneal cavity and a solution (dialysate) is infused through it. The dialysate remains in the peritoneal cavity for some hours (dwell time), and is then drained out before fresh dialysate is reinfused (the drainage-reinfusion procedure is called an exchange). Four exchanges are performed daily in continuous ambulatory peritoneal dialysis (CAPD); a mechanical device can also be used to infuse and drain the dialysate during the night, in combination with one or two day-time exchanges (automated peritoneal dialysis, APD) (Gokal and Mallick 1999).

Dialysis occurs during the stagnation time, when solutes and water move from peritoneal capillary blood to the dialysate, by diffusive and convective transports through the peritoneal membrane (PM). Whereas solutes diffuse according to their concentration gradient, fluid movements are osmotic-driven and made possible by the presence of an osmotic agent, i.e. glucose, in the dialysate. The PM consists of three anatomical barriers: the capillary wall, the interstitium and the mesothelium. The parameters influencing solutes and water movements are i) the exchange frequency and volume of the dialysate; ii) the peritoneal blood flow; iii) the effective peritoneal vascular surface area (EPSA); and iv) the PM permeability (Rippe and Krediet 1994).

The PM permeability can be evaluated with the peritoneal equilibration test (PET) which is based on the rate of small solutes transport during an exchange (Twardowski *et al.,* 1987). Patients with a high permeability membrane exhibit a rapid solute diffusion and a fast dissipation of the osmotic gradient (due to glucose diffusion from the dialysate to the blood), which might impair their ability to achieve a sustained ultrafiltration (UF). In contrast, patients with a low permeability membrane exhibit a decreased solute diffusion but keep a favourable osmotic gradient for UF (Leypoldt and Mistry 1994).

2. THE PM CAPILLARY ENDOTHELIUM CONTAINS ULTRA SMALL PORES SELECTIVE FOR WATER

Solutes transports during PD are best explained by the 3-pore model based on computer simulations (Rippe *et al.,* 1991). According to this model, the major transport barrier of the PM is the capillary endothelium, which contains three types of pores. The small pores (radius between 4 and 5 nm) are probably the clefts located between endothelial cells; they allow the diffusion of low-molecular weight solutes, including glucose and albumin. The large pores (radius > 15 nm), thought to correspond to the venular interendothelial gaps, allow the diffusion of larger solutes such as immunoglobulins. The existence of a third, ultrasmall type of pore (radius < 0.3 nm) which allows the diffusion of water but not that of solutes, is based not only on computer simulations, but also on clinical observations. Such ultrasmall pores explain indeed the dissociation between sodium and water transport observed during PD with hypertonic dwells: during the first hour of the dwell, the dialysate-over-plasma ratio of sodium falls markedly (that phenomenon is called "sodium sieving"), as a result of free water diffusion within the peritoneal cavity (Rippe and Krediet 1994, Twardowski *et al.,* 1987). These ultrasmall pores might account for the effectiveness of glucose as an osmotic agent in PD, despite its small size (radius 0.29 nm). The fact that 50 % of the UF during an hypertonic dwell occurs through the ultrasmall pores illustrates their major clinical importance in PD patients (Leypoldt and Mistry 1994).

3. AQP1 IS THE MOLECULAR COUNTERPART OF THE ULTRA SMALL PORE

In addition to its abundant expression in proximal convoluted tubules and descending thin limbs of Henle's loop of the kidney (Nielsen *et al.,* 1993a), aquaporin-1 (AQP1) is located in the apical and basolateral membranes of endothelial cells lining non-fenestrated capillaries in numerous tissues (Nielsen *et al.,* 1993b) including the PM (Carlsson *et al.,* 1996). Studies in rat and human PM have confirmed that AQP1 is located in endothelial cells lining capillaries, venules and small veins, but not in small-size arteries (Carlsson *et al.,* 1996, Pannekeet *et al.,* 1996, Devuyst *et al.,* 1998, Combet *et al.,* 1999).

Several lines of evidence support the hypothesis that AQP1 is the molecular counterpart of the ultrasmall pore of the PM. First, the distribution of AQP1 in the capillary endothelium is consistent with the predicted topology of the pore (Rippe *et al.,* 1991). Second, AQP1 is a constitutively expressed and specific pore for water, as demonstrated in *Xenopus* oocytes and proteoliposomes (Agre *et al.,* 1998). Third, ultrastructural informations about the pore formed by AQP1 (Walz *et al.,* 1997) fit the postulated ultrasmall pore size. Fourth, functional studies conducted in rats have shown that peritoneal water permeability is significantly inhibited by $HgCl_2$, a well-known (but aspecific) inhibitor of AQP1 (Carlsson *et al.,* 1996). Finally, elegant studies performed in knockout mice provided evidence that osmotically driven water transport across the PM is significantly decreased by AQP1 deletion, whereas it remains unchanged in case of AQP4 deletion (Yang *et al.,* 1999). It must be noted that staining for AQP1 has also been observed in rat and human mesothelium (Pannekeet *et al.,* 1996, Devuyst *et al.,* 1998), where its functional role remains obscure (Rippe and Krediet 1994).

Could aquaporins different from AQP1 play a role in PM permeability for water ? The only aquaporin that has been consistently located in the capillary endothelium of the PM is AQP1 (Carlsson *et al.,* 1996, Pannekeet *et al.,* 1996, Devuyst *et al.,* 1998, Combet *et al.,* 1999). The vasopressin-sensitivity of AQP2, as well as the $HgCl_2$-insensitivity of AQP4 render unlikely their participation in the water permeability of the PM (Devuyst *et al.,* 1998, Yang *et al.,* 1999). Expression of AQP3 in the PM has been suggested on the basis of a RT-PCR study (Akiba *et al.*, 1997), but could be explained by cross-contamination with erythrocytes (Roudier *et al.*, 1998). Furthermore, we were unable to document any expression of AQP3 and AQP4 in the capillary endothelium of rat peritoneum (Combet *et al.*, 1999). Thus, a significant intervention of an aquaporin different from AQP1 in water permeability of the PM appears unlikely.

4. AQP1 AND ENOS ARE DIFFERENTIALLY REGULATED IN CLINICAL CONDITIONS ASSOCIATED WITH A LOSS OF UF

A decrease in UF might occur in acute or chronic situations, such as peritonitis or long-term PD. The UF failure, which has severe clinical consequences for PD patients because of fluid overload (Gokal and Mallick 1999), is due to either an increased EPSA (which results in faster glucose absorption and dissipation of the osmotic gradient) or a functional alteration of the pore itself (Goffin *et al.*, 1999). In recent studies, we investigated the potential contributions of AQP1 and endothelial nitric oxide synthase (eNOS) in UF failure.

Because nitric oxide (NO) is a key mediator for vascular tone and permeability (Kone 1997), it has been suggested that NO is critical for regulating PM permeability. Indeed, addition of the NO donor nitroprusside to the dialysate increases EPSA and PM permeability (Nolph *et al.*, 1977), whereas addition of NOS inhibitors such as L-NAME increases UF in animal models (Breborowicz *et al.*, 1998). Furthermore, the three NOS isoforms – the neuronal NOS (nNOS, NOS1), the inducible NOS (iNOS, NOS2), and the endothelial NOS (eNOS, NOS3) - are expressed in human and rat PM, and a specific NOS activity has been assayed in the PM (Combet *et al.*, 2000).

Although both proteins are selectively expressed in endothelial cells, two major differences between AQP1 and eNOS were detected in the PM. First, eNOS is located in the endothelium lining all types of blood vessels including arterioles, whereas AQP1 is restricted to capillaries, venules and small veins (Devuyst *et al.*, 1998, Combet *et al.*, 1999). Second, the expression of eNOS is markedly increased in extracts from inflammatory PM, whereas AQP1 expression remains stable (Devuyst *et al.*, 1998). The differential regulation of AQP1 and eNOS was further illustrated in a long-term PD patient characterized by UF failure due to abolition of transcellular water permeability: despite these functional changes, both expression and localization of AQP1 were apparently normal in this patient, whereas eNOS was clearly upregulated (Goffin *et al.,* 1999).

A model of acute peritonitis in rats allowed us to further investigate expression of AQP1 and NOS isoforms, in parallel with permeability and structure of the PM and NOS activities (Combet *et al.,* 1999). Compared with controls, rats with peritonitis had an increased permeability for small solutes and a loss of UF (exactly like PD patients with acute peritonitis), but also a 10-fold increase in NOS activity, mediated by both iNOS and eNOS. In contrast, despite the loss of water permeability in presence of an effective osmotic gradient, there was neither quantitative nor qualitative evidence for a decrease in AQP1 expression in rats with peritonitis. These data suggest that increased NO level, secondary to NOS upregulation, is a major regulator of water permeability across the PM, because it induces a vasodilation (thus increasing EPSA and glucose reabsorption) and promotes angiogenesis in the PM (Combet *et al.,* 1999). However, these results might also suggest that increased NO could modify plasma membrane proteins such as AQP1 and interfere with their functions in endothelial cells (Goffin *et al.,* 1999). Such interaction, probably favoured by the co-localization of AQP1 and eNOS in caveolae (Carlsson *et al.,* 1996, Devuyst *et al.,* 1998), is also suggested by the intense reactivity for nitrotyrosine in the capillary endothelium of rats with peritonitis (Combet *et al.,* 1999).

5. FUNCTIONAL INTERACTION BETWEEN NO AND AQP1

To investigate if NO, either as peroxynitrite ($ONOO^-$) or free radical ($NO^{\cdot}$), might interferes with AQP1 function, preliminary experiments were performed to measure a potential inhibition of water permeability after exposure of human AQP1-expressing *Xenopus laevis* oocytes to NO donors. As previously reported (Abrami *et al.,* 1995), AQP1-expressing oocytes exhibited a 10-fold increase in water permeability compared to water- or non-injected oocytes. Incubation with 3-morpholinosydnonimine (SIN-1), that produces $ONOO^-$ and $NO^{\cdot}$, decreased significantly water permeability (39% inhibition, $N = 10$, $P = 0.001$). This inhibition was abolished by superoxide dismutase (SOD) or the reducing agent β-mercaptoethanol. Therefore, SIN-1 appears as an inhibitor of AQP1. As this inhibition is abolished by SOD, this effect is likely mediated by peroxynitrite. These data were confirmed by experiments with S-nitroso-N-acetylpenicillamine (SNAP) which, in oxidizing conditions, reproduced a significant inhibition of water permeability (28% inhibition, $N = 11$, $P < 0.02$). N-acetylpenicillamine (NAP), which does not produce NO, had no effect on water permeability of AQP1. These results confirm that NO species with oxidizing capacity, probably peroxynitrite, exert an inhibitory effect on AQP1-mediated water permeability.

6. CONCLUSIONS

Transcellular water permeability is an essential component of PD. Several lines of evidence suggest that AQP1 is the molecular counterpart of the ultrasmall pore located in the endothelium lining PM capillaries. Common clinical condition associated with loss of UF and absence of transcellular water permeability are characterized by an increased NOS activity in the PM, due to the upregulation of eNOS, whereas AQP1 expression remains apparently unchanged. Since preliminary data show that peroxynitrite is a potential inhibitor of AQP1-mediated permeability, future studies are required to determine the potential interactions between NO and AQP1 within endothelial cells.

ACKNOWLEDGMENTS

The work presented here was supported by the Fonds National de la Recherche Scientifique (9.4540.96), the Fonds de la Recherche Scientifique Médicale (3.4566.97), and Baxter Healthcare.

REFERENCES

Abrami L Tacnet F and Ripoche P 1995 Evidence for a glycerol pathway through aquaporin 1 (CHIP28) channels *Pflugers Arch* **430**: 447-458

Agre P Bonhivers M and Borgnia MJ 1998 The aquaporins blueprints for cellular plumbing systems *J Biol Chem* **273**: 14659-14662

Akiba T Ota T Fushimi K Tamura H Hata T Sasaki S and Marumo F 1997 Water channel AQP1 3 and 4 in the human peritoneum and peritoneum membrane *Adv Perit Dial* **13**: 3-6

Breborowicz A Wieczorowska-Tobis K Korybalska K Polubinska A Radkowski M and Oreopoulos DG 1998 The effect of a nitric oxide inhibitor (L-NAME) on peritoneal transport during dialysis in rats *Perit Dial Int* **18**: 188-192

Carlsson O Nielsen S Zakaria ER and Rippe B 1996 In vivo inhibition of transcellular water channels (aquaporin-1) during acute peritoneal dialysis in rats *Am J Physiol* **271**: H2254-H2262

Combet S Balligand J-L Lameire N Goffin E and Devuyst O 2000 A specific method for measurement of nitric oxide synthase enzymatic activity in peritoneal biopsies *Kidney Int* **57**: 332-338

Combet S van Landschoot M Moulin P Piech A Verbavatz J-M Goffin E Balligand J-L Lameire N and Devuyst O 1999 Regulation of aquaporin-1 and nitric oxide synthase isoforms in a rat model of acute peritonitis *J Am Soc Nephrol* **10**: 2185-2196

Devuyst O Nielsen S Cosyns J-P Smith BL Agre P Squifflet J-P Pouthier D and Goffin E 1998 Aquaporin-1 and endothelial nitric oxide synthase expression in capillary endothelia of human peritoneum *Am J Physiol* **275**: H234-H242

Goffin E Combet S Jamar F Cosyns J-P and Devuyst O 1999 Expression of aquaporin-1 in a long-term peritoneal dialysis patient with impaired transcellular water transport *Am J Kidney Dis* **33**: 383-388

Gokal R and Mallick NP 1999 Peritoneal dialysis *Lancet* **353**: 823-828

Kone BC 1997 Nitric oxide in renal health and disease *Am J Kidney Dis* 30: 311-333

Leypoldt JK and Mistry CD 1994 Ultrafiltration in peritoneal dialysis In *The Textbook of Peritoneal Dialysis* (R Gokal and KD Nolph eds) Kluwer Academic Publishers Dordrecht pp 135-160

Nielsen S Smith BL Christensen EI and Agre P 1993b Distribution of the aquaporin CHIP in secretory and resorptive epithelia and capillary endothelia *Proc Natl Acad Sci USA* **90**: 7275-7279

Nielsen S Smith BL Christensen EI Knepper MA and Agre P 1993a CHIP28 water channels are localized in constitutively water-permeable segments of the nephron *J Cell Biol* **120**: 371-383

Nolph KD Ghods A Brown PA Miller F Harris PD Pyle K and Popovich R 1977 Effects of nitroprusside on peritoneal mass transfer coefficients and microvascular physiology *ASAIO Trans* **23**: 210-218

Pannekeet MM Mulders JB Weening JJ Struijk DG Zweers MM and Krediet RT 1996 Demonstration of aquaporin-chip in peritoneal tissue of uremic and CAPD patients *Perit Dial Int* **16**: S54-S57

Rippe B and Krediet RT 1994 Peritoneal physiology - transport of solutes In *The Textbook of Peritoneal Dialysis* (R Gokal and KD Nolph eds) Kluwer Academic Publishers Dordrecht pp 69-113

Rippe B Stelin G and Haraldsson B 1991 Computer simulations of peritoneal fluid transport in CAPD *Kidney Int* **40**: 315-325

Roudier N Verbavatz J-M Maurel C Ripoche P and Tacnet F 1998 Evidence for the presence of aquaporin-3 in human red blood cells *J Biol Chem* **273**: 8407-8412

Twardowski ZJ Nolph KD Khanna R Prowant BF Ryan LP Moore HL and Nielsen MP 1987 Peritoneal equilibration test *Perit Dial Bull* **7**: 138-147

Walz T Hirai T Murata JB Heymann JB Mitsuoka K Fujiyoshi Y Smith BL Agre P and Engel A 1997 The three-dimensional structure of aquaporin-1 *Nature* **387**: 624-627

Yang B Folkesson HG Yang J Matthay MA Ma T and Verkman AS 1999 Reduced osmotic water permeability of the peritoneal barrier in aquaporin-1 knockout mice *Am J Physiol* **276**: C76-C81

2.4 Mammalian physiology

REGULATION OF RENAL AQUAPORINS AND SODIUM TRANSPORTERS DURING VASOPRESSIN-ESCAPE IN THE RAT

[1,3]Carolyn A. Ecelbarger, [1]Takashi Murase, [2]Søren Nielsen, [1]Ying Tian, [3]Chung-Lin Chou, [1]Beatriz Olson, [3]Mark A. Knepper and [1]Joseph G. Verbalis

[1]Division of Endocrinology and Metabolism, Department of Medicine, Bldg D, Rm 232, Georgetown University, 4000 Reservoir Rd, NW, Washington, DC 20007, [2] Department of Cell Biology, Institute of Anatomy, University of Aarhus, Aarhus C, Denmark, [3]Laboratory of Kidney and Electrolyte Metabolism, National Heart, Lung and Blood Institute, National Institutes of Health, Bethesda, MD 20892. E-mail: ecelbarc@gunet.georgetown.edu

1. INTRODUCTION

Several clinical conditions, e.g., congestive heart failure and cirrhosis are associated with inappropriately elevated vasopressin levels relative to serum osmolality (Bichet *et al.,* 1992). In addition, ectopic production of vasopressin-like substances (usually by tumors) can produce inappropriate antidiuresis. This disorder is known as the Syndrome of Inappropriate Antidiuretic Hormone (vasopressin) secretion (SIADH). As expected, inappropriately elevated plasma vasopressin levels will often result in water retention and hyponatremia. Eventually patients go through a physiological process known as "vasopressin-escape" in which they begin to excrete a higher volume of more dilute urine, despite continually high levels of circulating vasopressin. This escape process helps to reduce water retention and more symptomatic hyponatremia. Until recently little was known about the molecular mechanisms underlying escape.

2. AN ANIMAL MODEL FOR VASOPRESSIN ESCAPE

Previously, Verbalis (1994) developed and characterized a rat model of hyponatremia that closely resembled clinical SIADH. For the model, rats are implanted with osmotic mini-pumps to administer dDAVP (1-deamino-[8-D-argininine]-vasopressin), a V2-selective vasopressin agonist, while being offered a diet high in water content (liquid formula diet). The rats consume a high volume of water in order to get the calories they need. This model has virtually 0% mortality and low morbidity associated with it. The rats initially retain water and become rapidly hyponatremic. However, after about 2 days, they begin to "escape" by excreting significantly higher volumes of increasingly more dilute urine. For studies described in this chapter (Ecelbarger *et al.,* 1997a, b, 1998c, 1999, Murase *et al.,* 1998, Tian *et al.,* 2000), male Sprague-Dawley rats were infused, by minipump, with dDAVP for 5-11 total days. After a 4-day baseline period, rats were divided into two groups; and for 1-7 additional days, "control" rats received dDAVP, dry diet and ad libitum water while "water-loaded" rats received dDAVP and a high-water diet [liquid diet (Verbalis 1994, Ecelbarger *et al.,* 1997a, Murase *et al.,* 1998, Tian *et al.,* 2000) or a gelled-agar diet (Ecelbarger *et al.,* 1998b, c 1999)]. Water-loaded rats were offered 50-60 ml of water per day incorporated into the diet. The rats consumed most of the water on the first day and slightly lesser amounts on subsequent days.

3. DECREASED ABUNDANCE OF AQUAPORIN-2 CORRELATES WITH PHYSIOLOGICAL ESCAPE

A significant increase in urine volume and decrease in urine osmolality in the water-loaded rats was observed by the second day, indicating onset of vasopressin escape (Ecelbarger *et al.,* 1997a, b 1998, Murase *et al.,* 1998, Tian *et al.,* 2000). Furthermore, water-loaded rats developed a marked hyponatremia (plasma sodium range approximately 98-122 mmol/L) (Ecelbarger *et al.,* 1997a, c1999). The onset of escape coincided temporally with a marked decrease in renal aquaporin-2 protein as measured by semiquantitative immunoblotting (Figure 1A) and immunohistochemistry (Ecelbarger *et al.,* 1997a), as well as, decreased mRNA expression, as assessed by Northern blotting (Figure 1B) (Ecelbarger *et al.,* 1997a).

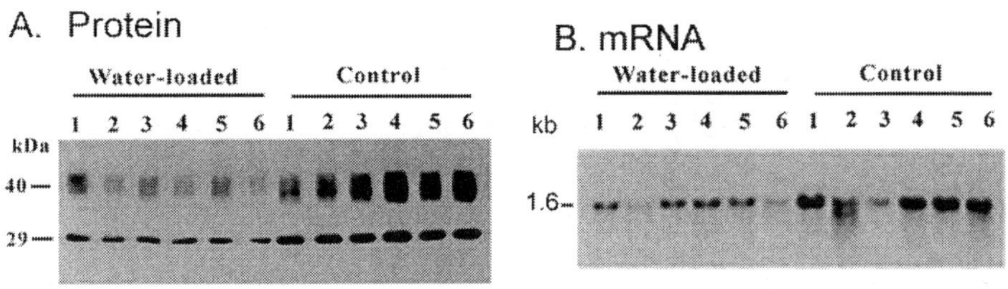

Figure 1. Aquaporin-2 expression in whole kidney homogenates after 3 days of water-loading during vasopressin escape. Each lane is loaded with a sample from a different animal n=6 rats/treatment. A. Aquaporin-2 protein blot is probed with anti-aquaporin-2 antibody L127 (Nielsen *et al.,* 1993). B. Aquaporin-2 mRNA (Ecelbarger *et al.,* 1997a). In both cases, a significant decrease in expression for aquaporin-2 protein or mRNA was observed (unpaired t-test, $p < 0.05$ considered significant).

Furthermore, immunohistocytochemical localization and differential centrifugation studies demonstrated that trafficking of aquaporin-2 to the plasma membrane remained intact during vasopressin escape (Ecelbarger *et al.,* 1997a). In contrast to aquaporin-2, there were no decreases in the renal expression of aquaporins 1, 3, or 4 (Ecelbarger *et al.,* 1997a). In fact, aquaporin-3 abundance was significantly increased in the water-loaded rats (Ecelbarger *et al.,* 1997a, c, 1999). These results suggest that escape from vasopressin-induced antidiuresis is attributable, at least in part, to a selective vasopressin-independent decrease in aquaporin-2 water channel expression in the renal collecting duct. In contrast, the escape phenomenon does not appear to be associated with an impairment of the short-term process by which vasopressin regulates collecting duct water permeability, namely vasopressin-induced trafficking of aquaporin-2 to the apical plasma membrane. Aquaporin-2 is the only water channel known to be present in the apical plasma membrane of the collecting duct principal cells. Since the apical plasma membrane is the rate-limiting barrier for transepithelial water transport (Flamion & Spring 1990), the fall in aquaporin-2 expression seen with vasopressin escape would be predicted to correspond to a marked decrease in collecting duct water permeability. In fact, this was found to be the case. In subsequent studies (Ecelbarger *et al.,* 1998b), we measured the osmotic water permeability of inner medullary collecting ducts (IMCDs) from 6 water-loaded rats and 6 controls using the isolated perfused tubule technique (Burg *et al.,* 1966). We found that, with 100 nM vasopressin in the peritubular bath, the osmotic water permeabilities of the IMCDs from the water-loaded rats were reduced on average to 46% of their corresponding controls (water-loaded, 240 ± 25 μm/s; control, 527 ± 63 μm/s; $P < 0.002$).

4. DECREASED EXPRESSION AND BINDING OF VASOPRESSIN V2 RECEPTORS DURING ESCAPE

The decrease in aquaporin-2 expression in the renal collecting ducts from the water-loaded i.e., "vasopressin-escape" rats is likely to be due in part to decreased signalling through the vasopressin V2 receptor. Tian *et al.*, (2000) found a marked reduction in vasopressin V2 receptor binding in the inner medulla of water-loaded (vasopressin-escape) rats (Figure 2).

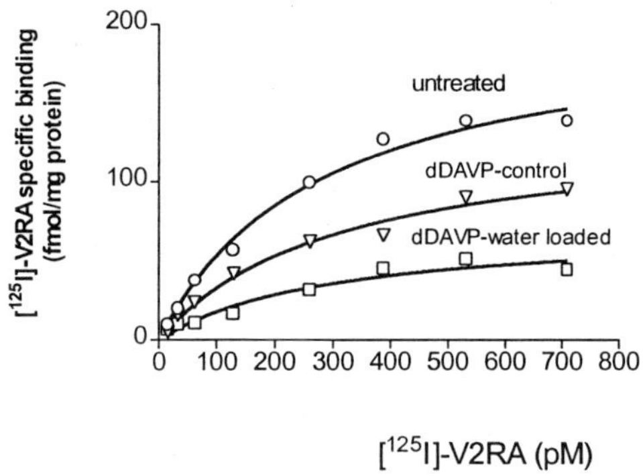

Figure 2. Effect of vasopressin escape on V2 vasopressin receptor binding in the inner medulla of rats after 3 days of water-loading (7 days of dDAVP infusion). Circles = untreated rats, triangles = control/dDAVP-treated, squares = water-loaded/dDAVP-treated (Tian *et al.*, 2000).

In these studies inner medullary homogenates were prepared from untreated rats, control and water-loaded rats. V2 binding capacity was assessed by incubating the samples with a radioiodinated V2 antagonist, $d(CH_2)_5[D\text{-}Ile^2,Ile^4,Tyr\text{-}NH_2^9]AVP$. DDAVP infusion alone decreased V2 receptor binding capacity (B_{max}) to 81% of untreated levels after 7 days of dDAVP infusion (n=8, $p<0.05$). Furthermore, by the third day of water-loading, vasopressin V_2 receptor binding decreased to 43% of the dDAVP-control rats (n=10, $p<0.01$). No differences were observed between the 3 treatment groups for binding affinity (Kd). Complementary Northern-blotting studies by Murase *et al.*, (1998) also demonstrated decreased expression of V2 receptor mRNA in the water-loaded rats.

In support of these findings, we demonstrated in parallel studies (Ecelbarger *et al.*, 1998b) a decrease in cyclic AMP generating capacity in

IMCD suspensions prepared from water-loaded rats (Figure 3). IMCD suspensions were prepared from 6 Water-Loaded rats and 6 corresponding Controls. An individual suspension was prepared for each animal, and 50-μl aliquots of each rat's suspension were incubated under eight conditions: 0, 0.1, 1.0, and 10 nM DDAVP in the presence or absence of 0.25 mM IBMX, a phosphodiesterase inhibitor. After 5 min at 37°C, the incubation was terminated by adding 200 μl of 10% trichloroacetic acid and placing the sample tube on ice. Cyclic AMP levels in the supernatant were determined by a radioimmunoassay. Protein content of each suspension aliquot was measured for normalization.

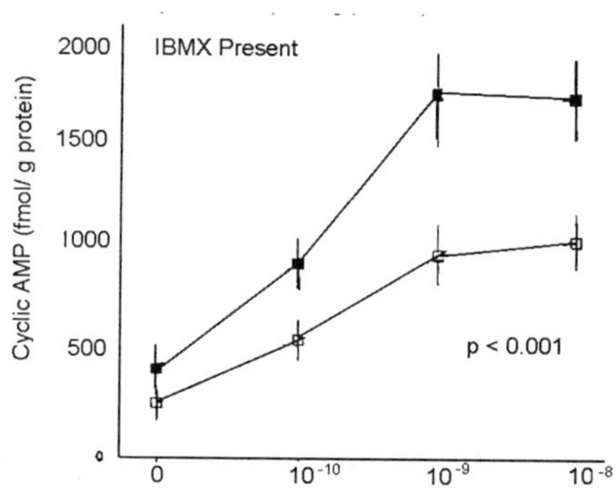

Figure 3. Cyclic AMP production during vasopressin escape in IMCD suspensions treated with IBMX after 3 days of water loading (Ecelbarger *et al.*, 1998b). Filled boxes = Controls, Open boxes = Water-Loaded.

Whether or not IBMX was included in the incubation solution, there was a significant reduction in the cyclic AMP accumulation in tubules from water-loaded (vasopressin-escape) rats. Because the difference between the treatments was no less apparent in the presence of the phosphodiesterase inhibitor, IBMX, we concluded that the collecting ducts from the vasopressin-escape rats have a decreased ability to produce cyclic AMP, i.e., a relative vasopressin resistance. Our observation of decreased cyclic AMP generating capacity of the inner medullary collecting duct from rats undergoing vasopressin escape, as well as, the reduced expression of aquaporin-2, might be explained by the measured reduction in V2 receptor expression and binding capacity. However, additional unknown mechanisms, both V2-receptor dependent or V2-receptor independent, may play a role.

More recent studies (Ecelbarger *et al.,* 1999c) have shown that several distal sodium reabsorptive mechanisms are upregulated during vasopressin escape. The abundances of the thiazide-sensitive NaCl cotransporter of the distal convoluted tubule, the α-subunit of the epithelial sodium channel (ENaC) of the collecting duct and the 70-kDa band of the γ-subunit of ENaC were all significantly increased in whole kidney homogenates from the water-loaded rats. No changes in abundance were observed for the β-subunit of the epithelial sodium channel. Similar protein changes have recently been associated with elevated aldosterone levels in rats (Kim *et al.,* 1998, Masilamani *et al.,* 1999). However, plasma aldosterone levels were significantly suppressed in this model (Ecelbarger *et al.,* 1999c). Nevertheless, these protein changes would be predicted to result in increased sodium reabsorptive capacity of the distal convoluted tubule and collecting duct during vasopressin escape, which could potentially ameliorate the resulting hyponatremia.

5. SUMMARY

Overall, the abundances of several critical renal transporters are regulated during the physiological phenomenon of vasopressin escape. Moreover, the down-regulation of aquaporin-2 protein is likely to be a primary means for the onset and maintenance of physiological vasopressin escape. A reduction in V2-receptor binding capacity may facilitate decreased cyclic AMP generation and decreased aquaporin-2 expression. Regulation of aquaporin-2 protein abundance, as well as, changes in sodium transporter expression would be predicted to modify the transport capacity of various affected renal tubules, such as the collecting duct, distal convoluted tubule and thick ascending limb. These adaptations likely are critical in allowing the animal to maintain salt and water homeostasis when vasopressin levels are inappropriately elevated.

ACKNOWLEDGEMENTS

This work was supported by the National Institutes of Diabetes and Kidney Diseases grants #DK-38094 (J. G. Verbalis, Y. Tian, T. Murase) and K01 #DK02672-01 (C. A. Ecelbarger), as well as by the intramural budget of the National Heart, Lung, and Blood Institute (C. A. Ecelbarger, M. A. Knepper, C.-L. Chou and B. R. Olson) Dr. S. Nielsen was supported by the

Danish Research Academy, the Novo Nordic Foundation, the Danish Medical Council, and the University of Aarhus Research Foundation. Dr. Murase was a Lilly International Fellow.

REFERENCES

Bichet DG Kluge R Howard RL and Schrier RW 1992 Hyponatremic states In The Kidney: Physiology and Pathophysiology (D W Seldin and G Giebisch eds) Raven New York p 1727-1751

Burg MB Grantham JJ Abramow M and Orloff J 1966 Preparation and study of fragments of single rabbit nephrons *Am J Physiol* **210**: 1293-1298

Ecelbarger CA Nielsen S Olson B Murase T Baker E A Knepper MA and Verbalis JG 1997 Role of renal aquaporins in escape from vasopressin-induced antidiuresis in rat *J Clin Invest* **99**: 1852-1863

Ecelbarger CA Chou C-L Lee AJ DiGiovanni SR Verbalis JG and Knepper MA 1998 Escape from vasopressin-induced antidiuresis: role of vasopressin resistance of the collecting duct *Am J Physiol* **274**: F1161-F1166

Ecelbarger CA Verbalis JG and Knepper MA 1999 Rapid up-regulation of thiazide-sensitive cotransporter and α-ENaC expression in a rat model of vasopressin escape [Abstract] *J Am Soc Nephrol* **10**: 14A

Flamion B and Spring KR 1990 Water permeability of apical and basolateral cell membranes of rat inner medullary collecting duct *Am J Physiol* **259**: F986-F999

Kim G-H Masilamani S Turner R Mitchell C Wade JB and Knepper MA 1998 The thiazide-sensitive Na-Cl cotransporter is an aldosterone-induced protein *Proc Natl Acad Sci USA* **95**: 14552-14557

Masilamani S Kim G-H Mitchell C Wade JB and Knepper MA 1999 Aldosterone-mediated regulation of ENaC α β and γ subunit proteins in rat kidney *J Clin Invest* **104**: R19-R23

Murase T Baker EA Tian Y Knepper MA and Verbalis JG 1998 Down-regulation of kidney AVP V2 receptor expression during renal escape from vasopressin-induced antidiuresis [Abstract] *J Am Soc Nephrol* **9**: 23A

Nielsen S DiGiovanni SR Christensen EI Knepper MA and Harris HW 1993 Cellular and subcellular immunolocalization of vasopressin-regulated water channel in rat kidney *Proc Natl Acad SciUSA* **90**: 1663-11667

Tian Y Sandberg K Murase T Baker EA Speth RC and Verbalis JG 2000 Vasopressin V2 receptor binding is down-regulated during renal escape from vasopressin-induced antidiuresis *Endocrinol* **141**: 307-314

Verbalis JG 1994 Pathogenesis of hyponatremia in an experimental model of the syndrome of inappropriate antidiuresis *Am J Physiol* **267**: R1617-R1625

2.5 Mammalian physiology

AQUAPORIN-8 AND -9: TISSUE LOCALIZATION

[1]Marie-Louise Elkjær, [1]Lene N. Nejsum, [4]Uffe B. Jensen, [2]Tae-Hwan Kwon, [3]Jørgen Frøkiær and [1]Søren Nielsen

[1]Department of Cell Biology, Institute of Anatomy, University of Aarhus, DK-8000 Aarhus, Denmark; [2]Department of Physiology, School of Medicine, Dongguk University, Kyungju, 780-714, Korea; [3]Institute of Experimental Clinical Research, University of Aarhus, DK-8000 Aarhus C, Denmark; [4]Department of Human Genetics, University of Aarhus, DK-8000 Aarhus, Denmark. E-mail: mle@ana.au.dk

1. INTRODUCTION

Transport of water and salt across cell membranes is of fundamental importance for all cellular function, and transport of water across epithelia in, among others, kidney, heart and lung is of great importance for the function of these organs. The molecular basis for transport of water across cell membranes was totally unknown until Peter Agre and collegues identified the first aquaporin waterchannel (Preston *et al.,* 1992) . Aquaporins are a family of homologous water transporting proteins expressed in various tissues. In general, they are expressed in plasma membranes of cells, that are involved in rapid fluid transport such as kidney tubules. The first aquaporin, aquaporin 1, was found to be a channel-forming integral membrane protein of 28 kDa (CHIP-28) expressed in mammalian red blood cells and kidney, and in other water-permeable epithelia (Agre *et al.,*. 1993) . Since then additionally 8 aquaporins in mammmals has been cloned. This review is focused on the two latest cloned rat aquaporins: aquaporin-8 and aquaporin-9.

2. AQUAPORIN 8

2.1 General characteristics of aquaporin 8

The aquaporin 8 gene was cloned in 1997 from rat (Ishibashi *et al.*, 1997, Koyoma *et al.*, 1997), and since then the cDNA has been isolated from other species including mouse and human (Ma *et al.*, 1997, Koyoma *et al.*, 1998, Calamita *et al.*, 1999). The rat AQP8 cDNA was found to encode a 263 amino acid protein (Ishibashi *et al.*, 1997, Koyoma *et al.*, 1997), while it in human and mouse was found to encode a 261 amino acid protein (Ma *et al.*, 1997, Koyoma *et al.*, 1998). It has amino acid sequence homology with other aquaporins (30-37%), and highest with a plant water-channel, AQP(–TIP (38-41%) (Ishibashi *et al.*, 1997, Koyoma *et al.*, 1997, Ma *et al.*, 1997, Koyoma *et al.*, 1998) . The human AQP8 is 74.9% identical to rat AQP8, and 76% identical to mouse AQP8 (Koyoma *et al.*, 1998). The human AQP8 gene has been localized to human chromosome 16p12 (Viggiano *et al.*, 1999). The transport selectivity of AQP8 is controversial. Urea transport is found in mouse AQP8 (Ma *et al.*, 1997), but not in rat AQP8 (Koyoma *et al.*, 1997). Ma *et al.*,. reported that mouse AQP8 is permeable to urea in addition to water, but not to glycerol (Ma *et al.*, 1997). The human AQP8 protein has been found to be a water-selective channel, with no permeability to urea, nor to glycerol (Koyoma *et al.*, 1998). The reason for these differences among species is still unclear. In rat and mouse the AQP8 protein was found to be mercury-sensitive (Ishibashi *et al.*, 1997, Koyoma *et al.*, 1997, Ma *et al.*, 1997).

2.2 Tissue localization of aquaporin 8

In rat and mouse AQP8 mRNA is expressed in testis, pancreas, salivary gland, liver and colon (Ishibashi *et al.*, 1997, Koyoma *et al.*, 1997, Ma *et al.*, 1997, Koyoma *et al.*, 1999). In addition rat AQP8 mRNA was localized in jejunum (Koyoma *et al.*, 1999). The cellular localization of AQP8 protein in rat was found to be in hepatocytes, acinal cells of pancreas and salivary gland, and in absorptive colonic epithelial cells (Koyoma *et al.*, 1997). In rat testes mRNA expression was reported to be in all stages of spermatogenesis from primary spermatocytes to spermatids in seminiferous tubules, suggesting a role in spermatogenesis (Ishibashi *et al.*, 1997). The mouse mRNA expression was also reported in heart, placenta, lung, kidney, diaphragm, spleen, stomach and brain (Ma *et al.*, 1997). Human aquaporin 8 protein has been localized in pancreas and colon, but interestingly it was found to be absent in testes, where rat AQP8 mRNA and protein has been shown to be abundant (Ishibashi *et al.*, 1997, Koyoma *et al.*, 1998). It has been shown by western blotting that the rat aquaporin 8 protein is localized in liver, testes, jejunum and colon, confirming the mRNA results reported (Elkjær *et al.*, 2000). Also it has been shown to be localized in epididymis. The cellular localization has been shown to be intracellular in hepatocytes, in the duct cells in epididymis, in the spermatogenic cells in testes and in the surface epithelia cells in colon and jejunum (Elkjær *et al.*, 2000). Aquaporin 8 mRNA has been reported to be present in mouse kidney (Ma *et al.*, 1997). The presence of AQP8 in kidney has been confirmed in rat, where the expression of AQP8 has been found as a 28 kDa protein in membrane fractions from total rat kidney and from all kidney zones (cortex, outer stripe and inner stripe of the outer medulla and inner medulla). The localization was mainly intracellular. The cellular localization of AQP8 was found to be in proximal tubule cells and in collecting ducts (Elkjær *et al.*, 2000).

3. AQUAPORIN 9

3.1 General characteristics of aquaporin 9

The rat aquaporin 9 gene and the human aquaporin 9 gene was cloned in 1998 (Ishibashi *et al.*, 1998, Tsukaguchi *et al.*, 1998), and later on by others (Ko *et al.*, 1999). AQP9 cDNA was found to encode a 295 amino acid protein in rat as well as in human (Ishibashi *et al.*, 1998, Tsukaguchi *et al.*, 1998, Ko *et al.*, 1999). Human and rat AQP9 has low to moderate homology with with AQP3 (48-49%) and AQP7 (45-47%) and other aquaporins (30%), suggesting that AQP3, AQP7 and AQP9 belong to a subfamily of the aquaporin family (Ishibashi *et al.*, 1998, Tsukaguchi *et al.*, 1998). Rat AQP9 has greatest amino-acid sequence identity with human AQP9 (75%) (Ko *et al.*, 1999). Rat aquaporin 9 was found to have high permeability to both water and solutes (carbamides, polyols, purines and pyremidines) (Tsukaguchi *et al.*, 1998), and consistent with this, another group found it to be permeable to water, urea and glycerol (Ko *et al.*, 1999). Both groups found rat AQP9 to be mercury chloride sensitive (Tsukaguchi *et al.*, 1998, Ko *et al.*, 1999). Conflicting results have been reported for human AQP9. One group found the permeability to be restricted to water and urea (Ishibashi *et al.*, 1998), while another group found it to allow passage of water and a wide varity of non-charged solutes including carbamides, polyols, purines, and pyrimidines in a phloretin- and mercurial-sensitive manner (Tsukaguchi *et al.*, 1999). Flourescence in situ hybridization has assigned the locus of human AQP9 to chromosome 15 q22.1-22.2 (Tsukaguchi *et al.*, 1999).

3.2 Tissue localization of aquaporin 9

In rat AQP9 mRNA has been found in rat liver, testes, brain (Ko *et al.*, 1999, Tsukaguchi *et al.*, 1998) and lung (Ko *et al.*, 1999). The subcellular localization in these tissues was shown by in situ hybridization to be in hepatocytes, in the inner surface of seminiferous tubules and in the interstitial Leydig cells (Tsukaguchi *et al.*, 1998). Also AQP9 was detected in astrocytes throughout the brain (Tsukaguchi *et al.*, 1998). Human AQP9 has been localized by northern blotting to peripheral leukocytes, liver, lung, spleen (Ishibashi *et al.*, 1998, Tsukaguchi *et al.*, 1999) and bone marrow (Tsukaguchi *et al.*, 1999).The mRNA localization of AQP9 in rat liver and testes, has been confirmed by immunoblotting and immunohistochemistry. Immunoblotting showed AQP9 to be a 27-28 kDa protein in rat liver, testes, spleen and epididymis (Elkjær *et al.*, 2000). The cellular localization was found by immunohistochemistry to be in the interstitial Leydig cells in testes, in the stereocilia in epididymis and in the plasmamembrane in the hepatocytes; mainly the sinusoidal surface (Elkjær *et al.*, 2000).

ACKNOWLEDGMENTS

The work cited from our laboratory was supported partly by grants from EU-Commission (TMR and Biotech programs), Karen Elise Jensen Foundation and Danish Research Council.

REFERENCES

Agre P Preston GM Smith BL Jung JS Raina S Moon C Guggino WB and Nielsen S 1993 Aquaporin CHIP: the archetypal molecular water channel *Am J Physiol* **265**: F463-F476

Calamita G Spalluto C Mazzone A Rocchi M and Svelto M 1999 Cloning structural organization and chromosomal localization of the mouse aquaporin-8 water channel gene (Aqp8) *Cytogenet Cell Genet* **85**: 237-241

Elkjær M-L Nejsum LN Jensen UB Kwon T-H Frøkier J and Nielsen S Localization of aquaporin-8 (AQP8) mRNA and protein in rat kidney and other organs as determined by RT-PCR immunoblotting and immunocytochemistry Abstract MIP2000 Göteborg Sweden

Elkjær M-L Nejsum LN Jensen UB Kwon T-H Frøkier J and Nielsen S Localization of aquaporin –9 protein in rat liver testes and epididymis as determined by immunoblotting and immunohistochemistry Abstract MIP2000 Goteborg Sweden

Ishibashi K Kuwahara M Gu Y Tanaka Y Marumo F and Sasaki S1998 Cloning and functional expression of a new aquaporin (AQP9) abundantly expressed in the peripheral leukocytes permeable to water and urea but not to glycerol *Biochem Biophys Res Commun* **244**: 268-274

Ishibashi K Kuwahara M Kageyama Y Tohsaka A Marumo F and Sasaki S 1997 Cloning and functional expression of a second new aquaporin abundantly expressed in testis *Biochem Biophys Res Commun* **237**: 714-718

Ko SB Uchida S Naruse S Kuwahara M Ishibashi K Marumo F Hayakawa T and Sasaki S 1999 Cloning and functional expression of rAOP9L a new member of aquaporin family from rat liver *Biochem Mol Biol Int* **47**: 309-318

Koyama N Ishibashi K Kuwahara M Inase N Ichioka M Sasaki S and Marumo F 1998 Cloning and functional expression of human aquaporin8 cDNA and analysis of its gene *Genomics* **54**: 169-172

Koyama Y Yamamoto T Kondo D Funaki H Yaoita E Kawasaki K Sato N Hatakeyama Kand Kihara I 1997 Molecular cloning of a new aquaporin from rat pancreas and liver *J Biol Chem* **272**: 30329-30333

Koyama Y Yamamoto T Tani T Nihei K Kondo D Funaki HYaoita E Kawasaki K Sato N Hatakeyama K and Kihara I 1999 Expression and localization of aquaporins in rat gastrointestinal tract *Am J Physiol* **276**: C621-C627

Ma T Yang B and Verkman AS 1997 Cloning of a novel water and urea-permeable aquaporin from mouse expressed strongly in colon placenta liver and heart *Biochem Biophys Res Commun* **240**: 324-328

Preston GM Carroll TP Guggino WB and Agre P 1992 Appearance of water channels in Xenopus oocytes expressing red cell CHIP28 protein *Science* **256**: 385-387

Tsukaguchi H Shayakul C Berger UV Mackenzie B Devidas S Guggino WB Van Hoek AN and Hediger MA 1998 Molecular characterization of a broad selectivity neutral solute channel *J Biol Chem* **273**: 24737-24743

Tsukaguchi H Weremowicz S Morton CC and Hediger MA 1999 Functional and molecular characterization of the human neutral solute channel aquaporin-9 *Am J Physiol* **277**: F685-F696

Viggiano L Rocchi M Svelto Mand Calamita G 1999 Assignment of the aquaporin-8 water channel gene (AQP8) to human chromosome 16p12 *Cytogenet Cell Genet* **84**: 208-210

2.6 Mammalian physiology

EXPRESSION OF THE EPITHELIAL NA$^+$ CHANNEL (ENAC) IN RAT KIDNEY COLOCALISES WITH AQUAPORIN-2 (AQP-2)

[1]Henrik Hager, [2]Tae-Hwan Kwon, [3]Mark Knepper, [4]Jørgen Frøkiær and [1]Søren Nielsen

[1]Department of Cell Biology, Institute of Anatomy, University of Aarhus, DK-8000 Aarhus, Denmark; [2]Department of Physiology, School of Medicine, Dongguk University, Kyungju, 780-714, Korea; [3]NHLBI, National Institutes of Health, Bethesda, MD 20814, USA; [3]Institute of Experimental Clinical Research, University of Aarhus, DK-8000 Aarhus C, Denmark. E-mail: hh@ana.au.dk

1. INTRODUCTION

The apical surface of many tight epithelia express Na$^+$-selective channels primarily characterized by the fact that they are selectively inhibited by submicromolar concentrations of the diuretic inhibitor amiloride (Garty and Palmer 1997). This channel is termed ENaC (epithelial Na$^+$ channel) and belongs to the DEGenerin/ENaC gene super family. The members of this family are implicated in several cellular functions, including mechanosensation, proprioseption, pain sensation, gametogenesis and sodium transport (Mano *et al.*, 1999). ENaC has been cloned from rat (Canessa *et al.*, 1995), human (Mc. Donald *et al.*, 1995) and other species. ENaC is composed of three homologous subunits α-, β- and γ–ENaC. The three subunits share limited sequence similarity (~30-40%), suggesting that they are derived from a common ancestral gene. The ENaC subunits have similar topologies with short intracellular NH_2 and COOH termini, two transmembrane domains, and a large cysteine rich extracellular loop. Putative functions for these regions have been obtained by analysing mutations found in genes coding for human ENaC subunits. The amiloride sensitivity and the selectivity of the channel has been shown to be dependent

Molecular Biology and Physiology of Water and Solute Transport
Edited by Hohmann and Nielsen, Kluwer Academic/Plenum Publishers, New York, 2000.

of a short segment proceeding the second membrane spanning domain suggesting an implication of this domain in the pore structure of ENaC (Schild *et al.,* 1997). A gating domain was identified within the highly conserved N-terminus whereas a well-conserved PPxY motif in the C-terminus promotes interaction with intracellular proteins e.g. Nedd4 (Abriel *et al.,* 1999) and proteins proteins in the clathrin mediated endocytic pathway (Shimkets *et al.,* 1997). The role of the extracellular loop is not well established (Firsov *et al.,* 1999). The expression of α–ENaC generates only very small Na^+ current, and the expression of either β–ENaC or γ–ENaC generates no current whereas the co expression of α–ENaC, β–ENaC and γ–ENaC in heterologous cells generates large Na^+ currents (Canessa *et al.,* 1995). Additional evidence of an interaction has come from the finding that the ENaC subunits can be co-immunoprecipitated (Firsov *et al.,* 1999). The functional ENaC is assembled with stoichiometry of 2α, 1β and 1γ (Firsov *et al.,* 1998).

2. EXPRESSION AND REGULATION OF ENAC

In various tight epithelia (see below) electrogenic entry of Na^+ from the lumen into the cells is mediated by the ENaC located in the apical membrane of the cell, and is the rate-limiting step of Na^+ absorption. On the basolateral side, the Na,K-ATPase actively transports sodium out of the cell into the extracellular compartment and provides the driving force for Na^+ absorption (Garty *et al.,* 1997). ENaC is located in the kidney, the colon, the urinary bladder epithelia, the lungs where it participates in fluid clearance at birth and during pulmonary oedema and in several other tissues including duct of the salivary and sweat glands.

In the kidney, ENaC subunits are expressed in principal cells of cortical (CCD), outer medullary (OMCD) collecting duct (CD) and connecting tubule (CNT) (Masilamani *et al.,* 1999). So far it is unknown whether ENaC is present in the inner medullary collecting duct (IMCD), known to display hormonal regulated Na-transport. A recent report confirmed the presence of all three subunits in principal cells of the CNT, CCD, and OMCD. Importantly also the IMCD cells exhibited significant labeling of α-, β- and γ-ENaC, with decreasing labeling along the axis of the CD, including the IMCD (Hager *et al.,* 2000) thus there is a high degree of colocalisation with aquaporin-2 (AQP-2) which is expressed in the same cells (Nielsen *et al.,* 1999). A marked labeling was also found of the surface epithelium covering the inner medulla (Hager *et al.,* 2000). ENaC participates in the fine-regulation of Na^+ excretion that occurs in the renal collecting duct under tight regulation by the hormones controlling the water and salt balance e.g. aldosterone and antidiuretic hormone (ADH or vasopressin).

Aldosterone has been shown to play a major role in the regulation of ENaC function, but the mechanisms involved are not completely understood (Garty 1994). Two potential mechanisms have been described:

1) Enhanced surface expression of ENaC via induced synthesis, decreased degradation and/or altered sorting into the apical membrane. In a recent study Masilamani *et al.,* showed that increased concentration of aldosterone induced either exogenous or by NaCl restriction, in rats, were associated with a multifold increase in α–ENaC protein whereas the abundance of β- and γ-ENaC remained unchanged indicating that α-ENaC is an aldosterone induced protein and the limiting step of ENaC expression (Masilamani *et al.,* 1999). It was furthermore shown that salt restriction caused a redistribution of all three subunits to the apical region of the collecting duct principal cells. Whether the increased abundance of α-ENaC, mediated by aldosterone, could be due to increased transcribtion of ENaC gene in collecting duct cells is controversial. Some investigators found no increase in mRNA level for any of the subunits in response to dietary sodium restriction (Renard *et al.,* 1995), whereas others found an increased level of α-ENaC mRNA in response to increased circulating aldosterone concentration (Stokes *et al.,* 1998). The abundance of α-ENaC protein could also be due, at least in part, to direct effect on protein synthesis. This is supported by work showing that the aldosterone response in A6 kidney cell at the protein level precedes the response at the mRNA level (May *et al.,* 1997). Finally, the increased abundance of α–ENaC protein could be due to increased half-life. It has been proposed that α-ENaC half life is regulated via ubiquitination by a ubiquintin ligase, Nedd4, that binds to PY regions of the COOH-terminus of α–ENaC (Staub *et al.,* 1997).

2) Stimulation of the channels already present in the plasma membrane or increased the open probability of the channels. In A6 cells aldosterone has been shown to increase the Na^+ influx by stimulation of the channels already present in the apical membrane possible by posttranslational modification (Becchetti *et al.,* 2000).

Vasopressin has been shown to increase both Na^{2+}- (Chen *et al.,* 1990) and water reabsorbtion (Grantham *et al.,* 1966) in kidney collecting ducts of rat and rabbit. vasopressin regulates water reabsorption through the vasopressin V2 receptor, and regulates the water permeability in the principal cells either by trafficking AQP-2 to the apical membrane (Nielsen *et al.,* 1995) or by increasing transcription DiGiovani *et al.,* 1994). The role of vasopressin in the regulation of Na^+ balance is not clear. In vitro, it has been shown that vasopressin increases sodium transport in perfused rat and rabbit collecting ducts (Chen *et al.*, 1990) and in several cell lines (Nakhoul *et al.*, 1998 and Canessa *et al.*, 1993). In vivo vasopressin is shown to increase the abundance of all three ENaC subunits in rat kidney (Ecelbarger *et al.*, 2000).

Future studies will be very important in defining whether ENaC is regulated by intracellular trafficking in response to vasopressin and/or aldosterone and how this relate to vasopressin regulation of AQP2. Moreover it will be essential to establish whether the stoichiometry of ENaC subunits are constant or a subject to regulation.

3. ALTERED REGULATION OF ENaC EXPRESSION IN HUMAN DISEASE

The function and regulation of ENaC is crucial, as indicated by the finding that a gene knockout of α–ENaC is lethal due to the inability of the α(-/-) mice to clear fluid from their lungs shortly after birth (Hummler *et al.*, 1996) whereas β- and γ-ENaC(-/-) exhibit significant defect in renal function leading to death within 48 hours. Moreover, several inherited human disorders were recently linked to mutations in or dysregulation of the ENaC subunits. These include the salt-wasting disease pseudohypoaldosteronism type 1 (Schild *et al.*, 1996), Liddle syndrome, an inherited autosomal dominant form of human hypertension (Botero-Velez *et al.*, 1994) and cystic fibrosis where abnormally high ENaC activity has been demonstrated (Mall *et al.*, 1999). Whether dysregulation of ENaC subunits also plays a part in the patophysiology in acquired diseases with altered Na^+ metabolism is yet unclear but is also a focus of future studies.

REFERENCES

Abriel H Loffing J Rebhun JF Pratt JH Schild L Horisberger JD Rotin D and Staub O 1999 Defective regulation of the epithelial Na+ channel by Nedd4 in Liddle's syndrome *J Clin Invest* **103**: 667-673

Becchetti A Kemendy AE Stockand JD Sariban-Sohraby S and Eaton DC 2000 Methylation Increases the Open Probability of ENaC in A6 Epithelia *J Biol Chem* in press

Botero-Velez M and Curtis JJ and Warnock DG 1994 Brief report: Liddle's syndrome revisited a disorder of sodium reabsorption in the distal tubule *N Engl J Med* **330**: 178-181

Canessa CM Horisberger JD Schild L Rossier BC 1995 Expression cloning of the epithelial sodium channel *Kidney Int* **48**: 950-955

Chen L Williams SK and Schafer JA 1990 Differences in synergistic actions of vasopressin and deoxycorticosterone in rat and rabbit CCD Am *J Physiol* **259**: F147-156

DiGiovanni SR Nielsen S Christensen EI and Knepper MA 1994 Regulation of collecting duct water channel expression by vasopressin in brattleboro rat *Proc Natl Acad Sci USA* **91**: 8984-8988

Firsov D Gautschi I Merillat AM Rossier BC and Schild L 1998 The heterotetrameric architecture of the epithelial sodium channel (ENaC) *EMBO J* **15**: 344-352

Firsov D Robert-Nicoud M Gruender S Schild L and Rossier BC 1999 Mutational analysis of cysteine-rich domains of the epithelium sodium channel (ENaC) Identification of cysteines essential for channel expression at the cell *JBiol Chem* **274**: 2743-2749

Garty H and Palmer LG 1997 Epithelial sodium channels: function structure and regulation *Physiol Rev* **77**: 359-396

Garty H Peterson-Yantorno K Asher C and Civian MM 1994 Effects of corticoid agonists and antagonists on apical Na+ permeability of toad urinary bladder *Am J Physiol* **266**: F108-116

Grantham JJ and Burg MB 1966 Effect of vasopressin and cyclic AMP on permeability of isolated collecting tubules *Am J Physiol* **211**: 255-259

Hager HT-H Kwon T-H Vinnikova AK Frokiaer J Knepper MA and S Nielsen S 2000 Immunocytochemical and immunoelectron microscopical (EM) localization of ENaC in rat kidney Abstract American Society of Nephrology

Hummler E Barker P Gatzy J Beermann F Verdumo C Schmidt A Boucher R and Rossier BC 1996 Early death due to defective neonatal lung liquid clearance in alpha-ENaC-deficient mice *Nat Genet* **12**: 325-328

Mall M Bleich M Kuehr J Brandis M Greger R and Kunzelmann K 1999 CFTR-mediated inhibition of epithelial Na+ conductance in human colon is defective in cystic fibrosis *Am J Physiol* **277**: G709-716

Mano I Driscoll M 1999 DEG/ENaC channels: a touchy superfamily that watches its salt *Bioessays* **21**: 568-578

Masilamani S Kim GH Mitchell C Wade JB and Knepper MA 1999 Aldosterone-mediated regulation of ENaC alpha beta and gamma subunit proteins in rat kidney *J Clin Invest* **104**: R19-23

May A Puoti A Gaeggeler HP Horisberger JD and Rossier BC 1997 Early effect of aldosterone on the rate of synthesis of the epithelial sodium channel alpha subunit in A6 renal cells *J Am Soc Nephrol* **8**: 1813-1822

McDonald FJ Price MP Snyder PM and Welsh MJ 1995 Cloning and expression of the beta- and gamma-subunits of the human epithelial sodium channel *Am J Physiol* **268**: C1157-1163

Nielsen S Chou CL Marples D Christensen EI Kishore BKand Knepper MA 1995 Vasopressin increases water permeability of kidney collecting duct by inducing translocation of aquaporin-CD water channels to plasma membrane *Proc Natl Acad Sci USA* **92**: 1013-1017

Nielsen S Kwon T-H Christensen BM Promeneur D Frøkiær J and Marples D 1999 Physiology and pathophysiology of renal aquaporins *J Am Soc Nephrol* **10**: 647-663

Renard S Voilley N Bassilana F Lazdunski M and Barbry P 1995 Localization and regulation by steroids of the alpha beta and gamma subunits of the amiloride-sensitive Na+ channel in colon lung and kidney *Pflugers Arch* **430**: 299-307

Schild L Schneeberger E Gautschi I and Firsov D 1997 Identification of amino acid residues in the alpha beta and gamma subunits of the epithelial sodium channel (ENaC) involved in amiloride block and ion permeation *J Gen Physiol* **109**: 15-26

Schild L 1996 The ENaC channel as the primary determinant of two human diseases: Liddle syndrome and pseudohypoaldosteronism *Nephrologie* **17**: 395-400

Shimkets RA Lifton RP and Canessa CM 1997 The activity of the epithelial sodium channel is regulated by clathrin-mediated endocytosis *J Biol Chem* **272**: 25537-25541

Staub O Gautschi I Ishikawa T Breitschopf K Ciechanover A Schild L and Rotin D 1997 Regulation of stability and function of the epithelial Na+ channel (ENaC) by ubiquitination *EMBO J* **16**: 6325-36

Stokes JB and Sigmund RD 1998 Regulation of rENaC mRNA by dietary NaCl and steroids: organ tissue and steroid heterogeneity *Am J Physiol* **274**: C1699-1707

2.7 Mammalian physiology

OXYTOCIN: ONE OF THE FACTORS FOR REGULATING AQP2 LOCALIZATION AND URINARY AQP2 EXCRETION

[1]Jin Suk Han, [2]Un Sil Jeon, [3]Kwon Wook Joo, [4]Ho Joon Chin, [5]Woosung Huh, [1]Jung Sang Lee, [6]Gheun-Ho Kim, [7]Jae-Ho Earm, [8]Jin Kim, [9]Søren Nielsen and [10]Mark A. Knepper

[1]Department of Internal Medicine, Seoul National University, Research Institute of Seoul National University Hospital, Seoul, Korea; [2]Gyeong Sang National University, Chinju, Korea; [3]Gachon Medical University, Incheon, Korea; [5]Cheju National University, Cheju, Korea; [5]Sungkyunkwan University, Seoul, Korea; [6]Hallym University, Seoul, Korea; [7]Chungbuk National University, Cheongju, Korea; [8]Department of Anatomy, Catholic University, Seoul, Korea; [9]Dept Cell Biology, Aarhus Univ, Aarhus, Denmark; [10]Laboratory of Kidney and Electrolyte Metabolism, NIH, Bethesda, U.S.A. E-mail: jshan@snu.ac.kr

1. INTRODUCTION

Aquaporin-2 (AQP2), an essential water channel for urinary concentration, is mainly located in the apical plasma membrane of collecting duct epithelial cells, and vasopressin is the major factor for regulation of AQP2 expression in the kidney (Nielsen *et al.*, 1993, Nielsen *et al.*, 1995, Ishibashi *et al.*, 1997). However, there has been some evidences of a moderate amount of basolateral localization of AQP2 at least in the inner medullary collecting duct (IMCD) (Nielsen *et al.*, 1993), and vasopressin-independent mechanisms of AQP2 regulation should exist (Marples *et al.*, 1995a, Ecelbarger *et al.*, 1997, Ecelbarger *et al.*, 1998, Marples *et al.*, 1998).

Oxytocin has been known as an antidiuretic hormone (Abdul-Karim and Assali 1961, Ahmed *et al.*, 1975, Han *et al.*, 1993, Chou *et al.*, 1995a, Chou *et al.*, 1995b) and may be involved in some part of vasopressin-independent urinary concentrating mechanisms. Here we present our studies undertaken to investigate whether the antidiuretic effect of oxytocin is exerted by regulating AQP2 protein in the kidney.

2. INTRACELLULAR LOCALIZATION OF AQUAPORIN-2 IN RAT KIDNEY

Oxytocin is a nonapeptide, secreted from posterior pituitary gland, and structurally very similar to vasopressin. Brattleboro rats were shown to concentrate their urine to hypertonic levels with severe dehydration, and plasma levels of oxytocin increased about six folds under this condition (Edwards and LaRochelle 1984). Knepper and his colleagues have documented that oxytocin can function as an antidiuretic hormone at physiologic plasma level, and the V_2 receptor mediates the antidiuretic action of oxytocin in rat IMCD (Han *et al.,* 1993, Chou *et al.,* 1995a, Chou *et al.,* 1995b).

We tested acute effect of exogenous oxytocin on AQP2 localiztion in the kidney using male Sprague-Dawley rats. Immunocytochemistry and semiquantitative immunoblot analysis were carried out using polyclonal peptide-derived antibodies to AQP2 (Nielsen *et al.,* 1993, Ecelbarger *et al.,* 1998). After a single intraperitoneal injection of 10 units of oxytocin, urine osmolality significantly increased compared with that of normal saline-injected control rats (1293 ± 118 *vs.* 737 ± 133 mOsm/kg H_2O, $P<0.05$) despite no significant changes in urine flow. Immunohistochemistry of IMCD revealed that AQP2 localization was shifted from diffuse cytoplasmic predominance in control rats to both apical and basolateral membranes in oxytocin-treated rats (Fig. 1). This pattern of AQP2 redistribution was noted in connecting tubules and outer medullary collecting ducts as well, although the tendency was somewhat weaker. In contrast, AQP2 was redistributed almost exclusively to apical plasma membranes of connecting tubule cells, principal cells and IMCD cells either with 48h-dehydration or after a single intraperitoneal injection of 5 units of vasopressin, as previously reported (Nielsen *et al.,* 1993, Marples *et al.,* 1995b).

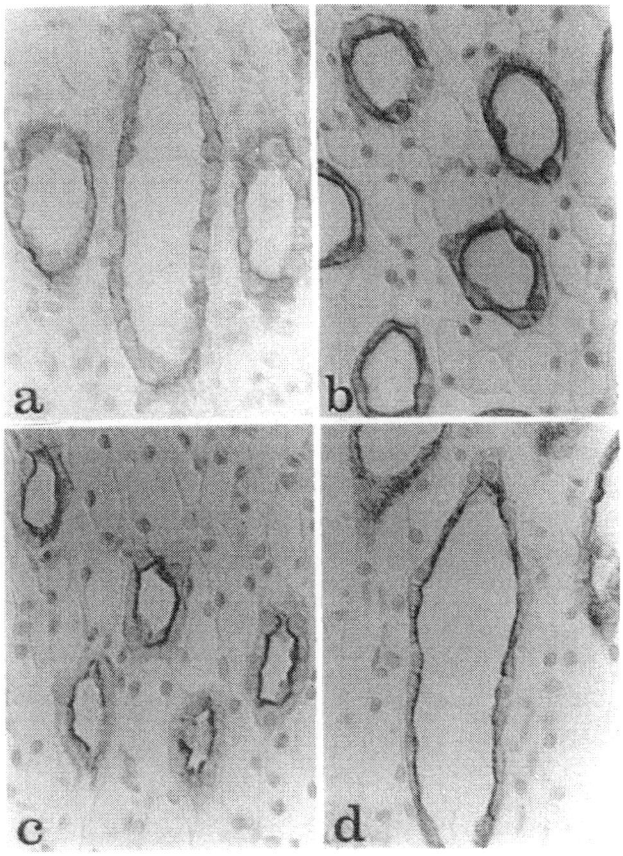

Figure 1. Immunocytochemical localization of AQP2 in rat kidney middle part of inner medulla (x 640). The IMCD cells exhibit predominant diffuse cytoplasmic labeling with some localization on apical and basolateral membranes in control (a). The AQP2 protein is strongly labeled on both basolateral and apical membranes in oxytocin-treated rat (b). Note more predominant labeling of basolateral membrane than that of apical membrane. In dehydrated rat (c) and vasopressin-treated rat (d), the IMCD cells exhibit strong apical labeling of AQP2 without basolateral localization.

We also performed semiquantitative immunoblot analysis from differential centrifugation of whole kidney homogenates to examine subcellular fractionation of AQP2 protein. The band density ratio of plasma membrane-rich fraction over cytoplasmic vesicle-rich fraction was higher in oxytocin-treated rats than in controls (3.64 ± 0.60 *vs* 1.09 ± 0.14, $p<0.05$, Fig. 2). This result appears compatible with our immunocytochemical finding that oxytocin induces redistribution of AQP2 protein from cytoplasmic vesicles to both apical and basolateral plasma membranes in rat kidney.

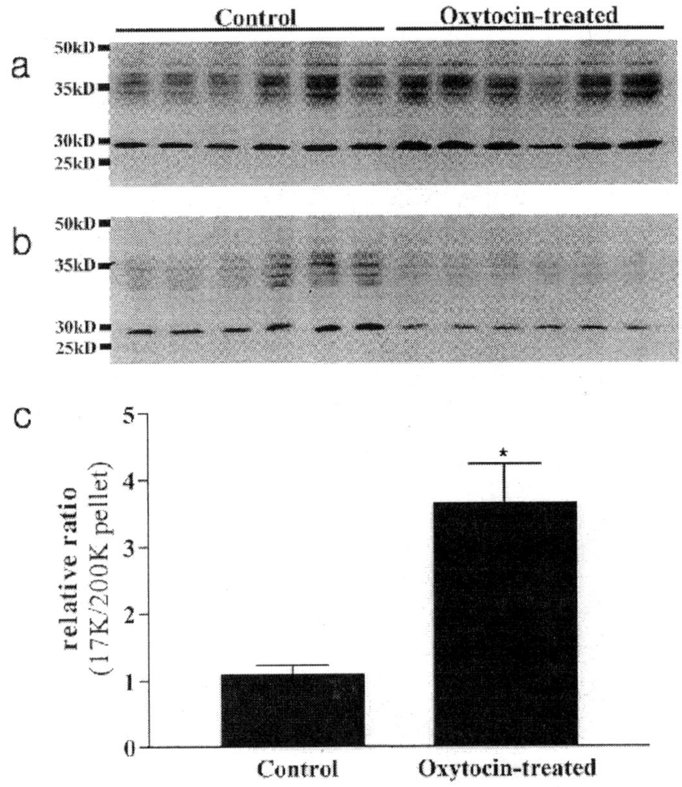

Figure 2. Immunoblots of AQP2 from plasma membrane-rich fractions(17K) and cytoplasmic vesicle-rich fractions(200K) of whole kidneys. Each lane was loaded with 20 μg of protein from a different rat. The AQP2 expression in plasma membrane-rich fraction is not different between control and oxytocin treatment (a). In contrast, the immunoblot of cytoplasmic vesicle-rich fraction reveals a lower AQP expression with oxytocin treatment (b). Therefore, the band density ratio of plasma membrane-rich fraction over cytoplasmic vesicle-rich fraction is increased by oxytocin treatment (c). * $P<0.05$ by Mann-Whitney U-test.

AQP3 and AQP4 are known as the major water channels on basolateral plasma membranes of collecting duct epithelial cells (Ishibashi *et al.,* 1997). Nielsen *et al.,* (1993) reported that AQP2 was also located on the basolateral plasma membrane of IMCD cells. However, the role(s) and regulatory mechanism(s) of basolaterally located AQP2 are not clear. In addition, how oxytocin stimulates basolateral targeting of AQP2 in renal epithelial cells is far from our knowledge at this time of writing. Recently, Terashima *et al.*, (1999) reported that oxytocin downregulated V_2 receptor mRNA and upregulated AQP2 mRNA expression in the collecting duct. They suggested that oxytocin upregulates AQP2 possibly through V_2 receptor.

However, our finding that oxytocin induces redistribution of AQP2 targeting from cytoplasm to both apical and basolateral plasma membranes of renal epithelial cells indicates that oxytocin might have a different way from vasopressin in the regulation of AQP2. Han *et al.*, (1993) demonstrated that oxytocin and vasopressin are involved in the regulation of water permeability in collecting duct through other receptor(s) as well as V_2 receptor, and it has been suggested that other than V_2 receptor may play a role in the regulation of AQP2 (Knepper 1997, Ecelbarger *et al.*, 1998, Marples *et al.*, 1998). Therefore, oxytocin may be different in the mechanism of regulation of AQP2 from vasopressin on the level of a receptor.

For this purpose, we examined the effect of V_2 antagonist pretreatment on the oxytocin-induced redistribution of AQP2 protein in rat kidney. The experimental rats (n=3) were fed a fixed amount of food slurry containing OPC-31260, a specific V_2 receptor antagonist, 20 mg/200 g BW/d for 3 days and treated with 10 units of oxytocin by a single intraperitoneal injection. Control rats (n=3) were fed the same amount of food slurry but without OPC-31260 and treated with the same dose of intraperitoneal oxytocin. Both groups had free access to drinking water. As expected, urine flow was higher in OPC-treated rats (24.2 ± 0.2 ml/d) than in controls(15.8 ± 0.8ml/day, $P<0.05$). Urine osmolality was lower in OPC-treated rats (554.3 ± 98.9 mOsm/kg) than in controls (924.7 ± 30.7 mOsm/kg). Immunocytochemistry revealed that AQP2 was found on both apical and basolateral membranes in oxytocin-treated control rats. This was the similar pattern of bipolarization observed in the previous experiment, although the general intensity of immunostaining was somewhat weaker. In OPC-31260 pre-treated rats, AQP2 labeling of basolateral plasma membrane was vanished and translocated to cytoplasm along the collecting duct. However, the apical staining of AQP2 was not affected by OPC-31260 pre-treatment. These results suggest that basolaterally targeted AQP2 induced by exogenous oxytocin may be mediated through V_2 receptor.

3. URINARY AQUAPORIN-2 EXCRETION FROM HUMAN

Previous studies have shown that AQP2 protein is measurable in the urine from human subjects by either RIA (Kanno *et al.*, 1995) or quantitative Western analysis (Elliot *et al.*, 1996). Urinary AQP2 may represent the altered levels of AQP2 in the apical plasma membrane of the collecting duct (Elliot *et al.*, 1996, Wen *et al.*, 1999), and changes in the urinary excretion of AQP2 can be used as an index of the action of vasopressin on the kidney (Kanno *et al.*, 1995).

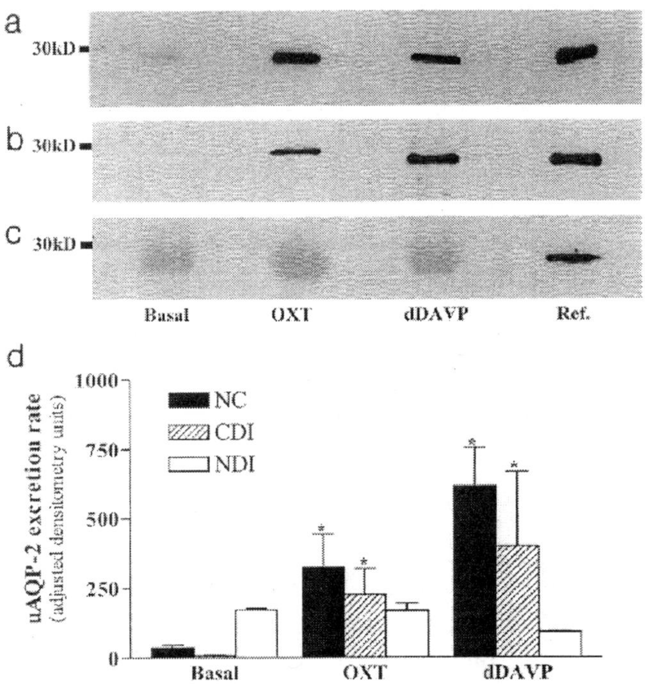

Figure 3. Immunoblots of urine AQP2 from a normal volunteers(a), a patient with CDI(b), a patient with NDI(c), and comparison of relative amount of urinary AQP2 excretion(d). Urinary excretion of AQP2 increased with treatment of oxytocin and dDAVP in normal volunteers and CDI patients. Each band density was corrected according to the density of reference sample. Band densities were adjusted by multiplicating the loaded sample amount-urine volume ratio(d). * P<0.05 compared with the basal data by Mann-Whitney U-test. Ref, Reference sample; Basal, Basal data; OXT, Data after oxytocin infusion; dDAVP, Data after the injection of dDAVP.

We investigated whether exogenous oxytocin affects urinary concentration in human and accompany changes in urinary AQP2 excretion. Intravenous oxytocin was administered at a rate of 20 mU/min over 2.5 hours in 10 healthy male volunteers, 7 patients with central diabetes insipidus(CDI) and 3 patients with nephrogenic diabetes insipidus (NDI). Two-hour urine was collected before and after the administration of oxytocin, and urinary AQP2 was measured by quantitative Western blot analysis. Urine volume and free-water clearance decreased significantly with oxytocin administration in normal volunteers and CDI patients but not in NDI patients. Urine osmolality increased with oxytocin administration in normal volunteers and CDI patients but not in NDI patients. Notably, urinary AQP2 excretion increased about 20-fold with oxytocin treatment in normal volunteers and CDI patients but not in NDI patients(Fig. 3). These results indicate that exogenous oxytocin has antidiuretic effect in human and its antidiuretic action may be mediated by AQP2 in the collecting duct.

4. CONCLUSION

We have demonstrated that localization of AQP2 protein is affected by short-term oxytocin administration in both rat and human, suggesting that oxytocin exerts antidiuretic action by regulating AQP2 targeting in a different way from vasopressin. We are left with the question about physiologic significance of basolaterally targeted AQP2 water channel along the connecting tubule and collecting duct. Further studies are required to elucidate intracellular pathways from oxytocin binding to AQP2 targeting in the kidney.

ACKNOWLEDGMENTS

This work is partly supported by the Grant 97-N1-02-04-A-05 from Ministry of Science and Technology of Korea and BK21 from Ministry of Education. We would like to thank Dr. Young Hee Kim, Ms. So Young Kim and Hanna Jung for their excellent technical assistance to our studies, and also to Drs. Yoon Sook Cho, Ki Young Na, Ji Eun Oh, Woo Kyoung Jung, Dae Youn Whang and Kang Suk Kim for their patients recruitment.

REFERENCES

Abdul-Karim R and Assali NS 1961 Renal function in human pregnancy V Effects of oxytocin on renal hemodynamics and water and electrolyte excretion *J Lab and Clin Med* **57**: 522-533

Ahmed AJ Clark EH and Jacobs HS 1975 Water intoxication associated with oxytocin infusion *Postgrad Med J* **51**: 249-252

Chou CL Digiovanni SR Luther A Lolait SJ and Knepper MA 1995a Oxytocin as an antidiuretic hormone II Role of V_2 vasopressin receptor *Am J Physiol* **269**: F78-85

Chou CL Diogiovanni SR Mejia R Nielsen S and Knepper MA 1995b Oxytocin as an antidiuretic hormone I Concentration dependent of action *Am J Physiol* **269**: F70-77

Ecelbarger CA Nielsen S Olson BR Murase T Baker EA and Knepper MA 1997 Role of renal aquaporins in escape from vasopressin-induced antidiuresis in rat *J Clin Invest* **99**: 1852-1863

Ecelbarger CA Chou CL Lee AJ DiGiovanni SR Verbalis JG and Knepper MA 1998 Escape from vasopressin induced antidiuresis: role of vasopressin resistance of the collecting duct *Am J Physiol* **274**: F1161-F1166

Edwards BR Jr and LaRochelle FT 1984 Antidiuretic effect of endogenous oxytocin in dehydrated Brattleboro homozygous rats *Am J Physiol* **247**: F453-465

Elliot S Goldsmith P Knepper M Haughey M and Olson B 1996 Urinary excretion of aquaporin-2 in humans: A potential marker of collecting duct responsiveness to vasopressin *J Am Soc Nephrol* **7**: 403-409

Han JS Maeda Y and Knepper MA 1993 Dual action of vasopressin and oxytocin in regulation of water permeability in terminal collecting duct *Am J Physiol* **265**: F26-F34

Ishibashi K Sasaki S Fushimi K Yamamoto T Kuwahara M and Marumo F 1997 Immunolocalization and effect of dehydration on AQP3 a basolateral water channel of kidney collecting ducts *Am J Physiol* **41**: F235-F241

Kanno K Sasaki S Hirata Y Ishkawa S Fushimi K Nakanishi S Bichet DG and Marumo F 1995 Urinary excretion of aquaporin-2 in patients with diabetes Insipidus *N Eng J Med* **332**: 1540-1545

Knepper MA 1997 Molecular physiology of urinary cencebtrating mechanism: regulation of aquaporin water channels by vasopressin *Am J Physiol* **272**: F3-F12

Marples D Christensen S Christensen EI Ottosen PD and Nielsen S 1995a Lithium-induced downregulation of aquaporin-2 water channel expression *J Clin Invest* **95**: 1838-1845

Marples D Knepper MA Christensen EI and Nielsen S 1995b Redistribution of aquaporin-2 water channels induced by vasopressin in rat kidney inner medullary collecting duct *Am J Physiol* **269**: C655-C664

Marples D Christensen BM Frokiaer J Knepper MA and Nielsen S 1998 Dehydration reverses vasopressin antagonist-induced diuresis and aquaporin-2 downregulation in rats *Am J Physiol* **275**: F400-F409

Nielsen S DiGiovanni SR Christensen EI and Knepper MA 1993 Cellular and subcellular immunolocalization of vasopressin-regulated water channel in rat kidney *Proc Natl Acad Sci USA* **90**: 11663-11667

Nielsen S Chou CL Marples D Christensen EI Kishore BK and Knepper MA 1995 Vasopressin increases water permeability of kidney collecting duct by inducing translocation of aquaporin-CD water channels to plasma membrane *Proc Natl Acad Sci USA* **92**:1013-1017

Terashimi Y Kondo K and Osio Y 1999 Administration of oxytocin affects vasopressin V_2 receptor and aquaporin-2 gene expression in the rat *Life Sci* **64**: 1447-1453

Wen H Frokiaer J Kwon TH and Nielsen S 1999 Urinary excretion of aquaporin-2 in rat is mediated by a vasopressin-dependnet apical pathway *J Am Soc Nephrol* **10**: 1416-1429

2.8 Mammalian physiology

FLUID TRANSPORT BY HUMAN NONPIGMENTED CILIARY EPITHELIAL MONOLAYER

[1]Zhiqiang Han and [1,2]Rajkumar V. Patil
Departments of [1]Ophthalmology and Visual Sciences, and [2]Molecular Biology and Pharmacology Washington University School of Medicine, St. Louis, Missouri 63110. E-mail: patil@vision.wustl.edu

1. INTRODUCTION

Ciliary epithelium is responsible for the secretion of aqueous humor into posterior chamber of the eye. Most of aqueous humor secretion is driven by the active transport of ions from plasma into the posterior chamber followed by rapid movement of water and water soluble substances (Jacob and Civan, 1996). Both the outer nonpigmented (NPE) and inner pigmented (PE) epithelial layers exhibit properties of transporting epithelia (Helbig *et al.*, 1988; Wolosin *et al.*, 1993). The NPE is thought to provide the direct driving force for aqueous humor formation. Na^+/K^+-ATPase which is located on the basolateral membranes of NPE drives sodium secretion and provides the main ion motive force for sodium dependent cotransporters (Coca-Prados and Lopez-Briones, 1987; Flugel and Lutjen-Drecoll, 1988; Okami *et al.*, 1989; Ghosh *et al.*, 1990). Recently, it has been proposed that the rate limiting step in active secretion of aqueous humor is a Cl^- transport via a DIDS sensitive, cAMP dependent chloride channel (Chen *et al.*, 1994; Jacob and Civan, 1996). Although the Na^+/K^+-ATPase active transport system is a primary transducer of cellular metabolic energy into fluid transport, it does not completely account for a variety of observations and functions that are prerequisites for the effective transport of fluid and electrolytes by the dual-layered ciliary epithelium. For example, cell volume regulation which is coupled to ion transport through parallel K^+ and, more importantly, Cl^-

Molecular Biology and Physiology of Water and Solute Transport
Edited by Hohmann and Nielsen, Kluwer Academic/Plenum Publishers, New York, 2000.

channel activity appear to be an intrinsically important aspects of normal aqueous secretion (Civan *et al.*, 1992; Yantorno *et al.*, 1992; Fischbarg, 1997). It has been suggested that the same mechanisms that underlie volume regulation may also subserve fluid secretion (Civan *et al.*, 1992).

Fluid transport is a major function of the ciliary epithelium. In the human eye, there are approximately four million nonpigmented ciliary epithelial cells with an estimated volume of 8 μl (Brubaker, 1998). Their function has been measured in vivo indirectly as aqueous flow by fluorophotometry (Jones and Maurice, 1966). It is estimated that over the course of a day, the ciliary epithelium secretes approximately 3.25 ml of aqueous humor (peak flow of 3 μl/min and trough flow of 1.5 μl/min i.e. average flow of 2.25 μl per min). It would be necessary for these cells to secrete nearly one third of their volume every minute at peak aqueous flow. Given the secretory capacity of this tissue, there have been tremendous efforts in identifying the basic transport process responsible for aqueous humor formation using transepithelial and intracellular electrical recording, radioactive and spectroscopic measurements, and patch clamping techniques. Studies of the intact excised ciliary epithelium has helped to reach the basic understanding of aqueous humor formation free from the complexities of uncontrolled variations in ciliary blood flow and outflow dynamics of the aqueous humor. However, there are no reported, convenient in vitro models for studying the fluid transport across ciliary epithelial cells and the mechanism by which rapid water exchange accompanies ion transport into the aqueous humor formation remains poorly understood. Ciliary epithelial cells have been cultured for several years, and characterized for the presence of several ion transporters and receptors (Wax and Patil, 1994; Ortego and Coca-Prados, 1999). The purpose of this report is to demonstrate that cultured monolayers of these cells can be used successfully to measure the osmotically directed transepithelial fluid movement in vitro.

2. METHODS

2.1 Cultured human NPE cell layers

The human NPE cells (passage # 8) were grown on 24.5 mm diameter transparent permeable, collagen-treated Transwell supports (Costar, Cambridge, MA) at a density of 1 X 10^6 per support in Dulbecco's modified Eagle medium (DMEM) supplemented with 10% fetal bovine serum (FBS), in a 5% CO_2 atmosphere at 37 °C. Transport experiments were performed within 1-7 days after confluence. Confluence of the cell culture was determined both by microscopically and by determining the specific resistance of each insert using an Endohm chamber and a resistance meter (both from WPI, Sarasota, FL). For 12 h prior to their beginning and throughout the experiments, the cell layers were bathed by the DMEM containing FBS without antibiotics. The basal side of the confluent NPE layer grown in this fashion is closer to the collagen support and the apical side is facing away from the support.

2.2 Measurement of Fluid Transport Rates

The NPE cells were grown to confluence on 24.5 mm diameter Transwell-Col culture chambers and 4 h prior to the experiments the apical medium was replaced by 0.4 ml of hyperosmotic DMEM medium (400 mosm i.e. 100 mM sucrose in DMEM medium), and the medium from the basolateral chamber was replaced by fresh isoosmotic (300 mosm) DMEM medium. Mercuric chloride was added to the top chamber to study its effect on the net fluid movement by NPE cells. After 4 h, the liquid on the apical side was carefully collected and its volume was measured with a calibrated pipette, and the net fluid movement was determined. The results are expressed as fluid flow in μl secreted/cm^2/hour and are the means + S.E. of three separate experiments performed in triplicate.

3. RESULTS AND DISCUSSION

In order to determine if the NPE cell layer transport fluid in vitro, we measured the transepithelial net fluid movement rate in these cells grown on collagen-coated permeable filters. Figure 1 shows the cells grown on the filters form a tight monolayer. In some experiments, the electrical resistance of the cells was measured to confirm the confluence of the cells. In one experiment, we determined fluid movement in response to differences in hydrostatic pressure between the inside and outside surfaces of Transwell membranes in the absence of a cellular monolayer. Volumes of fluid in the inner (0.4 ml) and outer (1.0 ml) chambers were selected which produced no net change in fluid movement across the supporting membrane in the absence of cells. To determine if evaporation occurred during the incubation period, samples of known osmolarity were incubated at no transmembrane hydrostatic pressure in Transwells containing no cells for 4 h, after which the osmolarity of the upper chamber and bottom chamber fluids was measured. There was no difference between the osmolarity of the top and bottom chamber fluid after 4 h.

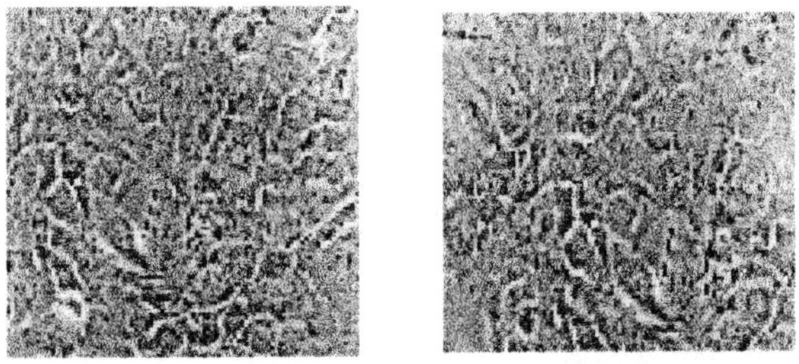

Figure 1. Photographs of nonpigmented ciliary epithelial monolayers cultured on Transwell-Col supports observed by phase contrast. The human NPE cells (passage # 8) were grown on 24.5 mm diameter transparent permeable, collagen-treated supports at a density of 1 X 106 per support in DMEM supplemented with 10% FBS, in a 5% CO2 atmosphere at 37 oC. The basal side of the confluent NPE layer grown in this fashion is closer to the collagen support and the apical side is facing away from the support. The cells were incubated with or without mercury chloride for 4 hour.

Figure 2 shows that NPE cells from human can be successfully used to measure the net fluid transport in vitro. Fluid movement was measured as described in 'Methods' section. After 4 hour, the liquid from top chamber was collected, its volume measured with a calibrated pipette, and the net fluid secretion rate was determined. Cultured NPE layers transported 1.96 $\pm$ 0.12 $\mu l.h^{-1}.cm^2$ (n = 3) fluid from their basal to their apical sides in response to osmotic gradient. Mercury chloride is an unspecific but potent blocker of Hg-sensitive aquaporins. Fluid transport by NPE monolayers was reduced by 25.5 and 61.2% in the presence of 50 and 100 μM $HgCl_2$. Mercury chloride was not harmful to the cells at this concentrations as the cells were intact after the treatment (see Figure 1). If mercury chloride was harmful to the cells then a toxic effect of mercuric chloride would have increased the leak and therefore there will be increased fluid movement from bottom chamber to top chamber. Instead, mercuric chloride caused decrease in fluid movement compared to the control, suggesting that it affected pathways for fluid movement, such as water channels. Typically, experiments were performed in three inserts on three different batches of cells. Bumetanide, a potent inhibitor of Na-K-Cl cotransport, had no effects on the fluid transport by these monolayers (data not shown). These results are consistent with the observation that bumetanide had no effect on the aqueous humor formation and intraocular pressure in organ-cultured human eyes (Uusitalo *et al.*, 1985; Gabelt *et al.*, 1997).

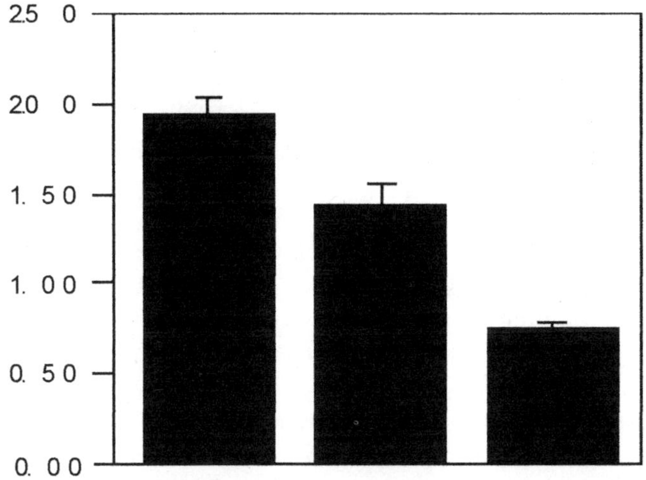

Figure 2. Net fluid transport by the human nonpigmented ciliary epithelial monolayers. The human NPE cells were grown to confluence on 24.5 mm diameter Transwell-Col culture chambers and the apical fluid was replaced by 0.4 ml of hyperosmotic medium (400 mosm i.e. 100 mM sucrose in medium), and the medium from the basolateral chamber was replaced by fresh isoosmotic (300 mosm) medium. Mercuric chloride was added to the apical side of layer (top chamber). After 4 h, the liquid on the apical side was collected and its volume was measured with a calibrated pipette, and the net fluid movement was determined. The results are expressed as fluid flow in ml secreted/cm2/hour and are the means + S.E. of three separate experiments performed in triplicate.

Since the fluid movement took place from apical to basal side of NPE cells in vivo we modified our experimental system to measure the net fluid secretion rates of NPE cells under more physiological conditions i.e. in the presence of a transepithelial basolateral to apical osmotic gradient by replacing the bottom chamber growth medium with hyperosmotic medium (400 mosm) and top chamber growth medium with isoosmotic (300 mosm) medium. After 4 h the amount of fluid from top chamber was collected and the net fluid movement from top chamber to bottom chamber was calculated. The net fluid movement rate by NPE cells under these conditions was similar to the fluid transport rate from basal to apical side (Table 1) and once again mercuric chloride reduced the net fluid movement by these cells.

Table 1. Net fluid movement across NPE monolayers (μl/cm²/h)

NPE cells	Hyperosmotic Top Chamber	Hyperosmotic Bottom Chamber
Control	1.96 ± 0.12	1.75 ± 0.14
50 μM Hg	1.46 ± 0.15	1.39 ± 0.12
100 μM Hg	0.76 ± 0.08	0.81 ± 0.14

We observed the fluid movement by nonpigmented ciliary epithelium alone without the intervention of second pigmented layer, which suggest the importance of this layer in the fluid transport by ciliary epithelium. Our observation is consistent with the recent report by Walker et al in which they observed cell volume changes with the movement of fluid from pigmented to nonpigmented layer (Walker *et al.*, 1999). In addition, present results indicate that cultures of the individual cell layers, like the one utilized here, would constitute a useful in vitro model to study the aqueous humor secretion by the ciliary epithelium. However, the rate of fluid transport by the pigmented and nonpigmented layer when coupled together could be more efficient since any one of these two layers independently transport fluid. Since this method is highly dependable and needs no specialized equipments it will be very useful in studying the effect of various other compounds such as cAMP, forskolin, and Atrial Natriuretic Peptide (ANP), which modulate the aqueous humor inflow, on the net fluid transport across the NPE cells. In summary, we have shown that human cultured NPE layer actively transports the fluid from basal to apical as well as apical to basal in response to osmotic gradient. Furthermore, these cells will provide a simple in vitro model in determining the role of various channel proteins such as aquaporins in the aqueous inflow pathway employing antisense technology and provide new insights into normal physiology and disease mechanisms, and may yield novel, more efficacious, and perhaps safer therapies to regulate fluid movement in glaucoma.

ACKNOWLEDGMENTS

Supported by NIH grants EY10423 (RVP), Core Grant, EY02687, and Unrestricted grant from Research to Prevent Blindness, Inc. RVP is a Research to Prevent Blindness Olga Keith Wiess Scholar

REFERENCES

Brubaker R 1998 Clinical Measurements of Aqueous Dynamics: Implications for Addrerssing Glaucoma In: The Eye's Aqueous Humor From Secretion to Glaucoma (M Civan Ed) Academic Press New York pp 234-284

Chen S Inoue R Inomata H and Ito Y 1994 Role of cyclic AMP-induced Cl conductance in aqueous humour formation by the dog ciliary epithelium *Br J Pharmacol* **112**: 1137-45

Civan MM Peterson-Yantorno K Coca-Prados M and Yantorno RE 1992 Regulatory volume decrease by cultured non-pigmented ciliary epithelial cells *Exp Eye Res* **54**: 181-191

Coca-Prados M and Lopez-Briones L 1987 Evidence that the a and a(+) isoforms of the catalytic subunit of NaK-ATPase reside in distinct ciliary epithelial cells of the mammalian eye *Biochem Biophys Res Comm* **145**: 460-466

Fischbarg J 1997 Mechanism of fluid transport across corneal endothelium and other epithelial layers: a possible explanation based on cyclic cell volume regulatory changes *Br J Ophthalmol* **81**: 85-89

Flugel C and Lutjen-Drecoll E 1988 Presence and distribution of Na^+/K^+-ATPase in the ciliary epithelium of the rabbit *Histochem* **88**: 613-621

Gabelt BT Wiederholt M Clark AF and Kaufman PL 1997 Anterior segment physiology after bumetanide inhibition of Na-K-Cl cotransport *Invest Ophthalmol Vis Sci* **38**: 1700-1707

Ghosh S Freitag AC Martin-Vasallo P and Coca-Prados M 1990 Cellular distribution and differential gene expression of the three a subunit isoforms of the NaK-ATPase in the ocular ciliary epithelium *J Biol Chem* **265**: 2935-2940

Helbig H Korbmacher C Berweck S Kuhner D and Wiederholt M 1988 Culture of ciliary body epithelia and studies of ion transport *Fortschr Ophthalmol* **85**: 42-45

Jacob TJ and Civan MM 1996 Role of ion channels in aqueous humor formation *Am J Physiol* **271**: C703-C720

Jones RF and Maurice DM 1966 New methods of measuring the rate of aqueous flow in man with fluorescein *Exp Eye Res* **5**: 208-220

Okami T Yamamoto A Omori K Akayama M Uyama M and Tashiro Y 1989 Quantitative immunocytochemical localization of Na^+K^+-ATPase in rat ciliary epithelial cells *J Histochem Cytochem* **37**: 1353-1361

Ortego J and Coca-Prados M 1999 Functional expression of components of the natriuretic peptide system in human ocular nonpigmented ciliary epithelial cells *Biochem Biophys Res Commun* **258**: 21-28

Uusitalo RJ Palkama A and Uusitalo H 1985 Stimulation of adenylate cyclase of ciliary processes in response to decreased inflow of aqueous humour *Acta Ophthalmol (Copenh)* **63**: 401-407

Walker VE Stelling JW Miley HE and Jacob TJ 1999 Effect of coupling on volume-regulatory response of ciliary epithelial cells suggests mechanism for secretion *Am J Physiol* **276**: C1432-C1438

Wax MB and Patil RV 1994 Immunoprecipitation of A1 adenosine receptor-GTP-binding protein complexes in ciliary epithelial cells *Invest Ophthalmol Vis Sci* **35**: 3057-3063

Wolosin JM Chen M Gordon RE Stegman Z and Butler GA 1993 Separation of the rabbit ciliary body epithelial layers in viable form: identification of differences in bicarbonate transport *Exp Eye Res* **56**: 401-409

Yantorno RE Carre DA Coca-Prados M Krupin T and Civan M 1992 Whole cell patch clamping of ciliary epithelial cells during anisosmotic swelling *Am J Physiol* **262**: C501-C509

2.9 Mammalian physiology

STRUCTURAL FEATURES OF BARRIER MEMBRANES

Warren G. Hill, John C. Mathai and Mark L. Zeidel
Renal-Electrolyte Division, Department of Medicine, University of Pittsburgh, Pittsburgh, Pennsylvania 15261, U.S.A.. E-mail: whill@pitt.edu

1. INTRODUCTION

Most cell membranes are surprisingly permeable to water and small uncharged solutes. This property is a consequence of both the fluid and anisotropic nature of the phospholipid bilayer. Water molecules are able to partition into the proximal region of the hydrocarbon core and then diffuse across the bilayer before dissolving in the aqueous environment on the other side. While this 'solubility-diffusion' hypothesis is well accepted, the mechanism of diffusion of water molecules across the bilayer is still a matter of intense speculation. One model hypothesises the existence of "mobile kinks" which initially provide a free volume space for a water molecule to occupy, and which then translocates the water molecule by virtue of consecutive *trans-gauche* isomerisations along the length of the acyl chain, combined with lateral diffusion of the phospholipid within the bilayer. The mobile kink is in effect a self-propagating "hole" which carries the water molecule across the membrane (Haines and Liebovitch, 1994). Another model (Deamer and Bramhall, 1986) postulates the presence of transient defects at the water/bilayer interface which allows a "strand" of water, perhaps only several molecules long, to penetrate into the bilayer. Individual water molecules then break off and can be translocated across the bilayer by virtue of further defects which arise from thermal fluctuations. While the precise mechanism may not be known, the functional consequences of high membrane permeability are that water and solutes will equilibrate across membranes in response to simple osmotic or concentration gradients. In

Molecular Biology and Physiology of Water and Solute Transport
Edited by Hohmann and Nielsen, Kluwer Academic/Plenum Publishers, New York, 2000.

contrast to this, there are several specialised cell membranes which exhibit exceptionally low water permeability. These are termed barrier membranes and are most often found at the apical pole of epithelial cells. Since epithelia represent the interface between organism and environment it is not surprising that special membranes have evolved to protect the organism by compartmentalising molecular and aqueous gradients. Good examples of this may be seen in the stomach, the urinary bladder and in the filtration organ *par excellence,* the kidney. Epithelia lining these structures must, for protective or functional purposes, effectively be impermeable. In this chapter we wish to present new information on three aspects of apical membrane structure that contribute to barrier function, and contrast these strategies with those employed by much simpler organisms – the archaebacteria – which manage to survive in extreme environments.

2. UNIQUE FEATURES OF THE APICAL MEMBRANE

2.1 Bilayer Asymmetry

The apical membrane of epithelial cells is a highly unusual membrane domain because the two leaflets which comprise the bilayer have different lipid compositions. The cell achieves this by lipid sorting in the trans-Golgi network and then packaging those lipids into transport vesicles destined for the apical membrane (Simons and van Meer, 1988). This results in asymmetric delivery of specific lipids to either the outer or the inner leaflet of the plasma membrane. Once lipids have been topologically oriented in this way, their asymmetric distribution is maintained by the presence of tight junctions which effectively “fence off” the outer leaflet lipids while leaving the inner to freely mix through lateral diffusion with the basolateral membrane (van Meer and Simons, 1986). Asymmetry may also be maintained through the activity of energy-dependent flippases which can move specific phospholipids from one leaflet to the other. While it is known that lipids such as the glycosphingolipids (GSLs) are enriched in the outer leaflet of epithelial apical membranes, the functional significance of such specific targeting has not been clearly established. Our working hypothesis was that the localisation of particular lipids in the outer leaflet might serve to reduce the fluidity of that leaflet and in doing so reduce the overall permeability of the membrane.

In an effort to model this phenomenon we conducted experiments with the rare earth metal praeseodymium (Pr^{3+}) and artificial membranes in the form of liposomes. Pr^{3+} binds to the phosphocholine head groups of dipalmitoylphosphatidylcholine (DPPC) liposomes and thereby rigidifies the outer leaflet of the bilayer. The reduction in fluidity at temperatures above the phase-transition (T_m) were confirmed by fluorescence anisotropy measurements using an anisotropy probe which specifically localised to the outer leaflet (DPH-HPC). Permeability measurements were performed using stopped-flow fluorimetry. One syringe containing vesicles or liposomes loaded with a fluorescent dye is rapidly mixed with a solution of different composition designed to create an extravesicular gradient. Water efflux in response to a rapidly imposed osmotic gradient can be measured as a function of vesicle shrinkage and dye self-quenching. Other similar assays exist for measuring solute efflux and NH_3 and proton influx. Liposomes which had a single leaflet rigidified showed reduced membrane permeability to water, small non-electrolytes and NH_3 (Negrete *et al.*, 1996a; Hill *et al.*, 1999). This demonstrated that bilayer asymmetry was a potential mechanism by which cells could erect a barrier membrane. A further observation to emerge from these studies was that each leaflet in a phospholipid bilayer offered an independent resistance to permeation. This novel insight indicates that the membrane can be thought of as two thin sheets of lipid just a single molecule deep and loosely associated with each other for energetic considerations, but with completely independent and additive biophysical properties. This relationship can be summarised by the following equation in which the resistance to permeation can be rewritten as the reciprocal of the permeability:

$$\frac{1}{P_{AB}} = \frac{1}{P_A} + \frac{1}{P_B} \quad \text{(Eq. 1)}$$

where P_{AB} is the permeability of a bilayer composed of leaflets A and B; P_A is the permeability of leaflet A and P_B is the permeability of leaflet B.

2.2 Lipid Composition

It has been apparent for many years that the apical membrane of epithelial cells has some unusual compositional features (for review see Simons and van Meer, 1988). Elegant experiments by van Meer and Simons (1982) demonstrated that enveloped viruses which budded from different membrane domains of MDCK cells had very different lipid compositions and that these lipids were a reflection of the apical or basolateral exit points. Further work from the same investigators demonstrated that the outer leaflet of the apical membrane of these cells was enriched in sphingomyelin and GSLs (van Meer *et al.,* 1987). The functional significance of localising specific lipids in this way is still unclear. However, Type 1 MDCK cells have been shown to have extremely low water permeability when grown as polarised cells on semi-permeable supports (Lavelle *et al.,* 1997). We decided to attempt to reconstitute the barrier properties of the MDCK apical membrane by making liposomes which had compositions equivalent to the outer leaflet and the inner leaflet. Using published analyses of the lipid composition of this membrane (Hansson *et al.,* 1986; van Meer and Simons, 1986) and by making a number of simplifying assumptions we made leaflet-specific liposomes as shown in Table 1. It is apparent that the outer leaflet is enriched in sphingolipids and cholesterol while the inner is predominantly composed of phosphatidylethanolamine and phosphatidylserine.

Table 1. Predicted composition of the MDCK cell apical membrane

Lipids[1]	Outer (Exofacial) Leaflet (mol%)	Inner (Cytoplasmic) Leaflet (mol %)
PE	-	40.1
PS	-	20.1
PI	-	2.5
Cholesterol	52.3	37.3
PC	8.6	-
SM	18.4	-
GSL	20.7	-

[1] PE, phosphatidylethanolamine; PS, phosphatidylserine; PI, phosphatidylinositol; PC, phosphatidylcholine; SM, sphingomyelin

Exofacial liposomes had approximately 20-fold lower water permeability than the cytoplasmic liposomes. When solute permeabilities were measured, the difference was as high as 90-fold for large solutes like glycerol. This clearly established that the exofacial leaflet was able to generate a considerable resistance to water and solute permeability. When we modified the composition of the outer leaflet liposomes by removing individual components, we found that SM and GSLs both contributed to the barrier. Their removal resulted in a modest (3 – 6-fold) increase in permeability. However, removal of cholesterol resulted in massive increases in permeability (up to 280-fold). This indicated a critical role for cholesterol, but more importantly implicated interactions between the sphingolipids and cholesterol as being necessary to build a watertight barrier at the outer leaflet. From eq. 1 we derived a theoretical permeability for our reconstituted apical membrane that matched very closely the experimentally measured permeabilities for this membrane as determined by others.

2.3 Integral Membrane Proteins

The mammalian urinary bladder fulfils an essential role in allowing the body to store and then void the urine which collects as a result of blood filtration in the kidneys. Given that the composition of urine may vary

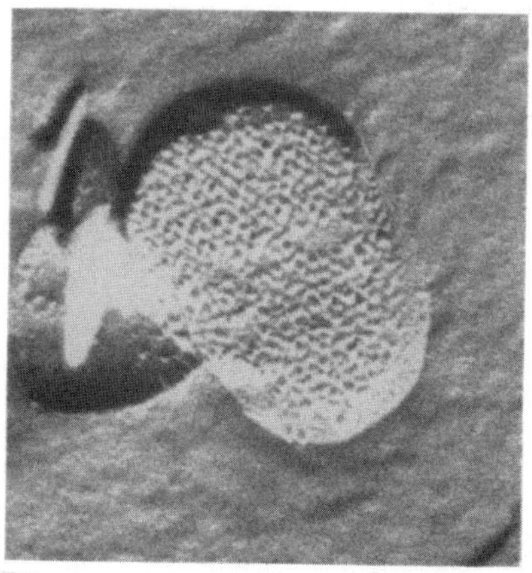

Figure 1. Uroplakin Proteoliposome

enormously (osmolalities may range from 50 – 1000 mOsm/kg and pH's from 4.5 – 9.0), the bladder's ability to maintain strikingly high chemical gradients when compared to the blood and surrounding interstitium, is remarkable. The features of the lumenal surface of the bladder which confer this remarkable impermeability are unknown, although it is now accepted that the barrier to permeation resides in the apical membrane of the superficial epithelial cells (Lewis and Hanrahan, 1990; Negrete *et al.*, 1996b).

Early electron microscopic analyses of mammalian bladder revealed the presence of a highly unusual hexagonal latticework that covered a large percentage of the lumenal surface area of the bladder (Caruthers and Bonneville, 1977; Stubbs *et al.*, 1979). These hexagonal particles caused an apparent thickening of the outer leaflet of the membrane and hence came to be termed the asymmetric unit membrane (AUM). Later studies revealed that the AUM was composed of proteins embedded into the outer leaflet of the apical membrane. The proteins consisted of four subunits termed uroplakins Ia, Ib, II and III, which interacted tightly with each other to form highly symmetric arrays (Wu and Sun, 1993; Wu *et al.*, 1994). The uroplakins are uniquely localised to the bladder and are a late stage marker of urothelial differentiation. Their function is currently unknown.

To examine whether uroplakins may be contributing to the barrier function of the bladder we obtained highly purified bovine uroplakins and reconstituted the protein into proteoliposomes. Freeze-fracture electron microscopy revealed a population of liposomes of uniform diameter (120-140 nm) containing variable amounts of membrane-embedded uroplakin. Figure 1 shows that some of the reconstituted vesicles had up to 100% of their surface area covered by protein plaques. When we tested the control and protein-embedded liposomes we detected a 40% reduction in water permeability with uroplakin present ($P < 0.05$). We estimate that areas of membrane covered with protein have at least a 10-fold lower permeability than naked membrane. This is the first time that integral membrane proteins have been shown to increase a membrane's resistance to the permeation of water and offers the first evidence for a potential function of uroplakins in the bladder.

3. ARCHAEA LIPIDS

Archaebacteria live in harsh environmental conditions with pH values as low as 2 to 3 and temperatures as high as 85 °C. One of the unique characteristics of this group is the presence of ether-linked lipids in place of the usual ester-linked lipids of the eubacteria and eukaryotes. In general these complex ether lipids fall into two major classes, diphytanyl glycerol diether and dibiphytanyl diglycerol tetraether lipids (Fig. 2). The hydrocarbon tail in some cases also contains a number of cyclopentane rings in the C_{40} chains. The interesting structure of these lipids led us to examine the permeability properties of liposomes composed of usual diester-linked lipids (but containing isoprenoid chain) and diether, tetraether-linked lipids. The water permeability of liposomes composed of ether-linked lipid was reduced by 30-40% compared to liposomes of ester-linked lipids. The permeability of solutes such as glycerol and urea decreased by 40% in liposomes composed of ether lipids. Strikingly, the proton permeability of ether lipids decreased by 90% whereas the ammonia permeability decreased by more than 95%. These results indicate a very tight packing of the ether lipids and the presence of a selective permeability barrier. In addition, permeability studies on bilayers composed of these lipids will allow us to define the quantitative contribution of each element of lipid structure to permeation of small molecules. Since some of these bacteria live in acidic conditions, the extremely low lipid permeability to protons would help sustain a proton gradient required for energy production (Elferink *et al.* 1994).

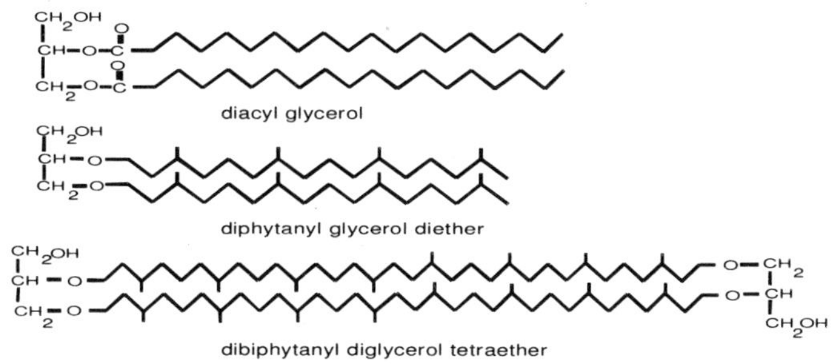

Figure 2. Unusual structure of archaea lipids

4. CONCLUSIONS

Barrier membranes are critical for the proper function of a number of organs, and the consequences of barrier breakdown may be seen in human diseases such as interstitial cystitis, carcinoma *in situ* and peptic ulcer disease. A greater knowledge of how epithelial tissues in particular are able to generate barrier membranes, is important not only for our understanding of normal physiological processes but also an understanding of how barriers may be compromised.

ACKNOWLEDGMENTS

We wish to thank Dr. Tung-Tien Sun and Dr. Xiangpong Kong for providing purified bovine uroplakins and Dr. Dennis Sprott for archaebacteria lipids. Thanks also to Dr. Donna Beer-Stolz for freeze-fracture EM. This work was supported by NIH grant DK43955.

REFERENCES

Caruthers JS and Bonneville MA 1977 Isolation and characterization of the urothelial lumenal plasma membrane *J Cell Biol* **73**: 382-99

Deamer DW and Bramhall J1986 Permeability of lipid bilayers to water and ionic solutes *Chem Phys Lipids* **40**: 167-88

Elferink MG de Wit JG Driessen AJ Konings WN 1994 Stability and proton-permeability of liposomes composed of archaeal tetraether lipids *Biochim Biophys Acta* **1193**: 247-54

Haines TH and Liebovitch LS A molecular mechanism for the transport of water across phospholipid bilayers In Permeability and Stability of Lipid Bilayers Disalvo EA Simon SA Ann Arbor: CRC Press 1994 pp123-136

Hansson GC Simons K and van Meer G 1986 Two strains of the Madin-Darby canine kidney (MDCK) cell line have distinct glycosphingolipid compositions *EMBO J* **5**: 483-489

Hill WG Rivers RL and Zeidel ML 1999 Role of leaflet asymmetry in the permeability of model biological membranes to protons solutes and gases *J Gen Physiol* **114**: 405-414

Lavelle JP Negrete HO Poland PA Kinlough CL Meyers SD Hughey RP and Zeidel ML 1997 Low permeabilities of MDCK cell monolayers: a model barrier epithelium *Am J Physiol* **273**: F67-75

Lewis SA and Hanrahan JW 1990 Physiological approaches for studying mammalian urinary bladder epithelium *Methods Enzymol* **192**: 632-650

Negrete HO Lavelle JP Berg J Lewis SA and Zeidel ML 1996a Permeability properties of the intact mammalian bladder epithelium *Am J Physio***271**: F886-894

Negrete HO Rivers RL Goughs AH Colombini M and Zeidel ML 1996b Individual leaflets of a membrane bilayer can independently regulate permeability *J Biol Chem***271**: 11627-11630

Simons K and van Meer G 1988 Lipid sorting in epithelial cells *Biochemistry* **27**: 6197-6202

Stubbs CD Ketterer B Hicks RM 1979 The isolation and analysis of the luminal plasma membrane of calf urinary bladder epithelium *Biochim Biophys Acta* **558**: 58-72

van Meer G and Simons K 1982 Viruses budding from either the apical or the basolateral plasma membrane domain of MDCK cells have unique phospholipid compositions *EMBO J* ;**1**: 847-852

van Meer G and Simons K 1986 The function of tight junctions in maintaining differences in lipid composition between the apical and basolateral cell surface domains of MDCK cells *EMBO J* **5**: 1455-1464

van Meer G Stelzer EHK Wijnaendts van Resandt RW and Simons K 1987 Sorting of sphingolipids in epithelial (Madin-Darby Canine Kidney) cells *J Cell Biol* **105**: 1623-1635

Wu XR Lin JH Walz T Haner M Yu J Aebi U et al 1994 Mammalian uroplakins A group of highly conserved urothelial differentiation-related membrane proteins *J Biol Chem* **269**: 13716-13724

Wu XR and Sun TT 1993 Molecular cloning of a 47 kDa tissue-specific and differentiation-dependent urothelial cell surface glycoprotein *J Cell Sci* **106**: 31-43

2.10 Mammalian physiology

MOLECULAR CLONING OF A NEW AQUAPORIN SUPERFAMILY IN MAMMALS

AQPX1 and AQPX2

Kenichi Ishibashi, Michio Kuwahara, Yukio Kageyama, Sei Sasaki, Makoto Suzuki and Masashi Imai

Department of Pharmacology, Jichi Medical School, Tochigi, 329-0498, Japan; Second Internal Medicine and Urology, Tokyo Medical & Dental University, Tokyo, Japan. E-mail: kishiba@jichi.ac.jp

1. INTRODUCTION

Currently, aquaporins (AQPs) are divided into two subfamilies: AQP1 group and AQP3 group. AQP1 group is water-selective AQP and has 6 members in mammals: AQP0, AQP1, AQP2, AQP4, AQP5, AQP6. They share similar genome structure with 4 exons. AQP3 group is non-selective AQP with permeability to urea and/or glycerol and has 3 members: AQP3, AQP7, AQP9. They share similar genome structure with 6 exons. This dichotomy left AQP8 unclassified. AQP8 is more similar to AQP1 group but the genome structure is unique and its function is controversial; water selective in two investigators and water and urea permeable in one investigator. However, all have highly conserved two NPA boxes that are thought to be critical for their function and belong to MIP family. As AQPs are ancient origin and important for the life as are potassium channels, much more diversity is expected. Unexpectedly, we identified AQPs with poorly conserved NPA boxes. As the homology with other AQPs are so low, these new AQPs were now named AQPX1 and AQPX2 following the recent nomenclature of GLUT family (glucose facilitator).

2. IDENTIFICATION OF AQPX FAMILY

2.1 AQPX1

Rat AQPX1 was isolated from rat testis cDNA library with open reading frame (ORF) of 271 amino acids (six transmembrane domains with single NPA box). AQPX1 has 26-29% identity with other AQPs (highest with AQP8). Northern blot (Fig. 1) revealed the AQPX1 expression (1.8 and 4.4 kb) highest in testis; moderately in thymus, kidney, liver and intestine; lower in brain and lung. AQP10 was expressed at Sertoli cells in testis (in situ hybridization) and PST in kidney (RT-PCR). In Xenopus oocyte expression study, AQP10 stimulated osmotic water permeability from 17 to 82 X 10-6 cm/s, which was partially inhibited by 0.3 mM HgCl2. However, glycerol transport (measured by 14C-glycerol flux) was not stimulated. We also cloned human and mouse orthologs with 81% and 92% identities (GeneBank accession numbers are AB023644, AB028146, AB028147, AB028148). In human, AQPX1 (1.8kb) was also expressed in adrenal gland, pancreas and placenta. The human genomic clone of AQPX1 (8kb) has three exons with similar exon/intron boundaries of carboxy-half of AQP8 and was mapped to chr.11.

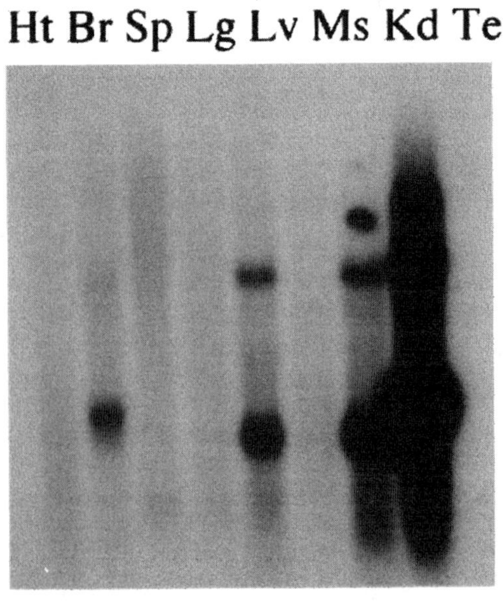

Figure 1. Northern blot analysis of AQPX1 in rat tissues (Clontech).

2.2 AQPX2

Human AQPX2 (accession number: AB040748) was isolated from human pancreas cDNA library with ORF of 295 amino acids (six transmembrane domains with single NPA box). AQPX2 has low homology with other AQPs (highest with AQPX1 at 32 % identity). In Northern blot, AQPX2 was expressed only in pancreas (1.5 kb). Even in human and mouse RNA Master Blots (Clontech) of 71 tissues, AQPX2 was expressed only in pancreas. Rat AQPX2 was also isolated with only 62% identity with human AQPX2 because of shorter carboxy-terminus of rat AQPX2 as is the case with AQP7. The human genomic clone has 5 exones and was mapped to chr. 2. The exon/intron boundaries of carboxy-halves of AQPX1 and AQPX2 were identical, suggesting the common origin of both clones.

3. DISCUSSION

AQPX1 was highly expressed in testis and more specifically at Sertoli cells. As these cells function as the barrier for blood-testis boundary, the role of AQPX1 may be the supplying water into seminiferous tubules whose luminal fluid is hypertonic to plasma. Recent cloning of AQPs from testis revealed unexpected complex water metabolism of spermatogenesis. AQP2, AQP7 and AQP8 are localized to sperms while AQP9 is localized to Leidig cells. The transports of water and/or small solutes through these five AQPs may play important roles in male reproduction physiology, which currently are little known. AQPX1 was also expressed other tissues albeit in lower amount. The roles in these tissues are also intriguing.

On the other hand, AQPX2 was selectively expressed in pancreas. The pancrease produce a large amount of digestive juice (~2 l a day) whose secretion is regulated by hormones. Currently, only known AQP expressed in pancreas is AQP8. However, AQP8 is localized to the apical membrane of acinar cells of pancreas and the main fluid of pancreatic juice is secreted by duct cells. AQPX2 is a good candidate for the water channel of pancreatic duct cells. As AQPX2 is only expressed in pancreas, it may have special roles in this organ as is the case with AQP2 in the kidney.

The physiological importance aside, both AQPX1 and AQPX2 have an interesting structural feature: poorly conserved NPA boxes especially at amino-halves. This structural feature is outstanding in huge family of MIP family where two NPA boxes are surprisingly highly conserved from bacteria to human. Both genes may have derived from a common ancestor gene as suggested by the conserved genome structures. The genome projects revealed some members in C. elegans and Drosophila but none in yeast and bacteria. Therefore, AQPX family may have been produced first at multicellular organisms. More members of this family are expected and they will expand the diversity of MIP family as is the case with another basic channel, potassium channel.

ACKNOWLEDGMENTS

The work from our laboratory was supported by grants from The Ichiro Kanehara Foundation, The Cell Science Research Foundation, The Ryoichi Naito Foundation for Medical Research, The Mochida Memorial Foundation for Medical and Pharmaceutical Research, and The Salt Science Research Foundation.

REFERENCES

Ishibashi K and Sasaki S 1997 Aquaporin water channels in mammals *Clin Exp Nephrol* **1**: 247-253

Ishibashi K and Sasaki S 1998 The dichotomy of MIP family suggests two separate origins of water channels *News Physiol Sci* **13**: 137-142

Ishibashi K Suzuki M and Imai M 1999 Molecular cloning of a first member of a new aquaporin (AQP) superfamily from rat testis *J Am Soc Nephrol* **10**: 16A

Sasaki S Ishibashi K and Marumo F 1998 Aquaporin-2 and -3: Representatives of two subgroups of the aquaporin family colocalized in the kidney collecting duct *Ann Rev Physiol* **60**: 199-220

2.11 Mammalian physiology

NEUROTRANSMITTERS REGULATE THE AMOUNT OF AQP5 IN THE APICAL PLASMA MEMBRANE VIA [Ca^{2+}]i IN PAROTID ACINAR CELLS

Yasuko Ishikawa and Hajime Ishida
Department of Pharmacology, Tokushima University School of Dentistry, 3 Kuramoto-cho, Tokushima, 770-8504, Japan. E-mail: isikawa@dent.tokushima-u.ac.jp

1. INTRODUCTION

The sympathetic or parasympathetic nerves innervate the salivary glands. In general, parasympathetic stimulation produces the high flow of saliva as a result of the activation of M3 muscarinic receptors on the salivary gland cells. In contrast, noradrenaline released from sympathetic nerve endings acts at both β- and α-adrenergic receptors in the salivary glands. Activation of b-adrenoceptors induces the secretion by exocytosis of salivary proteins contained in zymogen granules, whereas activation of a1-adrenoceptors induces the fluid secretion (Baum 1993, Turner 1993).

Fluid secretion from salivary gland cells actually consists of two processes: transcellular and paracellular transports. Stimulation of M3 muscarinic and a1-adrenergic receptors in the basolateral membrane of acinar cells induces activation of phospholipase C and a consequent increase in the cytosolic Ca^{2+} concentration ([Ca^{2+}]i), in turn, results in the opening of Ca^{2+} -activated K^+ channels and Cl^- channels in the basolateral membrane (BLM) and the apical plasma membrane (APM), respectively. Secretion of Cl^- results in increases in the paracellular transport of Na^+ and in the water secretion by way of leak intercellular junctions. The increase in the [Ca^{2+}]i also stimulates Na^+/H^+ exchange activity. A Cl^-/HCO_3^- exchanger acts in parallel with the Na^+/H^+ exchanger and stimulates HCO_3^--dependent water

Molecular Biology and Physiology of Water and Solute Transport
Edited by Hohmann and Nielsen, Kluwer Academic/Plenum Publishers, New York, 2000.

secretion. It is reported that the ability of HCO_3^- to support secretion is about the same as that of Cl^- when the two anions are present in the same concentrations. Salivary secretion is thus thought to be promoted osmotically in response to a transepithelial ion gradient generated by ion transport systems located in the APM and the BLM of the secretory cells (Baum 1993, Turner 1993).

Recently, the water-specific channels, aquaporins (AQPs), were cloned in the plasma membrane of mammalian cells. In salivary glands, the presence of AQP1 (He *et al.,* 1997), AQP4 (Ma *et al.,* 1999a), AQP5 (Raina *et al.,* 1995) and AQP8 (Koyama *et al.,* 1997) was reported.

1.1 AQP5 in salivary glands

The AQP5 cDNA was isolated from rat submandibular glands by Raina *et al.,* (1995). Ma *et al.,* (1999b) reported that in AQP5 knockout mice, pilocarpine-stimulated saliva production was reduced by more than 60 % compared with that in wild-type mice, suggesting that AQP5 plays an important role in salivary fluid secretion. The AQP5 protein is located in the APM of acinar cells and intercellular secretory canaliculi (Raina *et al.,* 1995, He *et al.,* 1997). However it remains to be established whether subcellular distribution of AQP5 in the cells changes by autonomic nerve stimulation. We have investigated by immunoblot analysis with anti-AQP5 antibody whether the translocation of AQP5 in the parotid glands is regulated by neurotransmitters (Ishikawa *et al.,* 1998, 1999).

2. NEUROTRANSMITTERES INDUCE THE TRAFFICKING OF AQP5 BETWEEN THE APMS AND ICMS IN RAT SALIVARY GLANDS

The treatment of rat parotid tissues with acetylcholine (ACh) or epinephrine (Epi) induces the translocation of AQP5 from the intracellular membrane (ICM) to the APM with a maximal at 15 sec or 1 min, respectively (Fig.1)(Ishikawa *et al.,* 1998, 1999). But the prolongation of the treatment of it with ACh for more than 5 min resulted in the converse translocation from the APM to the ICM (Fig 2).

ACh-induced increase in the amount of AQP5 in the APM was completely inhibited by atropine and p-F-HHSiD but not by methoctramine (Table 1), suggesting that ACh acted on M3 muscarinic receptors and induced the translocation of AQP5. This effect of Epi was mimicked by phenylephrine, but not by clonidine, dobutamine, or salbutamol, and Epi-induced increase in the amount of AQP5 was inhibited by phentolamine, but not by propranolol (Table 1), suggesting that Epi acted on a1-adrenergic receptors and increased the amount of AQP5 in the APM.

While, Matsuzaki *et al.*, (1999) reported that the stimulation of parotid gland cells with isoproterenol for 30 min resulted in the dispersal of AQP5 immunoreactivity into clusters of dots in the cytoplasm. This finding suggests that the activation of α1-adrenergic receptors induces the trafficking of AQP5 from the APM to the ICM in parotid glands.

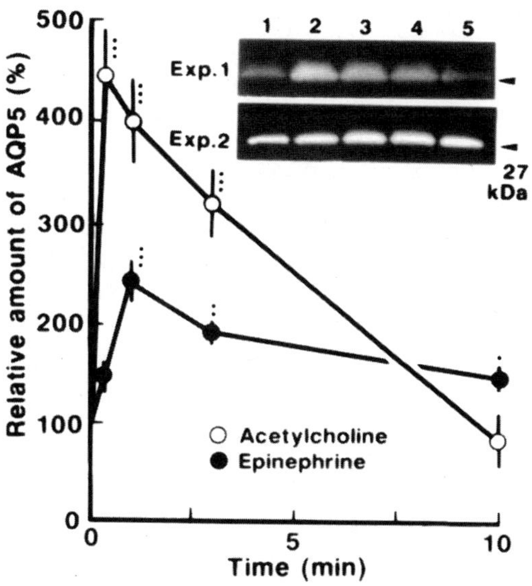

Figure 1. Effects of ACh and Epi on the amounts of AQP5 in the APM of rat parotid glands. Parotid tissues were incubated with 10 μM ACh and 1μM eserine (Exp.1, o) or 10 μM Epi (Exp.2, •) for 0 (lane 1), 0.25 (lane 2), 1 (lane 3), 3 (lane 4) and 10 (lane 5) min. After preparation of membranes, the APM (5 mg of protein) was separated on SDS-PAGE. The protein were transferred to a nitrocellulase membrane. A typical Western blot of AQP5 in the APM is shown (Exp. 1 and 2). Densitometric analysis was expressed as the relative intensity of chemiluminesence. The value was tested for statistical significance using Student's t-test. ***$p<0.001$, p**<0.01, *$p<0.05$.

Table 1. Effects of various cholinergic or adrenergic agents on the amount of AQP5 in the APM of rat parotid glands
The APMs (5 mg of protein) from parotid tissues incubated for 1 or 10 min in the absence or presence of the indicated agents were subjected to immunoblot analysis with antibody to AQP5. The amount of AQP5 in the APM was quantitated by densitometry and expressed as described in the legend for fig.1. Data are means ±s.e. of 3 to 6 independent experiments. Significantly different from the control value, ****p*<0.001, ***p*<0.01.

Agents (μM)	Relative amount of AQP5 (%)
None	100
ACh (10)	319.8 ± 14.3 ***
ACh (10) + atropine (100)	96.5 ± 2.7
ACh (10) + methoctramine (100)	261.8 ± 16.3***
ACh (10) + *p*-F-HHSiD (10)	120.8 ± 7.2
Epi (10)	233.8 ± 21.2**
Epi (10) + phentolamine (100)	124.9 ± 4.6
Epi (10) + propranolol (100)	210.0 ± 13.4**
Phenylephrine (10)	217.4 ± 14.7 **
Clonidine (10)	131.8 ± 8.4
Dobutamine (10)	129.3 ± 11.5
Salbutamol (10)	115.2 ± 7.4

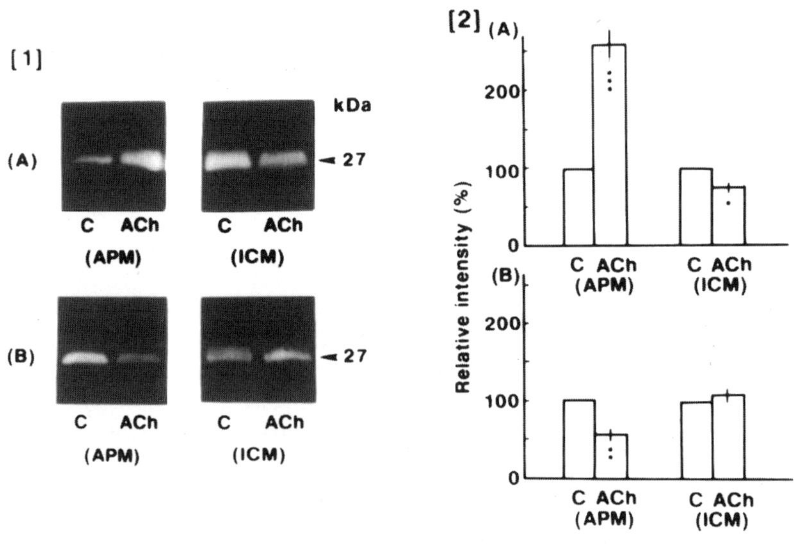

Figure 2. Effect of ACh on the amounts of AQP5 in the APM and the ICM of rat parotid tissues. Tissues were incubated with (ACh) or without (C) 10 μM ACh and 1 μM eserine for 1 (A) or 5 (B) min. The APM (10 mg of protein) and the ICM (40 mg of protein) were separated on SDS-PAGE. A typical Western blot of AQP5 is shown in [1]. Densitometric analysis was expressed as the relative intensity of chemiluminesence in [2]. The value was tested for statistical significance using Student's t-test. ***p<0.001, p**<0.01, *p<0.05

ACh or Epi induced a marked increase in the amount of AQP5 in the APM and an associated decrease in the abundance of this protein in the ICM (Fig 2). Immunofluoresence chemistry (He *et al.,* 1997, Matsuzaki *et al.,* 1999) and immunoblot analysis of subcellular fractions (Ishikawa *et al.,* 1998, 1999) revealed that AQP5 protein is not present in the BLM. Taken together, neurotransmitters regulate the trafficking of AQP5 between the APMs and ICMs.

3. [CA^{2+}]I REGULATES THE TRAFFICKING OF AQP5 IN RAT SALIVARY GLANDS

Stimulation of M3 muscarinic and a1-adrenergic receptors induces the activation of phospholipase C. The translocation of AQP5 to APM was profoundly decreased in the tissues treated with U-73122, an inhibitor of phospholipase C (Table 2). Dantrolene, an inhibitor of Ca^{2+}-induced Ca^{2+} release from the sarcoplasmic reticulum and unidentified Ca^{2+} stores, and TMB-8, an inhibitor of Ca^{2+} release from intracellular compartments, inhibited ACh- or Epi-induced increases in the amounts of AQP5 in the APM. These findings suggest that ACh and Epi induced trafficking of AQP5 from the ICM to the APM via the enhancement of $[Ca^{2+}]i$ in rat parotid gland cells.

Table 2. Effects of various agents on the amount of AQP5 in the APM
The APMs (5 mg of protein) from parotid tissues incubated for 1[§] or 10 min in the absence or presence of the indicated agents were subjected to immunoblot analysis with antibody to AQP5. The amount of AQP5 in the APM was quantitated by densitometry and expressed as described in the legend for fig.1. Data are means ± s.e. of 3 to 6 independent experiments. Significantly different from the control value, ***$p<0.001$, ** $p<0.01$.

Agents (µM)	Relative amount of AQP5 (%)
ACh (10)[§]	319.8±14.3 ***
ACh (10)	96.5±2.7
ACh (10) + TMB-8 (30)	81.8±8.3
Epi (10)	242.5±22.5**
Epi (10) + U-73122 (10)	81.3±1.8
Epi (10) + dantrolene (30)	125.0±12.0
Epi (10) + TMB-8 (30)	92.5± 8.9
Epi (10) + cytochalasin D (5)	120.0±3.0
Epi(10) + tubulozole C (5)	104.6±2.3

The calcium ionophore A23187 (10 mM) alone stimulates also the translocation of AQP5 between the APMs and ICMs in the presence of extracellular Ca^{2+} (Fig 3). These results suggest that the enhancement of $[Ca^{2+}]i$ triggers the trafficking of AQP5 in rat parotid cells.

4. CYTOSKELETON TAKES PART IN THE TRAFFICKING OF AQP5 IN RAT SALIVARY GLANDS

Membrane trafficking in polarized epithelial cells is thought to require both actin microfilaments and microtubules. Cytochalasin D, an actin-disrupting agent, and tubulozole-C, a microtubule-depolymerizing agent, also inhibited Epi- and A23187-induced increases in the amount of AQP5 in the APM (Fig 3). These results indicate that both actin filaments and microtubules are important for the Epi- and A23187-induced translocation of AQP5 to the APM via the enhancement of $[Ca^{2+}]i$ in rat parotid cells.

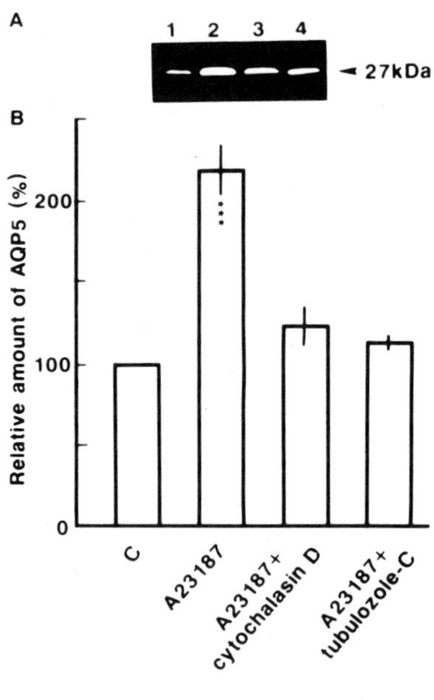

Figure 3. Effects of cytochalasin D and tubulozole-C on the A23187-indiced increases of AQP5 in the APM. (A) Tissues were pretreated without (lane 1 and 2) or with 5 μM cytochalasin D (lane 3) or 5 μM tubulozole-C (lane 4) and then incubated with (lane 2 to 4) or without (lane 1) 10 μM A23187 for 1 min. The APM was subjected to immunoblot analysis with antibodies to AQP5. (B) Densitometric analysis was expressed as the relative intensity of chemiluminesence. The value was tested for statistical significance using Student's t-test. ***$p<0.001$.

5. CONCLUSION

These results indicate that ACh and Epi act at M3 muscarinic and a1-adrenergic receptors, respectively, and induce translocation of AQP5 via $[Ca^{2+}]i$ elevation in response to IP3 and ryanodine receptor activation in rat parotid cells. In addition, cytoskeleton elements are shown to be involved in the translocation of AQP5 to APM in the cells.

ACKNOWLEDGMENTS

This work was supported in part by a Grand-in-Aid for Scientific Research from the Ministry of Education, Science, Sports, and Culture of Japan. We thank Yumiko Yoshinaga for help in preparing the manuscript.

REFERENCES

Baum BJ 1993 Principles of saliva secretion *Ann NY Acad Sci* **694**: 17-23

He X Tse CM Donowitz M Alper SL Gabriel SE and Baum BJ1997 Polarized distribution of key membrane transport proteins in rat submandibular gland *Pflugers Arch-Eur J Physiol* **433**: 260-268

Ishikawa Y Eguchi T Skowronski MT and Ishida H 1998 Acetylcholine acts on M3 muscarinic receptors and induces the translocation of aquaporin5 water channel via cytosolic Ca^{2+} elevation in rat parotid glands *Biochem Biophys Res Commun* **245**: 835-840

Ishikawa Y Skowronski MT Inoue N and Ishida H 1999 a1-Adrenoceptor-induced trafficking of aquaporin-5 to the apical plasma membrane of rat parotid cells *Biochem Biophys Res Commun* **265**: 94-100

Koyama Y Yamamoto T Kondo D Funaki H Yaoita E Kawasaki K Sato N Hatakeyama K and Kihara I 1997 Molecular cloning of a new aquaporin from rat pancreas *J Biol Chem* **272**: 30329-30333

Ma T and Verkman AS 1999 Aquaporin water channels in gastrointestinal physiology *J Physiol* **517**: 317-326

Ma T Song Y Gillespie A Carlson EJ Epstein CJ and Verkman AS 1999 Defective secretion of saliva in transgenic mice lacking aquaporin-5 water channels *J Biol Chem* **274**: 20071-20074

Matsuzaki T Suzuki T Koyama H Tanaka S and Takata K 1999 Aquaporin-5 (AQP5) a water channel protein in the rat salivary and lacrimal glands: immunolocalization and effect of secretory stimulation *Cell Tissue Res* **295**: 513-521

Raina S Preston GM Guggino WB and Agre P 1995 Molecular cloning and characterization of an aquaporin cDNA from salivary lacrimal and respiratory tissues *J Biol Chem* **270**: 1908-1912

Turner BJ 1993 Mechanisms of fluid secretion *Ann NY Acad Sci* **694**: 24- 35

2.12 Mammalian physiology

PERSISTENT INCREASE IN THE AMOUNT OF AQP5 IN THE APICAL PLASMA MEMBRANE OF RAT PAROTID ACINAR CELLS INDUCED BY THE MUSCARINIC RECEPTOR AGONIST SNI-2011

Yasuko Ishikawa and Hajime Ishida
Department of Pharmacology, Tokushima University School of Dentistry, 3 Kuramoto-cho, Tokushima, 770-8504, Japan. E-mail: isikawa@dent.tokushima-u.ac.jp

1. INTRODUCTION

SNI-2011, (±)-cis-2-methylspiro(1,3-oxathiolane-5, 3'-quinuclidine) hydrochloride hemihydrate, is a structurally rigid analog of acetylcholine (ACh) that has been described as for muscarinic M1 (M1) receptors selective agonist (Fisher *et al.*, 1991; Ono *et al.*, 1988 ; Iga *et al.*, 1996). SNI-2011 was shown to be a selective agonist for M1 and M3 receptors (Hargreaves *et al.*, 1992). The intravenous administration of SNI-2011 induces a long-lasting salivation in rats and dogs (Iwabuchi and Masuhara, 1994; Masunaga *et al.*, 1997).

Stimulation of muscarinic receptors in rat salivary glands with their respective agonists generates inositol 1,4,5-trisphosphate (IP3) and diacylglycerol (DAG) as a result of the activation of a heterotrimetic guanine nucleotide-binding protein (Gq) and phospholipase C-β (Dai *et al.*, 1991; Ehlert *et al.*, 1996). The generation of IP3 leads to an increase in the cytosolic concentration of Ca^{2+} ($[Ca^{2+}]i$) (Ishikawa *et al.*, 1988; Dai *et al.*, 1991). Type III RyRs (Ry3Rs) have also been identified in microsomal membranes of mouse parotid acini (DiJulio *et al.*, 1997) and other mammalian cells (Berridge, 1993). These receptors contribute to changes in $[Ca^{2+}]i$ induced by cell stimulation, and the function is inhibited by dantrolane (Ward *et al.*, 1986). In contrast, DAG activates protein kinase C (PKC) in the presence of Ca^{2+} (Nishizuka, 1986). An increase in $[Ca^{2+}]i$ in rat salivary gland cells results in the rapid secretion of saliva (Foskett and Melvin, 1989; Nauntofte, 1992). Aquaporins (AQPs), which form channels that selectively transport water across the plasma membrane, have been cloned from a variety of mammalian tissues (King and Agre, 1996). With the use of the specific antiserum to the AQP5, we recently detected AQP5 in apical plasma membranes (APMs) and intracellular membranes (ICMs) of rat parotid tissues. Furthermore, ACh, acting at M3 receptors and by increasing $[Ca^{2+}]i$, induced the translocation of AQP5 from ICMs to APMs of the parotid cells (Ishikawa *et al.*, 1998).

We have now investigated the effect of SNI-2011 on the translocation of AQP5 in rat parotid cells and compared it with that of ACh.

2. MECHANISMS UNDERLYING THE CHANGES IN THE SUBCELLULAR DISTRIBUTION OF AQP5 IN RAT PAROTID TISSUES

2.1 Changes in the subcellular distribution of AQP5 in rat parotid cells induced by SNI-2011

We first prepared APMs, basolateral membranes (BLMs), and ICMs from rat parotid tissues and subjected each preparation to immunoblot analysis with the specific antiserum to AQP5. The antiserum recognized a single protein, with a mobility corresponding to the predicted molecular mass of AQP5 (27 kDa) in both APM and ICM fractions, but it did not detect it in BLM fraction (Ishikawa *et al.*, 1998, 1999). The immunoreactive band in APM and ICM fractions was not detected with AQP5 antiserum that had been subjected to preabsorption with excess immunizing peptide. These results indicate that AQP5 is associated with APMs and ICMs, but not with BLMs, in rat parotid cells (Fig 1A).

Rat parotid tissues were incubated in the Krebs Ringer Tris medium containing 10mM SNI-2011 for 5 min at 37°C to investigate the effect of this drug on the subcellular distribution of AQP5. After the incubation, the tissue were immediately homogenized and APM, ICM, and BLM fractions were then prepared. Immunoblot analysis with the antiserum to AQP5 revealed that SNI-2011 induced approximately threefold increase in the amount of AQP5 in the APM and a corresponding decrease in the amount of this protein in the ICM fraction (Fig 1B). Treatment of the tissues with SNI-2011 did not affect the yields of the APM, BLM, and ICM fractions nor did it affect the immunoblot analysis per se, as was the case for treatment of tissue with ACh (Ishikawa *et al.*, 1998). Thus, exposure of rat parotid tissues to SNI-2011 induced the translocation of AQP5 from ICMs to APMs. Incubation of the tissues with various concentrations of SNI-2011 for 10 min at 37°C revealed that SNI-2011 increased the amount of AQP5 in APMs in a dose-dependent manner. The effect of SNI-2011 was significant at 1 nM and maximal (335.8±16.6% of control) at 100 nM, with a median effective concentration (EC50) of 2.60±0.58 nM.

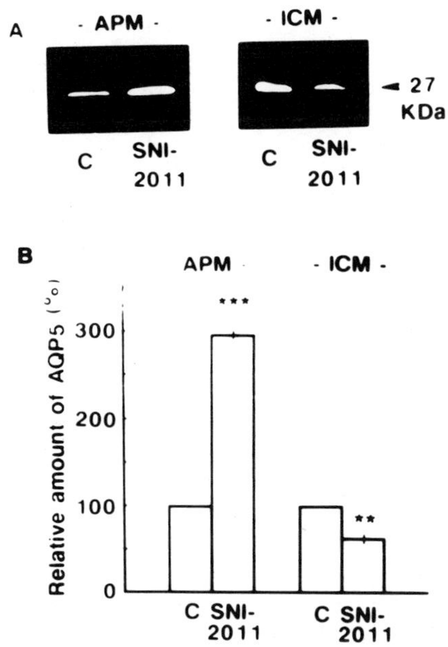

Figure 1. Effect of SNI-2011 on the amount of AQP5 in the APM and the ICM. A: APMs (5 mg of protein) and ICMs (40 mg of protein) from rat parotid tissues incubated in the absence (C) or presence (SNI-2011) of 10 μM SNI-2011 for 10 min were subjected to immunoblot analysis with antibodies to AQP5. B: Immuboblots shown in A were subjected to densitomeric analysis, and the amount of AQP5 in each fraction was expressed as described in fig.1. Data are means ± s.e. of 3 independent experiments. **$P<0.01$, ***$P<0.001$ versus the value for control cells.

The time courses of the effects of 10 μM SNI-2011 and 10 μM ACh on the translocation of AQP5 to APMs were also examined. The effect of SNI-2011 was apparent at 1 min and maximal at 10 min, and the amount of AQP5 in APMs remained increased after incubation with SNI-2011 for 30 min (Fig 2). In contrast, although ACh also induced a rapid and marked increase in the amount of AQP5 in APMs with the effect maximal at 15 s, the abundance of AQP5 in this fraction had returned to control values after incubation of tissues with ACh for 10 min (Fig 2). These findings demontrate that a long-lasting increase in the amount of AQP5 in APMs is induced by SNI-2011.

2.2 Action at M_3 muscarinic receptors

To determine the subclass of M receptors that mediate SNI-2011-induced increase in the amount of AQP5 in APMs of rat parotid cells, we investigated the effects of atropine and p-F-HHSiD, the latter a specific antagonist of M_3 receptors. The presence of 100 μM atropine or 10 μM p-F-HHSiD markedly inhibited the increase in the amount of AQP5 in APMs induced by 10 μM SNI-2011 for 10 min (Table 1). These results indicate that SNI-2011 acts at M_3 receptors in rat parotid tissue to induce the translocation of AQP5 to APMs.

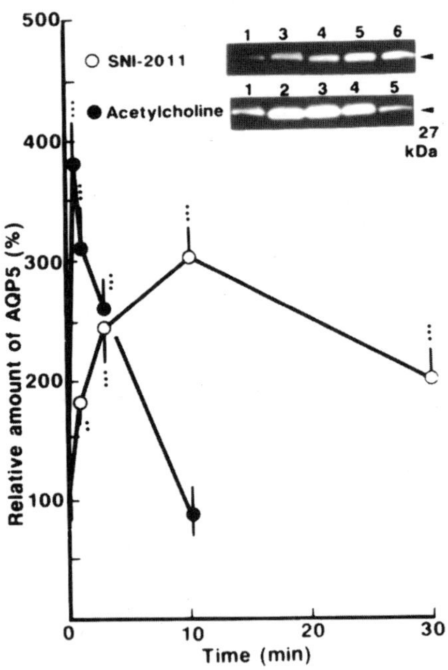

Figure 2. Time courses for the effects of SNI-2011 and ACh on the amount of AQP5 in the APM. The APMs (5 mg of protein) from rat parotid tissues incubated with SNI-2011 (10 μM), or ACh (10 μM) plus eserine (10 μM) for 0, 0.25, 1, 3, 10, or 30 min (corresponding to lanes 1 to 6, respectively, in the inset) were subjected to immunoblot analysis with antibody to AQP5. Data were quantitated by densitometry and are means ± s.e. of 3 to 5 independent experiments. **P<0.01, ***P<0.001 versus the value for control cells.

2.3 Stimulation of PLC activity and generation of IP_3

To investigate the signaling pathway by which SNI-2011 triggers the translocation of AQP5 from ICMs to APMs of rat parotid cells, we first investigated the role of phospholipase C (PLC) on this effect. U73122 (10 μM) inhibited SNI-2011-induced increase in the amount of AQP5 in APMs of rat parotid tissues. (Table 1). Treatments of the tissues with H-7 (300 mM) and GF109203X (10 μM), inhibitors of PKC, and phorbol 12-myristate 13-acetate (PMA, 1 μM), an activator of PKC, had no effect on the translocation of AQP5 induced by SNI-2011. These observations suggest that SNI-2011 activates PLC, but not PKC and contributes to effect of this drug on translocation of AQP5. We investigated whether SNI-2011 stimulated IP_3 production in rat parotid tissues. Incubation of tissues with 10 μM SNI-2011 resulted in an increase in the intracellular concentration of IP_3; the effect was maximal at 10 min (3.20±0.07 and 9.16±1.42 pmol/mg protein in the control and SNI-2011-treated tissues, respectively) and remained apparent at 30 min (3.48±0.34 and 5.78±0.92 pmol/mg protein in the control and SNI-2011-treated tissues, respectively).

Table 1. Effects of various agents on the amount of AQP5 in the APM

Agents (mM)	Relative amount of AQP5 (%)
None	100
SNI-2011 (10)	301.8 ± 9.2***
SNI-2011 (10) + atropine (100)	100.4 ± 5.1
SNI-2011 (10) + *p*-F-HHSiD (10)	141.1 ± 14.3
SNI-2011 (10) + U73122 (10)	95.6 ± 2.4
SNI-2011 (10) + TMB-8 (30)	94.9 ± 1.6
SNI-2011 (10) + dantrolene (30)	140.7 ± 9.6
SNI-2011 (10) + H-7 (300)	254.4 ± 4.8***
SNI-2011 (10) + GF 109203X (10)	305.5 ± 14.4***
SNI-2011 (10) + Tubulozole-C (5)	102.6 ± 1.0
SNI-2011 (10) + Cytochalasin D (5)	124.0 ± 0.1
PMA (1)	90.7 ± 3.9
A23187[§] (10)	234.5 ± 2.3***

The APMs (5 mg of protein) from parotid tissues incubated for 1[§] or 10 min in the absence or presence of the indicated agents were subjected to immunoblot analysis with antibody to AQP5. The amount of AQP5 in the APM was quantitated by densitometry and expressed as described in the legend for Fig 1. Data are means ± s.e. of 3 to 6 independent experiments. Significantly different from the control value, ***p<0.001.

2.4 Induction of generation of the oscillation of $[Ca^{2+}]i$

TMB-8, an inhibitor of Ca^{2+} release from intracellular stores, was used to investigate the importance of an increase in $[Ca^{2+}]i$ in SNI-2011-induced translocation of AQP5 to APMs. Incubation of the tissues with 30 μM TMB-8 markedly inhibited the increase in the amount of AQP5 in APMs induced by either 10 μM SNI-2011 (Table 1), indicating that the increase in the $[Ca^{2+}]i$ associated with the stimulation of M_3 receptors by SNI-2011 was involved in the translocation of AQP5 to APMs from ICMs in rat parotid tissues. Dantrolene, a hydantoin derivative, inhibits Ca^{2+} induced Ca^{2+} release from the sarcoplasmic reticulum of skeletal muscle and from unidentified Ca^{2+} stores in other cell types (Ohta *et al.*, 1990). In the present study, treatment of rat parotid tissues with 30 μM dantrolene inhibited by 30 % the translocation of AQP5 from ICMs to APMs induced by 10 μM SNI-2011 (Table 1). These observations suggest that an increase in $[Ca^{2+}]i$, at least part of which results from activation of the Ry3R, contributes to the translocation of AQP5 induced by SNI-2011. This suggestion is supported by the observation that A23187, a Ca^{2+} ionophore, also induces a rapid and marked increase in the amount of AQP5 in APMs (Table 1).

It has been proposed that an initial release of Ca^{2+} from the IP_3-sensitive pool acts as a primer for further Ca^{2+} release from ryanodine-sensitive pool producing oscillations (Berridge, 1993). The characteristics of oscillations of $[Ca^{2+}]i$ generated by full M agonists differ from those induced by partial M agonists. ACh evoked a transient elevation of $[Ca^{2+}]i$ lasting about 2-5 min (Tepkin *et al.*, 1992). Furthermore, the concentration of IP_3 is 4-fold higher than that of control at 15 s after addition of ACh. In contrast, it is 3-fold higher than that of control at 10 min after the addition of SNI-2011. As shown in fig. 3, SNI-2011 induced persistent oscillation of $[Ca^{2+}]i$ in the presence of extracellular Ca^{2+}.

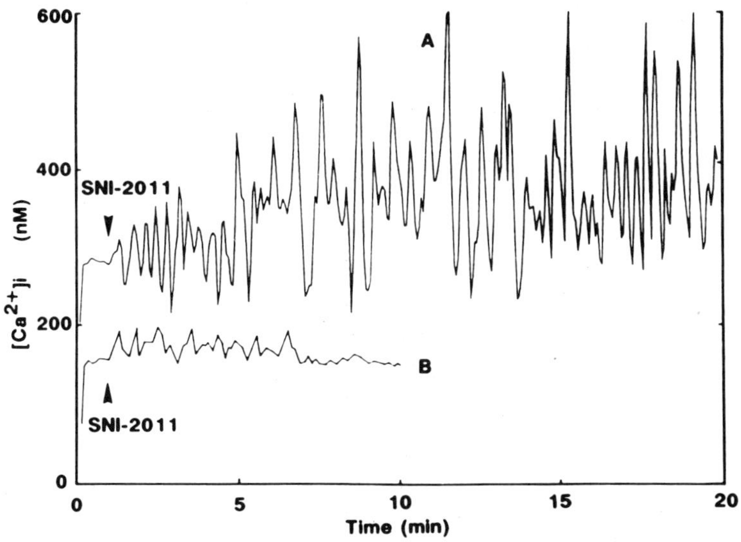

Figure 3. Effect of SNI-2011 on $[Ca^{2+}]i$ in rat parotid acinar cells. Isolated acinar cells were loaded with fura-2/AM and then exposed to 10 μM SNI-2011 in the presence (A) or absence (B) of 1 mM $CaCl_2$. The $[Ca^{2+}]i$ was calculated from fluorescence measurements as described in Methods. SNI-2011 was added at the time indicated by the arrow.

2.5 Role of the cytoskeleton in AQP5 trafficking

Cytochalasin (5 μM) and tubulozole C (5 μM) inhibited strongly SNI-2011-induced increase in the amount of AQP5 in APM, showing that both microfilaments and microtubules were necessary for the AQP5 trafficking (Table 1). These findings support that AQP5 traffics between APMs and the cytoplasm in parotid gland cells. However, the mechanism by which AQP5 interacts with microfilaments and microtubules remains unknown.

3. CONCLUSION

SNI-2011 acts at M_3 receptors in rat parotid cells to induce a long-lasting translocation of AQP5 from ICMs to APMs. This effect appears to be mediated by the generation of IP_3 and the consequent increase in $[Ca^{2+}]i$.

ACKNOWLEDGMENT

We thank Mrs. Yumiko Yoshinaga for excellent secretarial assistance.

REFERENCES

Berridge MJ 1993 Cell signalling A tale of two messengers *Nature* **365**: 388-389

Dai Y Ambudkar IS Horn VJ Yeh C-K Kousve-lari EE Wall SJ Li M Yasuda RP Wolfe BB and Baum BJ 1991 Evidence that M_3 muscarinic receptors in rat parotid gland couple to two second messenger systems *Am J Physiol* **261**: C1063-C1073

DiJulio DH Watson EL Pessah IN Jacobson KL Ott SM Buck ED and Singh JC 1997 Ryanodine receptor type III(Ry3R) identification in mouse parotid acini Properties and modulation of [3H]ryanodine-binding sites *J Biol Chem* **272**: 15687-15696

Ehlert FJ Griffin MT and Glidden PF 1996 The interaction of the enantiomers of aceclidine with subtypes of the muscarinic receptor *J Pharmacol Exp Ther* **279**: 1335- 1344

Fisher A Brandeis R Karton I Pittel Z Gurwitz D Haring R Sapir M Levy A and Heldman E 1991 (±)-cis-2-Methyl-spiro(13-oxathiolane-5 3') quinuclidine an M1 selective cholinergic agonist attenuates cognitive dysfunctions in an animal model of Alzheimer's disease *J Pharmacol Exp Ther* **257**: 392-403

Foskett JK and Melvin JE 1989 Activation of salivary secretion: Coupling of cell volume and $[Ca^{2+}]i$ in single cells *Science* **244**: 1582-1585

Hergreaves RJ McKnight AT Scholey K Newbetrry NR Street LJ Hutson PH Semark JE Harley EA Patel S and Freedman SB 1992 L-689660 a novel cholinomimetic with functional selectivity for M1 and M3 muscarinic receptors *Br J Pharmacol* **107**: 494-501

Iga Y Arisawa H Ise M Yoshida H and Takeshita Y 1996 Modulation of rhythmical slow activity long-term potentiation and memory by muscarinic receptor agonists *Eur J Pharmacol* **308**: 13-19

Ishikawa Y Skowronski MT Inoue N and Ishida H 1999 α1-Adrenoceptor-induced trafficking of aquaporin-5 to the apical plasma membrane of rat parotid cells *Biochem Biophys Res Commum* **265**: 94-100

Ishikawa Y Eguchi T Skowronski MT and Ishida H 1998 Acetylcholine acts on M_3 muscarinic receptors and induces the translocation of aquaporin5 water channel via cytosolic Ca^{2+} elevation in rat parotid glands *Biochem Biophys Res Commum* **245**: 835-840

Ishikawa Y Gee MV Ambudkar IS Bodner L Baum BJ and Roth GS 1988 Age-related impairment in rat parotid cell a1-adrenergic action at the level of inositol trisphosphate responsiveness *Biochim Biophys Acta* **968**: 203-210

Iwabuchi Y and Masuhara T 1994 Sialogogic activities of SNI-2011compared with those of pilocarpine and McN-A-343 in rat salivary glands: Identification of a potential therapeutic agent for treatment of Sjorgen's syndrome *Gen Pharmacol* **25**: 123-129

King LS and Agre P 1996 Pathophysiology of the aquaporin water channels *Annu Rev Physiol* **58**: 619-648

Masunaga H Ogawa H Uematsu Y Tomizuka T Yasuda H and Takeshita Y 1997 Long-lasting salivation induced by a novel muscarinic receptor agonist SNI-2011 in rats and dogs *Eur J Pharmacol* **339**: 1-9

Nauntofte B 1992 Regulation of electrolyte and fluid secretion in salivary acinar cells *Am J Physiol* **263**: G823-G837

Nishizuka Y 1986 Studies and perspectives of protein kinase C *Science* **233**: 305-312
Ohta T Ito S and Ohga A 1990 Inhibitory action of dantrolene on Ca-induced Ca^{2+} release from sarcoplasmic reticulum in guinea pig skeletal muscle *Eur J Pharmacol* **178**: 11-19
Ono S Saito Y Ohgane N Kawanishi G and Mizobe F 1988 Heterogeneity of muscarinic autoreceptors and heteroreceptors in the rat brain: Effects of a novel M1 agonist AF102B *Eur J Pharmacol* **155**: 77-84
Tepikin AV Voronina SG Gallacher DV and Petersen OH 1992 Acetylcholine- evoked increase in the cytoplasmic Ca^{2+} concentration and Ca^{2+} extrusion measured simultaneously in single mouse pancreatic acinar cells *JBiolChem* **267**: 3569-3572
Wade A Chaffman MO and Sorkin EM 1986 Dantrolene A review of its pharmacodynamic and pharmacokinetic properties and therapeutic use in malignant hyperthermia the neuroleptic malignant syndrome and an update of its use in muscle spasticity *Drugs* **32**: 130-168

2.13 Mammalian physiology

CYCLIC AMP-MEDIATED AQUAPORIN-2 TRANSLOCATION: IDENTIFICATION OF PROTEIN KINASE A ANCHORING PROTEINS AND THE ROLE OF THE SMALL GTPases OF THE RHO FAMILY

AKAPs and Rho proteins are required for AQP2 translocation

[1]Enno Klussmann, [1]Eberhard Krause, [1]Burkhard Wiesner, [1]Michael Beyermann, [1]Bayram Edemir, [1]Christoph Bouchaala, [1]Dorothea Lorenz, [1]Kenan Maric, [2]Grazia Tamma, [2]Giovanna Valenti and [1,3]Walter Rosenthal
[1]Forschungsinstitut für Molekulare Pharmakologie, Alfred-Kowalke-Str. 4, D-10315 Berlin, Germany; [2]Dipartimento di Fisiologia Generale ed Ambientale, Universität Bari, Italy; [3]Freie Universität Berlin, Institut für Pharmakologie, Thielallee 67-73, D-14195 Berlin, Germany. E-mail: klussmann@fmp-berlin.de

1. AKAPS IN AQP2 TRANSLOCATION

1.1 INTRODUCTION

The antidiuretic hormone arginine-vasopressin (AVP) regulates water reabsorption in inner medullary collecting duct (IMCD) principal cells by inducing the translocation of water channels (AQP2) from intracellular vesicles into the apical cell membranes (for review see: Klussmann *et al.*, 2000). AVP initiates the activation of protein kinase A (PKA) by stimulation of the G_s/adenylyl cyclase system. Using a recently established primary culture model of rat IMCD cells (Maric *et al.*, 1998) it was shown that the tethering of PKA by protein kinase A anchoring proteins (AKAPs; Scott *et al.*, 2000) is a prerequisite for the translocation of AQP2 (Klussmann *et al.*, 1999). In addition, in subcellular fractions from primary cultured rat IMCD cells enriched for AQP2-bearing vesicles or cell membranes various AKAPs

Molecular Biology and Physiology of Water and Solute Transport
Edited by Hohmann and Nielsen, Kluwer Academic/Plenum Publishers, New York, 2000.

were detected. Two strategies have been applied to identify AKAPs potentially involved in AQP2 translocation.

1.2 IDENTIFICATION OF AKAPs BY 2D GEL ELECTROPHORESIS AND MALDI MASS SPECTROMETRY

In the first approach AQP2-bearing vesicles from rat kidney inner medullae were isolated essentially as described (Harris *et al.*, 1994). Ficoll (18%) rather than Percoll was used in the density gradient centrifugation step. The proteins from purified AQP2-bearing vesicles (300-800 µg) were separated by a 2D gel electrophoresis system (Hartinger *et al.*, 1996), in which proteins are separated according to their molecular weight in both dimensions. In the first dimension (7.5 % acrylamide) the cationic detergent 16-BAC (benzyldimethyl-n-hexadecylammonium chloride) is present. The second dimension is a conventional SDS-PAGE (10 % acrylamide). The second dimension gels were either stained with Coomassie Blue (Fig. 1A) or blotted onto nitrocellulose membrane. To identify bands corresponding to AKAPs the membrane was overlaid with radioactively labelled PKA regulatory RII-subunits (RII-overlay technique; Bregman *et al.*, 1989).

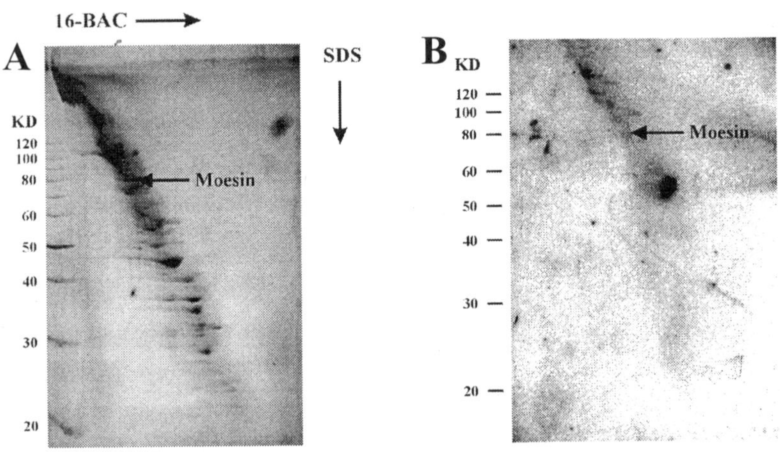

Figure 1. A: Coomassie Blue stained second dimension 16-BAC 2D electrophoresis gel. Proteins (300 µg) of purified AQP2-bearing vesicles were separated in the first dimension in the presence of the cationic detergent 16-BAC according to their molecular weight (7.5 % acrylamide). The second dimension was a conventional SDS-PAGE (10 % acrylamide). B: RII-overlay. AQP2-bearing vesicles were separated by 16-BAC 2D electrophoresis and blotted onto nitrocellulose. For detection of AKAPs the membrane was overlaid with radioactively labelled RII-subunits. The bands were visualized by autoradiography. The arrows in A and B indicate the spot that was localized in a Coomassie Blue stained gel after RII-overlay and identified as the AKAP moesin by mass spectrometry (Fig. 2).

Various bands between 55 and more than 120 kD in size were visualized by autoradiography (Fig. 1B). Corresponding bands on a Coomassie Blue stained gel were excised and analysed by peptide mass fingerprinting using MALDI mass spectrometry. The arrow (Figs. 1A and B) indicates the spot that was identified as the AKAP moesin (Fig. 2). The functional involvement of moesin in AQP2 translocation is currently being investigated.

A

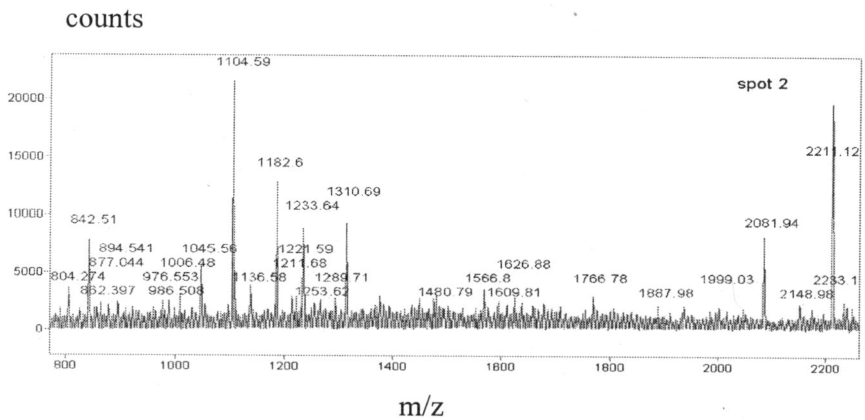

B

Min # Peptides to Match	Peptide Mass Tolerance (+/-)	Peptide Masses	Digest Used	Max # Missed Cleavages	Cysteines Modified by
7	0.050 Da	monoisotopic	Trypsin	1	carbamidomethylation

Detailed Results

1. 7/24 matches (29%). (67739.3 Da) Acc. # 2218139. RATTUS NORVEGICUS. (AF004811) moesin

Data submitted	MH+ matched	Delta Da	Fragment start	end	Peptide Sequenze
862.3967	862.3695	-0.0272	166	171	DQWEER
894.5410	894.5453	-0.0043	101	107	LFFLQVK
976.5531	976.5468	0.0063	28	35	QLFDQVVK
1104.5930	1104.5842	0.0088	238	246	IGFPWSEIR
1182.5980	1182.5948	0.0032	264	273	APDFVFYAPR
1233.6380	1233.6129	0.0251	172	180	IQVWHEEHR
1310.6930	1310.6897	-0.0033	263	273	KAPDFVFYAPR

The matched peptides cover 8% (50/577 AA's) of the protein.

Figure 2. Identification of Coomassie Blue stained proteins by MALDI mass spectrometry. The protein spot indicated by the arrow in Fig. 1 was excised from a Coomassie Blue stained second dimension gel (Fig. 1A) and "in-gel" trypsin digested. The peptides were eluted and analysed by mass spectrometry (Voyager-DE STR, Perceptive Biosystems, Wiesbaden, Germany). A: mass spectrum obtained from the analysis of the spot indicated in Fig. 1A and B as moesin. B: Peptide search parameters, detailed results and summary obtained from a search in the MS-FIT (UCSF, CA, USA) database.

1.3 IDENTIFICATION OF AKAPs BY cDNA EXPRESSION LIBRARY SCREENING

In a second approach cDNA expression library screening with PKA RII subunits led to the isolation of one clone which represents the rat homologue of AKAP18. Northern blot analysis demonstrated the presence of at least two different splice variants each with stronger expression in renal inner medulla, the main site of AQP2 expression, than in residual kidney tissue. An antiserum was raised against a peptide derived from an AKAP18 sequence which is common to all splice variants. Subsequent Western blot and immunofluorescence microscopic analysis of primary cultured IMCD cells with affinity-purified antibodies from this serum showed that AKAP18 variants are localized intracellularly and at the cell membrane. Immunofluorescence microscopic studies of sections of renal inner medulla demonstrated that these variants (as AQP2) are expressed mainly in cells lining the collecting ducts (most likely principal cells). Taken together these data indicate a possible involvement of AKAP18 in AQP2 translocation.

2. THE INVOLVEMENT OF THE SMALL GTPASES OF THE RHO FAMILY IN AQP2 TRANSLOCATION

The involvement of the small GTPases of the Rho family (Rho, Rac and Cdc42) was investigated by incubation of primary IMCD cells with bacterial toxins (Aktories 1997). *Clostridium difficile* toxin B inhibits all small G proteins of the Rho family, while cytotoxic necrotizing factor I derived from pathogenic *E. coli* activates Rho, Rac and Cdc42. Both toxins - to varying degrees - induced translocation of AQP2 in the absence of AVP, which is still active after toxin treatment. These data indicate the functional involvement of Rho proteins in AQP2 translocation. They also point to the presence of different pools of vesicles which may depend on different signals for their release: one pool may dependent on activation of Rho proteins while another requires inactivation of these proteins. Identification of the responsible Rho family member is attempted by the use of C3 toxin, derived from *Clostridium botulinum* (Barth *et al.,* 1998) and dominant negative or constitutively active mutants of RhoA, Cdc42 and Rac1.

REFERENCES

Aktories K 1997 Rho proteins: targets for bacterial toxins *Trends Microbiol* **5**: 282-288

Barth H Hofmann F Olenik C Just I and Aktories K 1998 The N-terminal part of the enzyme component (C2I) of the binary Clostridium botulinum C2 toxin interacts with the binding component C2II and functions as a carrier system for a Rho ADP-ribosylating C3-like fusion toxin *Infect Immun* **66**: 1364-1369

Bregman DB Bhattacharyya N and Rubin CS 1989 High affinity binding protein for the regulatory subunit of cAMP-dependent protein kinase II-B *J Biol Chem* **264**: 4648-4656

Harris Jr HW Zeidel ML Jo I and Hammond TG 1994 Characterization of purified endosomes containing the antidiuretic hormone-sensitive water channel from rat renal papilla *J Biol Chem* **269**: 11993-12000

Hartinger J Stenius K Högemann D and Jahn R 1996 16-BAC/SDS-PAGE: Two-dimensional gel electrophoresis system suitable for the separation of integral membrane proteins *Anal Biochem* **240**: 126-133

Klussmann E Maric K Wiesner B Beyermann M and Rosenthal W 1999 Protein kinase A anchoring proteins are required for vasopressin-mediated translocation of aquaporin-2 into cell membranes of renal principal cells *J Biol Chem* **274**: 4934-4938

Klussmann E Maric K and Rosenthal W 2000 The mechanisms of aquaporin control in the renal collecting duct *Rev Physiol Biochem Pharmacol* **141**: in press

Maric K Oksche A and Rosenthal W 1998 Aquaporin-2 expression in primary cultured rat inner medullary collecting duct cells *Am J Physiol* **275**: F796-801

Scott JD Dell'Aqua ML Fraser ID Tavalin SJ and Lester LB 2000 Coordination of cAMP signaling events through PKA anchoring *Adv Pharmacol* **47**: 175-207

2.14 Mammalian physiology

GASTROINTESTINAL PHENOTYPE OF AQUAPORIN KNOCKOUT MICE

Tonghui Ma and Alan S. Verkman
Cardiovascular Research Institute, Box 0521, University of California, SanFrancisco, CA 94143-0521, USA. E-mail: tonghui@itsa.ucsf.edu

1. INTRODUCTION

Fluid transport is a major function of the gastrointestinal (GI) tract. In humans, more than 9 litres/day of fluid are absorbed or secreted across epithelia and endothelia in salivary gland, stomach, the hepatobiliary tract, pancreas, small intestine and colon (Figure 1). There is increasing evidence that aquaporin water channels (AQPs) are involved in GI fluid transport and digestive function. At least seven aquaporins are expressed in various epithelia and endothelia in the GI system. There are functional data that some GI cell types expressing aquaporins have high or regulated water permeability. Phenotypic analysis of transgenic knockout mice lacking various aquaporins has provided direct evidence that aquaporins are involved in saliva secretion, dietary fat processing, and colon fluid transport. This brief chapter summarizes recent progress on the GI phenotype of aquaporin knockout mice.

Molecular Biology and Physiology of Water and Solute Transport
Edited by Hohmann and Nielsen, Kluwer Academic/Plenum Publishers, New York, 2000.

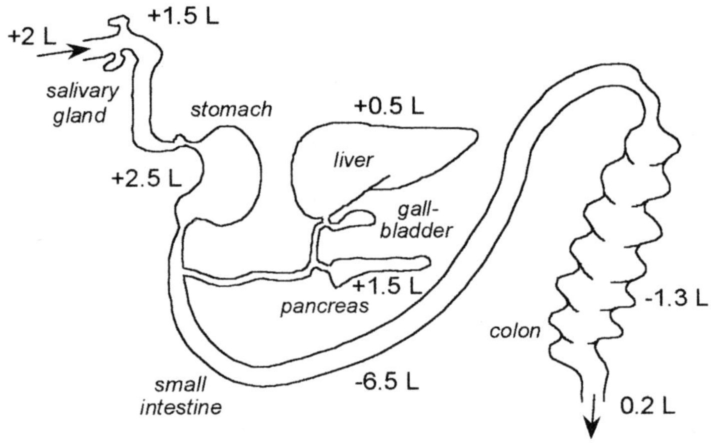

Figure 1. Fluid transport in gastrointestinal system, showing the amount of fluid secreted into and absorbed by various GI organs

2. AQUAPORIN EXPRESSION IN GASTROINTESTINAL SYSTEM

So far, at least seven AQPs are expressed in various tissues in the GI tract: AQP1 in the intrahepatic bile duct epithelium and intestinal lacteals, AQP4 in gastric parietal cells and colon surface epithelium, AQP3 in ileum and colonic epithelium, AQP5 in salivary serous gland, AQP7 in small intestine, AQP8 in liver, pancreas, salivary gland, and colon, and AQP9 in hepatocytes (reviewed in Ma and Verkman, 1999). These sites of aquaporin expression suggest a role of AQPs in fluid transport and digestive/absorptive functions in the GI tract.

3. AQUAPORIN KNOCKOUT MICE

Transgenic mice lacking aquaporins 1, 3, 4, and 5 have been generated in our laboratory by homologous replacement gene targeting Ma *et al.,* 1997, 1998, 1999, 2000). Initial phenotypic analysis of the knockout mice indicated a important role of these AQPs in various physiological functions (reviewed in Verkman *et al.,* 2000). AQP1 null mice manifest impaired renal concentrating function related to reduced proximal tubule water permeability and defective countercurrent multiplication. AQP3 null mice have nephrogenic diabetes insipidus (NDI) related to reduced collecting duct basolateral membrane water permeability. AQP4 plays a key role in brain edema. In the GI system, AQP5 facilitates near-isosmolar fluid transport in salivary serous glandular epithelium. The role of AQP1 in dietary fat processing and of AQP4 in stomach acid secretion and colonic fluid transport were recently investigated as described below.

4. AQUAPORINS IN GASTROINTESTINAL PHYSIOLOGY

4.1 AQP5 in saliva secretion

AQP5 is a water-selective channel expressed in the apical membrane of epithelial cells of serous acini in salivary gland. AQP5 null mice were generated by targeted gene disruption. The genotype distribution of offspring from fonuder AQP5 heterozygous mice was 237 : 217 : 101 wildtype : heterozygous : knockout, indicating impaired prenatal survival of the null mice. The knockout mice had grossly normal appearance, but grew ~20% slower than litter-matched wildtype mice when placed on solid food after weaning. Pilocarpine-stimulated saliva production was reduced by more than 60% in AQP5 knockout mice. Compared with the saliva from wildtype mice, the saliva from AQP5 null mice was hypertonic (420 mOsm) and dramatically more viscous (Figure 2A) with higher salt concentration (Figure 2B). Amylase and protein secretion reflecting salivary mucous gland function were not affected by AQP5 deletion. These results implicate a key role for AQP5 in saliva fluid secretion and provide direct evidence that high epithelial cell membrane water permeability is required for active, near-isosmolar fluid transport.

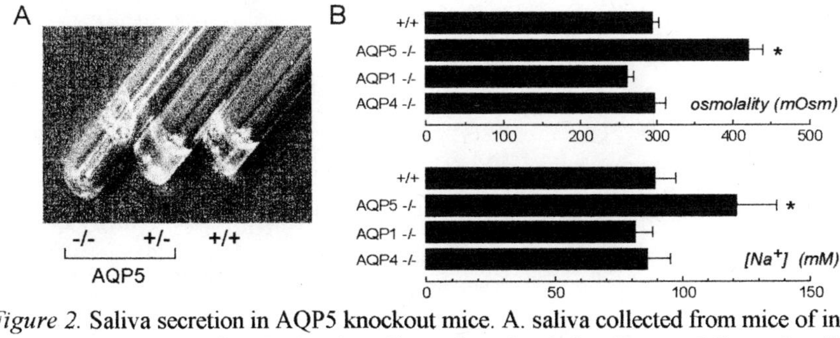

Figure 2. Saliva secretion in AQP5 knockout mice. A. saliva collected from mice of indicated genotype over the first 5 min after pilocarpine stimulation. B. osmolality and sodium concentration of saliva in A. *$p < 0.01$.

4.2 AQP1 in dietary fat processing

Immunocytochemistry showed expression of AQP1 water channels at sites involved in dietary fat processing, including intrahepatic cholangiocytes, gallbladder, pancreatic microvascular endothelium, and intestinal lacteals. To determine whether AQP1 has a role in dietary fat digestion and/or absorption, mice were placed on a diet containing 50% animal fat. Whereas wildtype [+/+] mice (3-3.5 weeks of age, 10-12 grams) gained 49 $\pm$ 5 % (SE, n=50) body weight in 8 days and [+/-] mice gained 46 $\pm$ 4 %, AQP1 [-/-] mice gained only 4 $\pm$ 3 %; weights became similar after return to a 6 % fat diet for 6 days (Figure 3A). The [-/-] mice on a high fat diet acquired an oily appearance, developed steatorrhea, and were relatively hypotriglyceridemic. Addition of pancreatic enzymes (Pancrease™) to the high fat diet partially corrected the decreased weight gain in [-/-] mice (23 $\pm$ 4 % weight increase on day 8). Absorption of ^{14}C-oleic acid from small intestine was not affected by AQP1 deletion as determined blood radioactivity after duodenal infusion. Fecal lipase activity was ten-fold greater in [-/-] than [+/+] mice on low and high fat diets (Figure 3B). Fluid collections done in older mice (which are less sensitive to a high fat diet) by ductal cannulation showed three-fold increased pancreatic fluid flow in response to secretin/CCK, but volumes, pH and amylase activities were affected little by AQP1 deletion. Bile flow rates and bile salt concentrations were not affected by AQP1 deletion. Together these results establish a dietary fat misprocessing defect in AQP1 null mice.

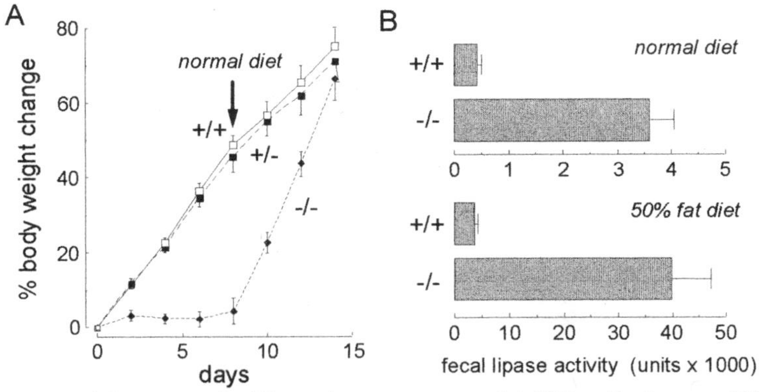

Figure 3. A. weight curves of wildtype, heterozygous, and AQP1 null mice on a 50% fat diet. A total of 60 mice were studied in each group in 8 independent sets of measurements. Initial mouse weight was 10-12 grams. The mice were switched to a normal diet on day 8. B. Fecal lipase activity. Stool samples were collected from weight matched AQP1 [-/-] and [+/+] mice (10-12 g) before and after three days on high fat diet.

4.3 AQP4 in gastric acid secretion

Immunocytochemistry using anti-AQP4 antibodies showed strong expression of AQP4 protein at the basolateral membrane of gastric parietal cells in wildtype [+/+] mice. The involvement of AQP4 in gastric acid secretion was studied using transgenic null [-/-] mice deficient in AQP4 protein (Wang *et al.*, 2000a). The null mice had grossly normal growth and appearance, and showed no differences in gastric morphology by light microscopy. Gastric acid secretion was measured in anesthetized mice in which the stomach was lumenally perfused at 0.3 ml/min with 0.9% NaCl containing [^{14}C] PEG as a volume marker. Collected effluent was assayed for titratable acid content and [^{14}C] PEG radioactivity. After a 45 min baseline perfusion, acid secretion was stimulated by intravenous pentagastrin (200 μg/kg/hr) for one hour or intravenous histamine (0.23 mg/kg) plus intralumenal carbachol (20 mg/l). As shown in Figure 4, baseline gastric acid secretion (SE, n=25) was 0.06±0.03 mEq/15 min in [+/+] and 0.03±0.02 mEq/15 min in [-/-] mice. Pentagastrin stimulated acid secretion was 0.59±0.14 mEq/15 min in [+/+] and 0.70±0.15 mEq/15 min in [-/-] mice. Histamine/carbachol stimulated acid secretion was 7.0±1.9 mEq/15 min in [+/+] and 8.0±1.8 mEq/15 min in [-/-] mice. In addition, there was no effect of AQP4 deletion on gastric fluid secretion, gastric pH, or fasting serum gastrin concentrations. These results provide direct evidence against a role of AQP4 in gastric acid secretion.

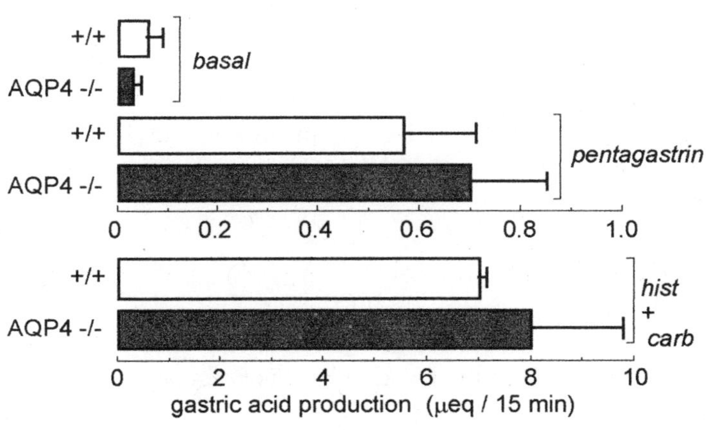

Figure 4. Basal and stimulated gastric acid secretion in AQP4 knockout mice

4.4 AQP4 in colon water transport and fecal dehydration

Transgenic null mice were used to test the hypothesis that AQP4 is involved in colon water transport and fecal dehydration (Wang *et al.*, 2000b). AQP4 was immunolocalized to the basolateral membrane of colonic surface epithelium of wildtype [+/+] mice and was absent in AQP4 null [-/-] mice. The transepithelial osmotic water permeability coefficient (P_f) of the *in vivo* perfused colon was measured using ^{14}C-polyethylene glycol as a volume marker. P_f in [+/+] mice was 0.016±0.002 cm/s and independent of osmotic gradient magnitude and direction (-40 to 280 mosM), as well the solute used to induce osmosis (180-1100 Da). P_f was independent of lumenal perfusion flow rate (0.9 to 6.8 μL/s) but was increased 2.5-fold by amphotericin B, suggesting minimal unstirred layer effects. P_f of proximal colon was greater than that of distal colon (0.020±0.004 vs. 0.009±0.003 cm/s, $P<0.01$). P_f was significantly lower in [-/-] mice when measured in full-length colon (0.009±0.002 cm/s, $P<0.05$) and proximal colon (0.013±0.002 cm/s, $P<0.05$), but not in distal colon. Apparent diffusional water permeabilities in colons from [+/+] vs. [-/-] mice were not different (P_d = 8.8±0.7 x 10^{-5} vs. 9.6±0.7 x 10^{-5} cm/s). There was no difference in water content of cecal stool from wildtype vs. AQP4 null mice (0.80±0.01 vs. 0.81±0.01, mean ± SE), but there was slightly higher water content in defecated stool from [-/-] mice (0.68±0.01 vs. 0.65±0.01, $P<0.05$). Despite the differences in water permeability with AQP4 deletion, colonic secretory function was not impaired in a model of theophylline-induced secretion (50±9 vs. 51±8 μL/min/g). These results provide evidence that transcellular water transport through AQP4 water channels in colonic epithelium facilitates transepithelial osmotic water permeability, but has little or no affect on colonic fluid secretion or fecal dehydration.

ACKNOWLEGMENTS

The work was supported by NIH grants DK35124, HL60288, DK59198 and DK43840, and grants from the Cystic Fibrosis Foundation.

REFERENCES

Frigeri A Gropper M Turck CW and Verkman AS 1995 Immunolocalization of the mercurial-insensitive water channel and glycerol intrinsic protein in epithelial cell plasma membranes *Proc Natl Acad Sci USA* **92**: 4328-4331

Ma T Yang B Gillespie A Carlson EJ Epstein CJ and Verkman AS 1997 Generation and phenotype of a transgenic knockout mouse lacking the mercurial-insensitive water channel aquaporin-4 *J Clin Invest* **100**: 957-962

Ma T Yang B Gillespie A CarlsonEJ Epstein CJ and Verkman AS 1998 Severely impaired urinary concentrating ability in transgenic mice lacking aquaporin-1 water channels *J Biol Chem* **273**: 4296-4299

Ma T Song Y Gillespie A Carlson EJ Epstein CJ and Verkman AS 1999 Defective secretion of saliva in transgenic mice lacking aquaporin-5 water channels *J Biol Chem* **274**: 20071-20074

Ma T and Verkman AS 1999 Aquaporin water channels in gastrointestinal physiology *J Physiol (Lond)* **517**: 317-326

Ma T Song Y Yang B Gillespie A Carlson EJ Epstein CJ and Verkman AS 2000 Nephrogenic diabetes insipidus in mice lacking aquaporin-3 water channels *Proc Natl Acad Sci USA* **97**: 4786-4791

Ma T WangKS Song Y Yang B Jayaraman S Bastidas JA and Verkman AS 2000 Defective dietary fat processing in transgenic mice lacking aquaporin-1 water channels *Am J Physiol* In press

Nielsen S Smith BI Christensen EI and Agre P 1993 Distribution of the aquaporin CHIP in secretory and resorptive epithelia and capillary endothelia *Proc Natl Acad Sci USA* **90**: 7275-7279

Verkman A Yang SB Song Y Manley G and Ma T 2000 Role of water channels in fluid transport studied by phenotype analysis of aquaporin knockout mice *Exp Physiol* In press

Wang KS Komar AR Ma T Filiz F McLeroy J Hoda K Verkman AS and Bastidas JA 2000 Gastric acid secretion in aquaporin-4 knockout mice *Am J Physiol* In press

Wang KS Ma T Filiz F Verkman AS and Bastidas JA 2000 Colon water transport in transgenic mice lacking aquaporin-4 water channels *Am J Physiol* In press

2.15 Mammalian physiology

TWO DISTINCT SIGNALS DETERMINE THE BASOLATERAL TARGETING OF AQP4 IN THE RENAL EPITHELIAL CELL LINE MDCK

Ricardo Madrid, Sophie Le Maout, Marie-Bénédicte Barrault, and Jean Mérot
DBCM/SBCe, CEA Saclay, 91191 Gif/Yvette, France. E-mail: ricardo@ferre.saclay.cea.fr

1. INTRODUCTION

To fulfill their physiological role in water handling by the kidney, aquaporins are differentially targeted to the apical and the basolateral membranes of epithelial kidney cells. For example, whereas AQP1 exhibits an un-polarized distribution in the proximal tubule cells (Sabolic *et al.,* 1992, Nielsen *et al.,* 1993), AQP2 is specifically targeted to the apical membranes of the principal cells in the collecting duct (Fushimi *et al.,* 1993, Nielsen *et al.,* 1993). On the other hand, AQP3 and AQP4 are exclusively localized in the basolateral membranes of these cells (Ecelbarger *et al.,* 1995, Frigeri *et al.,* 1995). Although a great deal of information about the cellular distribution of aquaporins in the body has emerged from previous studies, the mechanisms underlying their targeting in polarized epithelial cells remain poorly understood.

To answer this question we analyzed the expression and the targeting of AQP4 in a stably transfected epithelial model system, the MDCK cells. In a first step, we introduced a VSV tag on the extreme amino-terminus of the protein to enable the analysis of deletions in its C-terminal part which constitutes the epitope of AQP4 specific antibody. Confocal analysis showed that the VSV tag did not modify the basolateral targeting of AQP4. Then, several deletion mutants were established (Table 1) and analyzed by confocal microscopy using anti-VSV antibody.

The deletion of the last 42 a.a of the protein did not significantly modify basolateral AQP4 expression. However, larger deletion (52 a.a in AQP4-272 stop) resulted in an intracellular localization of AQP4 presumably the Golgi apparatus. On the other hand, the intra-molecular deletion of 31 a.a (from Q272 to I302) resulted in the complete re-routing of AQP4 to the apical membranes of the cells, whereas the deletion of either 15 or 21 a.a (AQP4 Δ272-286 and Δ282-302, respectively) did not modify the basolateral targeting of AQP4.

Together, our results show that AQP4 is basolateraly targeted in MDCK cells and that two independent signals, located in the cytoplasmic carboxy terminus of the protein, determine its basolateral localization.

2. MATERIALS AND METHODS

2.1 cDNA constructs and mutagenesis

A VSV epitope-tagged-AQP4 was generated by introducing the nucleotide sequence encoding the eleven a.a of the VSV tag just after the initiation codon of AQP4 cDNA (Le Maout *et al.*, 1997). Deletion mutants were constructed by PCR based mutagenesis. All the constructs were verified by sequencing. Wild type AQP4 and VSV-AQP4 cDNAs were cloned into the PCB6 expression vector containing the G418 resistance gene.

2.2 Transfection and immunofluorescence

Stably transfected AQP4-MDCK clonal cell lines were isolated in a selection medium containing 500μg/ml G418. For immunofluorescence experiments, the cells were grown on permeable supports until they formed a confluent monolayer. They were fixed with 3% paraformaldehyde and permeabilized with 0.1% Triton X-100 in PBS saline. They were then incubated with the P5D4 antibody (1:1000) directed against the VSV epitope or the AQP4 specific antibody (1/100) raised against the C-terminus of the protein. The cells were observed with a Biorad confocal microscope.

Table 1. Carboxy terminal amino acid sequences (single letter code) of wild type and mutant AQP4. From alanine A.269 to the carboxy terminal valine (V.323)

AA QQTKGSYMEV EDNRSQVETE DLILKPGVVH VIDIDRGDEK KGKDSSGEVL SSV	AQP4 Wild
AA QQTKGSYMEV EDNRSQVETE DLILKPGVVH VIDIDRGDEK KGKDSSGEVL ---	321 Stop
AA QQTKGSYMEV EDNRSQVETE DLILKPGVVH VI----------------------	303 Stop
AA QQTKGSYMEV E---------------------------------------	282 Stop
AA Q--	272 Stop
AA Q-------------------------------DIDRGDEK KGKDSSGEVL SSV	Δ272-302
AA QQTKGSYMEV E---------------------DIDRGDEK KGKDSSGEVL SSV	Δ282-302
AA Q--------- ------VETE DLILKPGVVH VIDIDRGDEK KGKDSSGEVL SSV	Δ272–286

3. RESULTS AND DISCUSSION

Indirect immunofluorescence analysis, using a specific anti-AQP4 antibody, revealed that AQP4 was exclusively detected in the basolateral membranes of the cells. Using a focal plan passing through the middle of the cell layer (Fig 1, left) a cobble stone pattern of fluorescence, characteristic of a basolateral localisation, was observed. Conversely, a strong fluorescence lining the cell contacts was clearly visible with a focal plan perpendicular to the cells (Fig 1, left bottom image). These observations are in accordance with previously reported localisation of AQP4 in original epithelial cells from the kidney and the intestinal tract (Terris *et al.*, 1995, Koyama *et al.*, 1999) and indicates that MDCK cells represent a valuable cell model system to analyse AQP4 targeting mechanisms in polarized epithelial cells. Figure 1 also shows that the VSV-AQP4 protein, detected with the anti-VSV antibody, was also targeted to the basolateral membranes of the cells. Together with previous studies this indicated that the N-terminal addition of a small epitope to aquaporin sequences did not introduce important structural modifications that could interfere with their targeting or their function (Kamsteeg *et al.*, 1999).

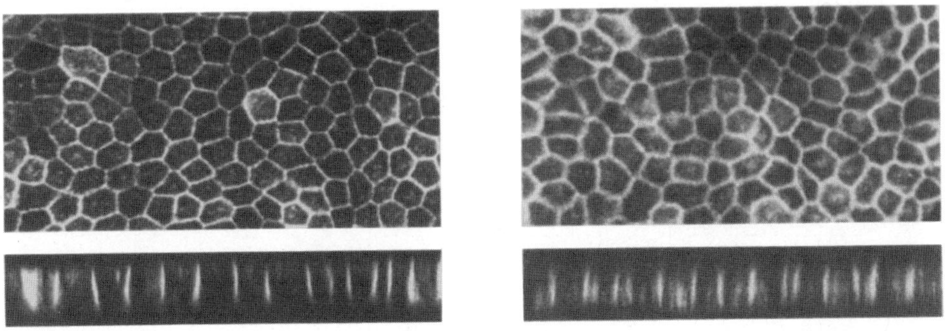

Figure 1. Confocal images showing the immunofluorescence pattern observed on AQP4-MDCK (left) and VSV-AQP4-MDCK (right) using anti-AQP4 and anti-VSV antibodies respectively. Upper images used a focal plan parallel to the cell layer whereas in the lower XZ images the focal plan was perpendicular to the cells. Apical membranes are facing the top of the XZ images.

Several studies identified basolateral targeting signals in the cytoplasmic carboxy-terminal domains of membrane proteins. We thus focused our attention on AQP4 carboxy-terminus and introduced several deletions in this region of the protein (see Table 1). Deletion of the last 3, 21 or 42 amino acids (AQP4-321, 303 and 282-Stop, respectively) did not abolish basolateral expression of AQP4 (Fig. 2, left panel). On the other hand, larger deletions (AQP4-272 Stop) resulted in a profound modification of AQP4 expression. As illustrated in figure 2 (right panel), the basolateral plasma membrane expression was no longer detectable and all the fluorescence was concentrated in intracellular compartments presumably the Golgi and the lysosome.

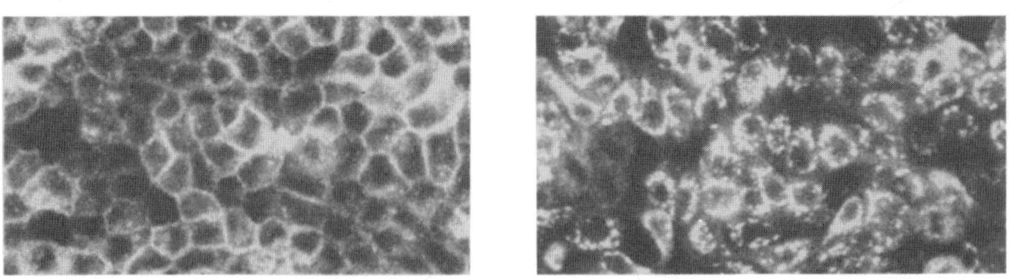

Figure 2. AQP4 basolateral membrane expression was observable until the last 42 amino acids were deleted (AQP4-282 Stop, left panel). Additional deletions resulted in an intracellular localisation of the protein (AQP4-272 Stop, right panel).

We reasoned that the intracellular localisation of AQP4-272 Stop might be due to improper folding or assembly of AQP4 sub-units when the carboxy-terminus of the protein is too short. We thus introduced intra-molecular deletions in AQP4 cytoplasmic domain (Table 1).

As illustrated on figure 3 (right panel), the deletion of 31 amino acids between D272 and I302 resulted in the apical expression of AQP4.

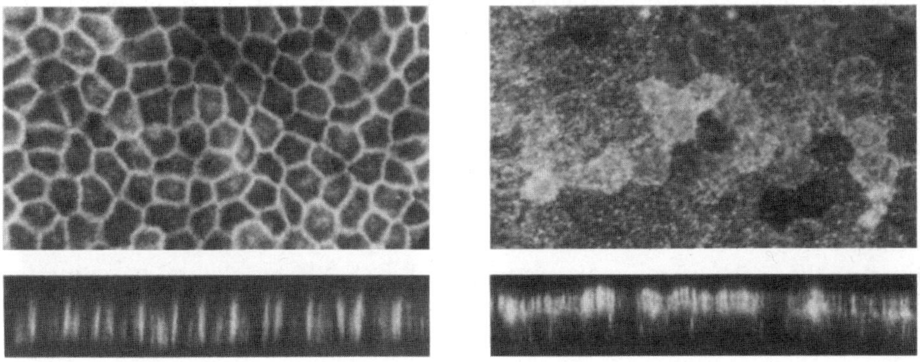

Figure 3. Confocal sections made at the level of the apical membranes of VSV-AQP4 (left) and VSV-AQP4-Δ272-302 (right). In XY sections the fluorescence was observed over the entire surface of the cells for AQP4 -Δ272-302 (upper right). The apical labelling was clearly demonstrated by XZ sectioning (lower right).

We concluded that this amino acid stretch contained AQP4 basolateral targeting determinant. To define more precisely the sorting signal, two additional mutants were constructed. As illustrated in figure 4, the re-addition of either the first 10 a.a or the last 16 a.a (see Table 1) of the basolateral sorting signal to the apically targeted AQP4-Δ272-302 restored the basolateral targeting of AQP4.

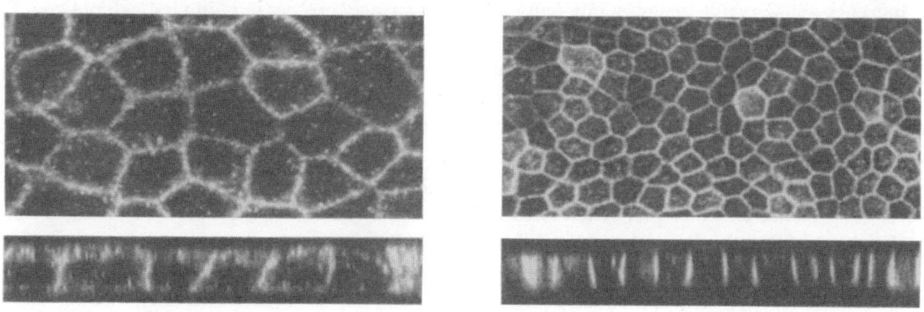

Figure 4. The 10 a.a between Q272 and E281 (AQP4 Δ282-302, left) and the 16 a.a between V287 and I302 (AQP4 Δ272-286, right) are necessary and sufficient to re-route the apically targeted mutant AQP4 Δ272-302 (Fig 3) to the basolateral membranes of MDCK cells.

4. CONCLUSIONS

Together with previous studies (Le Maout *et al.,* 1997, Kamsteeg *et al.,* 1999), our results show that MDCK cells represent a valuable cellular model system to analyse the mechanisms of aquaporins targeting in polarized epithelial cells. Using engineered epitope

tagged AQP4 and mutagenesis we showed that AQP4 basolateral targeting is dependent on two independent signals contained in the cytoplasmic carboxy-terminal domain of the protein. Interestingly, these two peptidic sequences contain putative tyrosine-based (YXXΦ; 277YMEV280) and double leucine (292LI293) signals that have been shown to be critical sorting determinants of basolateraly targeted proteins (Marks *et al.,* 1997). Whether these specific signals are involved in AQP4 basolateral targeting is now under study.

ACKNOWLEDGMENTS

This work was supported by the European TMR Project #FMRX-CT97-0128.

REFERENCES

Ecelbarger CA Terris J Frindt G Echevarria M Marples D Nielsen S and Knepper MA 1995 Aquaporin-3 water channel localization and regulation in rat kidney *Am J Physiol* **269**: F663-72

Frigeri A Gropper MA Turck CW and Verkman AS 1995 Immunolocalization of the mercurial-insensitive water channel and glycerol intrinsic protein in epithelial cell plasma membranes *Proc Natl Acad Sci U S A* **92**: 4328-4331

Fushimi K Uchida S Hara Y Hirata Y Marumo F and Sasaki S 1993 Cloning and expression of apical membrane water channel of rat kidney collecting tubule *Nature* 361 549-52

Kamsteeg EJ Wormhoudt T Rijss JPL Van OS C anf Deen P 1999 An impaired routing of wild-type aquaporin-2 after tetramerization with an aquaporin-2 mutant explains dominant nephrogenic diabetes insipidus *EMBO J* **18**: 2294-2400

Koyama Y Yamamoto T Tani T Nihei K Kondo D Funaki H Kawasaki K Sato N Hatakeyama K Kihara I 1999 Expression and localization of aquaporins in rat gastrointestinal tract *Am J Physiol;* **276**: C621-627

Le Maout S Brejon M Olsen O Merot J and Welling PA 1997 Immunolocalization of the mercurial-insensitive water channel and glycerol intrinsic protein in epithelial cell plasma membranes *Proc Natl Acad Sci U S A* **94**: 13329-13334

Marks MS Ohno H Kirchausen T and BonifacinoJ 1997 protein sorting by tyrosine based signals: adapting to the Ys and wherefores *Trends in Cell Biol* **7**: 124-128

Nielsen S DiGiovanni SR Christensen EI Knepper MA and Harris HW 1993 Cellular and subcellular immunolocalization of vasopressin-regulated water channel in rat kidney *Proc Natl Acad Sci U S A* **90**: 11663-11667

Nielsen S Smith BL Christensen EI Knepper MA and Agre P 1993 CHIP28 water channels are localized in constitutively water-permeable segments of the nephron *J Cell Biol* **120**: 371-383

Sabolic I Valenti G Verbavatz JM Van Hoek AN Verkman AS Ausiello DA and Brown D 1992 Localization of the CHIP28 water channel in rat kidney *Am J Physiol* **263**: C1225-1233

Terris J Ecelbarger CA Marples D Knepper MA Nielsen S 1995 Distribution of aquaporin-4 water channel expression within rat kidney *Am J Physiol* **269**: F775-785

2.16 Mammalian physiology

WATER CHANNEL PROTEIN, AQUAPORIN 3, IN EPITHELIAL CELLS

Toshiyuki Matsuzaki, Takeshi Suzuki and Kuniaki Takata
Laboratory of Molecular and Cellular Morphology, Institute for Molecular and Cellular Regulation, Gunma University, Maebashi, Gunma 371-8512, Japan.
E-mail: matoshi@med.gunma-u.ac.jp

1. INTRODUCTION

Water channel protein, aquaporin 3 (AQP3), was originally isolated from the kidney and is localized at the basolateral plasma membranes of renal collecting duct principal cells. AQP3, together with antidiuretic hormone (ADH)-sensitive water channel protein, AQP2, in apical aspects, plays an important role in the transcellular water reabsorption through the collecting duct epithelial cell layer (Fushimi *et al.*, 1993, Ishibashi *et al.*, 1994; Fig 1). Although many works on AQP3 have been done in the kidney, AQP3 in extrarenal organs is poorly understood. We surveyed the localization of AQP3 in rat extrarenal epithelial tissues. We discuss here possible roles of AQP3 in these epithelial tissues.

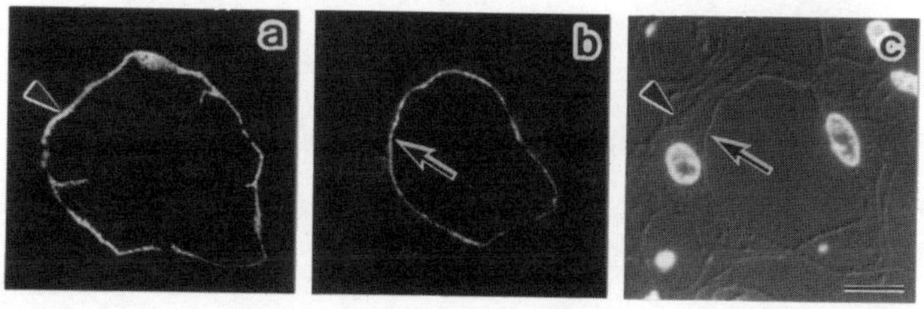

Figure 1 Immunofluorescence localization of AQP3 and AQP2 in the rat kidney collecting duct. a: AQP3 is localized at basolateral membrane (arrowhead). b: AQP2 is localized at apical membrane (arrow). c: Corresponding Nomarski differential interference-contrast image counterstained with DAPI for nuclei. Bar: 10 µm.

Molecular Biology and Physiology of Water and Solute Transport
Edited by Hohmann and Nielsen, Kluwer Academic/Plenum Publishers, New York, 2000.

2. LOCALIZATION OF AQP3 IN EPITHELIAL TISSUES

The expression of AQP3 and the localization of the AQP3 protein have been reported sporadically (Frigeri *et al.,* 1995a, b, Hamann *et al.,* 1998, Nielsen *et al.,* 1997). We carried out a comprehensive survey and localized AQP3 protein in rat epithelial tissues in detail (Matsuzaki *et al.,* 1999). AQP3 was present in the epithelia of the urinary tract, digestive tract, respiratory tract, and the epidermis of the skin. In the transitional epithelia of the renal pelvis, ureter, urinary bladder, and urethra, AQP3 was localized at cell membranes of basal and intermediate layers but not in the large superficial cells (Fig 2). In the stratified epithelia of the skin and the upper digestive tract from the oral cavity to the forestomach (nonglandular portion), AQP3 was localized at cell membranes of the basal and intermediate layers but not in the surface layer (Fig 3). In the simple epithelia of the fundic stomach, AQP3 was restricted to the surface mucous cells. In the ileum, AQP3 was localized at basolateral membranes of absorptive epithelial cells in the villous tips. In the distal colon and rectum, AQP3 was localized at the basolateral membranes of epithelial cells directly

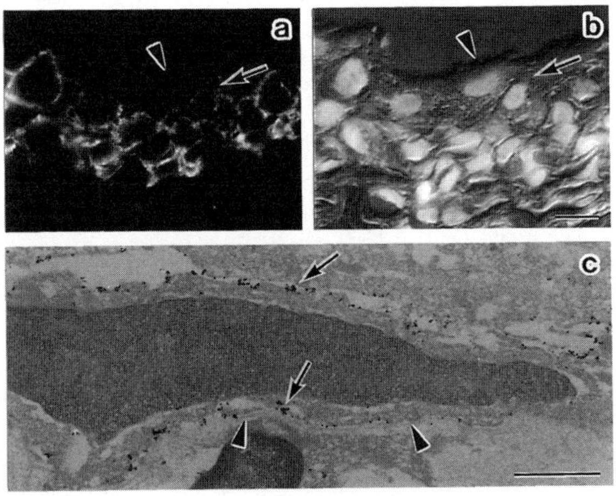

Figure 2 Immunofluorescence localization (a, b) and ultrastructural localization (c) of AQP3 in the rat urinary bladder. a: AQP3 is localized at cell membrane of basal and intermediate cell layers (arrow) but not in large superficial cells (arrowhead). b: Corresponding Nomarski differential interference-contrast image counterstained with DAPI for nuclei. Bar: 10 µm. c: AQP3 is localized along the plasma membrane of the basal and intermediate cells (arrows). Arrowheads indicate basal lamina. Bar: 1 µm. (Reproduced, with permission, from Matsuzaki T, Suzuki T, Koyama H, Tanaka S, and Takata K, Water channel protein AQP3 is present in epithelia exposed to the environment of possible water loss *J Histochem Cytochem* **47**: 1275-1286, 1999).

facing the lumen and at the neck of the crypts. In the pseudostratified ciliated epithelia of the respiratory tract from the nasal cavity to the intrapulmonary bronchi, AQP3 was localized at cell membranes of basal cells and at basolateral membranes of some ciliated cells. In summary AQP3 was present in the surface-covering epithelial cells, and it was localized at plasma membranes not directly facing the lumina.

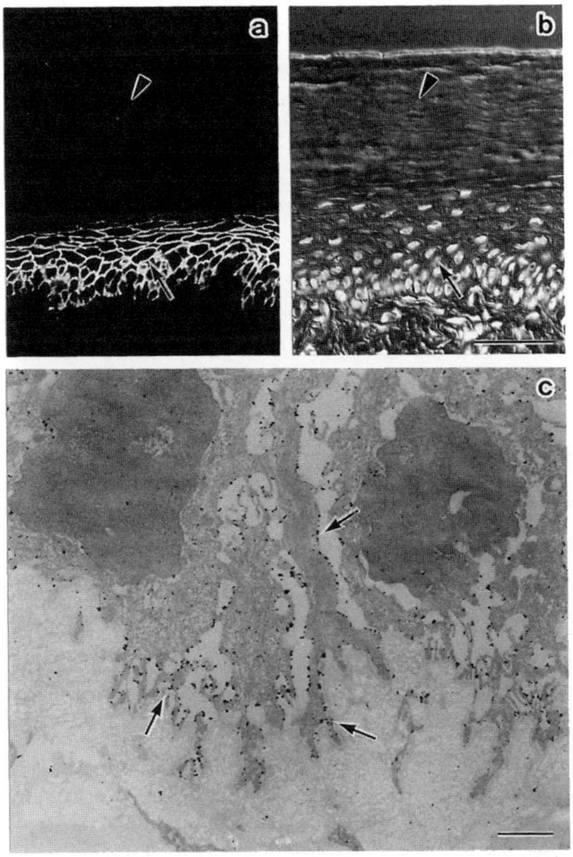

Figure 3 Immunofluorescence localization (a, b) and ultrastructural localization (c) of AQP3 in the rat skin. a: AQP3 is localized at cell membrane of basal and intermediate cell layers (arrow). Labeling intensity gradually decreases toward the surface from the basal cells, and no label for AQP3 is seen in the stratum corneum (arrowhead). b: Corresponding Nomarski differential interference-contrast image counterstained with DAPI for nuclei. Bar: 50 µm. c: AQP3 is localized along the plasma membrane of the basal cells of the epidermis (arrows). Bar: 1 µm. (Reproduced, with permission, from Matsuzaki T, Suzuki T, Koyama H, Tanaka S, and Takata K, Water channel protein AQP3 is present in epithelia exposed to the environment of possible water loss *J Histochem Cytochem* **47**: 1275-1286, 1999).

3. DEVELOPMENTAL CHANGES OF AQP3 EXPRESSION IN THE SKIN

Fetus rats have a low risk for the water loss from the skin because they are surrounded by the amniotic fluid. Upon delivery, newborn rats are suddenly exposed to the air and face a risk of dehydration. Examination of AQP3 expression in the skin development revealed that its expression in the epidermis commenced late in the fetal life toward the birth. The beginning of AQP3 expression seems to be a part of the preparatory steps toward their transformation from the life in the water to the life in the land.

4. POSSIBLE ROLE OF AQP3 IN EPITHELIAL CELLS

In the kidney collecting ducts, water is reabsorbed from the lumen to the interstitium. AQP3 is also present in the urinary tract downstream of the collecting ducts. However, AQP3 does not seem to be responsible for water reabsorption from the lumen because it is not present in the surface-covering large superficial cells of transitional epithelia. In addition, it has been shown that apical membranes of the superficial cells have low water-permeability. Therefore, we suggest that water movement in the urinary epithelia seems to be directed from subepithelial side to epithelial cells. In the skin AQP3-localization pattern is the same as that in the urinary epithelia, that is, AQP3 is not present at surface layers of epidermis. Here again it is suggested that water movement seems to occur from the dermis to the epidermis. Then what is the role of AQP3 in these epithelia? AQP3-positive epithelia are exposed to severe environments. For example, urinary epithelia are exposed to the hypertonic urine, epidermis to the air, and so on. These harsh and hostile environment results in a possibility of water loss. We therefore suggest that AQP3 present in epithelia exposed to a possible water loss serves to supply water to epithelial cells from subepithelial side to prevent their dehydration. Developmental changes of AQP3 expression in the epidermis, that is, AQP3 expression in the epidermis commences late in the fetal life toward the birth, also support the above role of AQP3.

5. AQP3 EXPRESSION IN MDCK CELLS

We examined the relationship between the AQP3 expression and the osmolality of the extracellular environment in Madin-Darby canine kidney

(MDCK) cultured cells. AQP3 expression was increased by the treatment with hypertonic media containing raffinose or NaCl. AQP3 was localized at basolateral plasma membranes in these cells. These results show that the AQP3 expression is controlled by the extracellular fluid osmolality.

6. CONCLUSION

AQP3 is present in surface-covering epithelia exposed to an environment of possible water loss and seems to have a possible role of providing these epithelial cells with water from the subepithelial side to protect them against dehydration. The expression of AQP3 seems to be controlled by the extracellular fluid osmolality.

ACKNOWLEDGMENTS

We thank H. Koyama and S. Matsuzaki for assistance. This work was supported in part by Grants-in-Aid for Scientific Research from the Ministry of Education, Science, Sports, and Culture of Japan. T. Matsuzaki is a JSPS Research Fellow.

REFERENCES

Frigeri A Gropper MA Turck CW and Verkman AS 1995a Immunolocalization of the mercurial-insensitive water channel and glycerol intrinsic protein in epithelial cell plasma membranes *Proc Natl Acad Sci USA* **92**: 4328-4331

Frigeri A Gropper MA Umenishi F Kawashima M Brown D and Verkman AS 1995b Localization of MIWC and GLIP water channel homologs in neuromuscular epithelial and glandular tissues *J Cell Sci* **108**: 2993-3002

Fushimi K Uchida S Hara Y Hirata Y Marumo F and Sasaki S 1993 Cloning and expression of apical membrane water channel of rat kidney collecting tubule *Nature* **361**: 549-552

Hamann S Zeuthen T Cour ML Nagelhus EA Ottersen OP Agre P and Nielsen S 1998 Aquaporins in complex tissues: distribution of aquaporins 1-5 in human and rat eye *Am J Physiol* **274** (*Cell Physiol* **43**): C1332-C1345

Ishibashi K Sasaki S Fushimi K Uchida S Kuwahara M Saito H Furukawa T Nakajima K Yamaguchi Y Gojobori T and Marumo F 1994 Molecular cloning and expression of a member of the aquaporin family with permeability to glycerol and urea in addition to water expressed at the basolateral membrane of kidney collecting duct cells *Proc Natl Acad Sci USA* **91**: 6269-6273

Matsuzaki T Suzuki T Koyama H Tanaka S and Takata K 1999 Water channel protein AQP3 is present in epithelia exposed to the environment of possible water loss *J Histochem Cytochem* **47**: 1275-1286

Nielsen S King LS Christensen BM and Agre P 1997 Aquaporins in complex tissues II Subcellular distribution in respiratory and glandular tissues of rat *Am J Physiol* **273** (*Cell Physiol* **42**): C1549-C1561

2.17 Mammalian physiology

REGULATION OF AQUAPORIN-2 WATER CHANNEL TRAFFICKING BY PHOSPHORYLATION: LESSONS FROM TRANSFECTED EPITHELIAL CELLS

Role of phosphorylation in AQP2 trafficking

[1]Giuseppe Procino, [1]Monica Carmosino, [2]Lorenzo A. Pinna, [3]Ildo Nicoletti, [4]Søren Nielsen, [1]Maria Svelto and [1]Giovanna Valenti

[1] Dipartimento di Fisiologia Generale ed Ambientale, University of Bari, Italy; [2]Dipartimento di Chimica Biologica, University of Padova, Italy; [3].Dipartimento di Medicina Clinica e Sperimentale, University of Perugia, Italy [4]; Department of Cell Biology, Institute of Anatomy, University of Aarhus, Denmark. E-mail: g.valenti@bilogia.uniba.it

1. INTRODUCTION

Vasopressin plays a key role in the modulation of water permeability in mammalian kidney. The ability of vasopressin to increase water transport is mediated by translocation of the specific vasopressin-sensitive water channel AQP2, from intracellular vesicles to the apical membrane. The initial mechanism of vasopressin action involves the hormone binding to the V2 receptor located on the basolateral cell surface of collecting duct principal cells and subsequent activation of cAMP-dependent PKA. It has been shown that a cAMP-dependent phosphorylation of AQP2 occurs at a serine level (ser-256) located in the cytoplasmic C-terminus. PKA-mediated phosphorylation of AQP2 has been proposed to be essential in regulating AQP2-containing vesicle exocytosis. Many progress in our understanding of the cellular biology of AQP2 recycling come from the establishment of cell colture models in which the regulated pathway of AQP2 insertion into the plasma membrane is retained. In cultured renal CD8 cells transfected with AQP2, vasopressin stimulation caused AQP2 redistribution from an

Molecular Biology and Physiology of Water and Solute Transport
Edited by Hohmann and Nielsen, Kluwer Academic/Plenum Publishers, New York, 2000.

intracellular site to the apical membrane increasing the water permeability coefficient (P_f) by approximately 5-fold (Valenti *et al.*, 1996, Valenti *et al.*, 1998).

2. OKADAIC ACID STIMULATES THE OSMOTIC WATER PERMEABILITY RESPONSE BY INDUCING AQP2 TRANSLOCATION

Recent finding from our laboratory (Valenti *et al*, 2000), demonstrate that in CD8 cells treated with the phosphatase inhibitor okadaic acid, phosphorylation of AQP2 is not required for its translocation toward the apical membrane. Okadaic acid is a tumor promoter which inhibits the activity of protein phosphatases 1 and 2A in intact cells. In CD8 cells both forskolin alone and okadaic acid alone increased the osmotic water permeability coefficient P_f by about 4-5 fold. In intact cells, in vivo phosphorylation studies reveal that forskolin stimulation results in a threefold increase in AQP2 phosphorylation. In contrast, 1μM okadaic acid treatment promotes only a 60% increase in AQP2 phosphorylation which is abolished when this treatment is performed in the presence of 1μM H89, a specific Protein kinase A (PKA) inhibitor (Fig. 1). Nevertheless, in this latter condition, confocal microscopy analysis reveals that AQP2 translocates and fuses to the apical membrane. (Fig. 1). Okadaic acid-induced AQP2 translocation is dose dependent having its maximal effect at a concentration of 1 μM. Therefore, okadaic acid exerts a full forskolin-like effect independent from AQP2 phosphorylation. The observation that okadaic acid stimulates water transport and AQP2 translocation in the absence of any measurable increase in AQP2 phosphorylation, clearly supports the presence of an alternative AQP2 phosphorylation-independent pathway to stimulate AQP2 sorting to the apical membrane. It is conceivable that different intracellular pools of AQP2-bearing vesicles under different regulatory mechanisms might exist in renal collecting duct epithelial cells. This hypothesis is also sustained by the observation of the partial additive effect of okadaic acid and forskolin in increasing the P_f. Confocal laser microscopy analysis suggests that cytoskeletal elements might be involved in okadaic acid action. Therefore a possible explanation for the effect of okadaic acid is that, by altering the phosphorylation state of unidentified protein/s, okadaic acid induces cytoskeletal rearrangements that promote vesicle fusion with the apical plasma membrane. In this respect it is known that synaptic vesicles are organized into two distinct functional pools, a large reserve pool in which vesicles are restrained by the actin cytoskeleton and a releasable pool in which vesicles approach the presynaptic membrane and fuse upon stimulation. The observed actin rearrangement induced in CD8 cells by okadaic acid treatment, in the absence of forskolin stimulation, might explain AQP2 translocation in renal cells as well. Visualization of actin network with phalloidin-FITC in CD8 cells double labeled with AQP2 and treated with either forskolin or okadaic acid would favour this hypothesis. Compared with untreated cells, okadaic acid treated cells show very few organized actin filaments and a massive fusion of AQP2 on the apical surface. Okadaic acid effect on actin network is still visible in H89-pretreated CD8 cells, a condition in which AQP2 translocates toward the apical membrane. In contrast, forskolin stimulation does not induce apparent modification in the actin network organization in H89-pretreated CD8 cells, and does not promote AQP2 redistribution as well as AQP2 phosphorylation. This suggests that okadaic acid acts primarily on the cellular machinery that facilitates the movement of AQP2 to the cell surface rather than directly on AQP2 protein itself. Moreover, these data indicate that a reorganization of apical actin network might be a prerequisite for promoting redistribution of AQP2-containing vesicles either for forskolin and for okadaic acid effects.

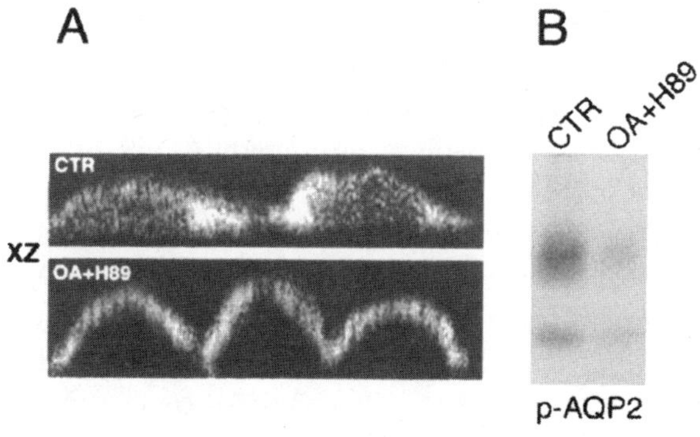

Figure 1. (A) Confocal xz reconstruction of AQP2 staining in CD8 cells. (B) Autoradiography of phosphorylated AQP2 (p-AQP2) immunoprecipitated from CD8 cells.

3. AQP2 PHOSPHORILATION BY THE GOLGI CASEIN KINASE MIGHT REGULATE THE TRAFFICKING WITHIN THE GOLGI NETWORK

Besides short term regulation of AQP2 trafficking by PKA, recent finding from our group (Procino *et al.*, 2000) indicate that AQP2 is phosphorylated at ser-256 by a Golgi-resident kinase, termed Golgi casein kinase (G-CK). *In vitro* phosphorylation experiments in CD8 cells reveal that during AQP2 maturation in the Golgi apparatus, AQP2, as well as a peptide reproducing the sequence around ser-256 is readily phosphorylated by purified G-CK. This suggests a potential role for G-CK in regulating constitutive AQP2 trafficking within the Golgi apparatus. Preincubation with 30μM H89 or a 20°C temperature shift result in blocking of newly synthesized AQP2 in the TGN as confirmed by immunolocalization experiments. In both conditions, the block is completely reversible by returning to basal conditions. Phosphorylation experiments in intact cells reveal that these treatments cause a significant rise in the phosphorylation level of AQP2, compared to untreated cells, even in a condition in which PKA is strongly inhibited. This evidence suggests that AQP2 is phosphorylated in the Golgi apparatus by a H89-insensitive kinase. Immunofluorescence and immunoblotting experiments using an antibody specifically recognizing ser-256-phosphorylated AQP2, demonstrate that AQP2 stacked in the Golgi apparatus is phosphorylated at ser-256 (Fig. 2). G-CK phosphorylates seryl residues specified by an acidic aminoacid at position +2. This consensus sequence occurs around AQP2 ser-256 (...RRQSVEL...) making it a potential substrate for G-CK while the same site is also a good target for PKA by virtue of the two arginyl residues at positions -2 and -3. In vitro phosphorylation studies of AQP2 immunoprecipitated from a Golgi-enriched fraction demonstrate that Mn^{2+}, a G-CK activator, increases the amount of phosphorylated AQP2 (p-AQP2) even in the presence of 30 μM H89, thus excluding any contribution of PKA in the phosphorylation state of AQP2. Synthetic peptide reproducing the AQP2 phosphorylation site (VRRRQSVELH), is phosphorylated in vitro by both PKA and G-CK. A derivative in which the glutamyl residue at position +2 relative to serine is replaced by alanine, fails to be phosphorylated by G-CK whereas still acts as a PKA substrate. It appears clear that ser-256 in AQP2 is phosphorylated by two different kinases: by a Golgi-resident kinase, likely the G-CK, during transition through the Golgi apparatus, and by PKA. We speculate that: 1) AQP2 phosphorylation by G-CK might be of physiological significance as a regulatory signal for AQP2 trafficking within the Golgi network, 2) AQP2 phosphorylation by cAMP dependent PKA would then represent the regulatory signal for vesicle targeting and fusion.

Interestingly, an autosomal dominant form of inheritance NDI caused by a mutation of the glu-258 in the AQP2, affects an aminoacid that is irrelevant to PKA targeting, while it is essential to make ser-256 a good target for G-CK. Previous studies have shown that when expressed in oocytes, mutated AQP2 was retained in the Golgi. For this case of NDI in which the phenotype is caused by an impaired routing of AQP2, the possible involvement of the G-CK might give some hints.

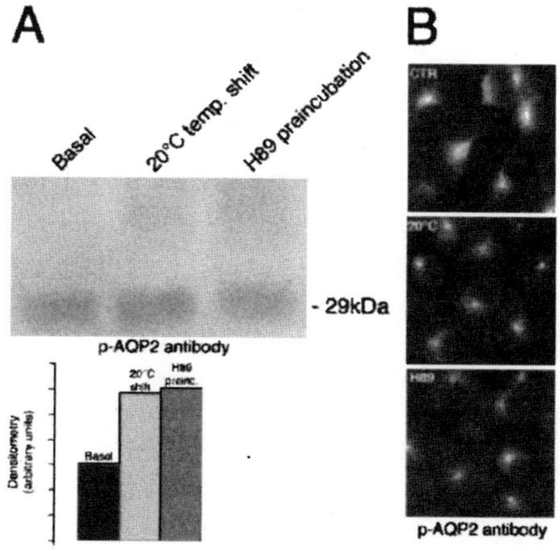

Figure 2. (A) Immunoblotting of immunoprecipitated AQP2 from CD8 cells revealed with anti-phosphorylated AQP2 antibody (p-AQP2 antibody). (B) Immunofluorescence detection of p-AQP2 in CD8 cells.

4. CONCLUSIONS

In addition to phosphorylation of AQP2 at ser-256, alteration in phosphorylation of other unidentified proteins appear to regulate vasopressin-induced AQP2 trafficking. Moreover, besides PKA, AQP2 water channel seems to be phosphorylated by other kinases. More extensive studies are needed for determination of which proteins are phosphorylated by PKA and what role these proteins play in regulating renal epithelial water permeability.

REFERENCES

Procino G Marin O Brunati AM Contri A Pinna LA Svelto M and Valenti G 1999 Potential role for the Golgi casein kinase in regulating aquaporin 2 trafficking within the Golgi network 32nd Annual Meetiting ASN A0118 (abstract)

Valenti G Frigeri A Ronco PM D'Ettorre C Svelto M 1996 Expression and functional analysis of water channels in a stably AQP2-transfected human collecting duct cell line (HCD) *J Biol Chem* **271**: 24365-24370

Valenti G Procino G Carmosino M Frigeri A Mannucci R Nicoletti I Svelto M 2000 The phosphatase inhibitor okadaic acid stimulates the osmotic water permeability response by inducing AQP2 translocation in renal collecting duct cells *J Cell Sci* **113**: 1985-1992

Valenti G Procino G Libenhoff U Frigeri A Benedetti PA Anhert-Hilger G Nurnberg B Svelto M and Rosenthal W 1998 A heterotrimeric G-protein of the Gi family is required for cAMP-triggered trafficking of AQP2 in kidney epithelial cells *J Biol Chem* **273**: 22627-22634

2.18 Mammalian physiology

FUNCTIONAL EXPRESSION OF AQP3 IN HUMAN EPIDERMIS AND KERATINOCYTE CELL CULTURES

[1]Rachid Sougrat, [2]Maryse Morand, [2]Catherine Gondran, [2]Frédéric Bonté, [2]Marc Dumas and [1]Jean-Marc Verbavatz

[1]Service de Biologie Cellulaire, CEA/Saclay, F-91191 Gif-sur-Yvette, [2]Laboratoire Biologie, Parfums Christian Dior, F-45804 Saint Jean de Braye, FRANCE. e-mail: rsougrat@cea.fr

1. INTRODUCTION

Skin is the largest organ of the body and acts as the first line of defense against the outside environment. The skin surface area of a male adult of 70 kg is approximately 1.80 m^2. Skin is comprised of three distinct layers: the epidermis, the dermis and the subcutaneous fat. The epidermis is a multi-stratified epithelium, the apical part of which is the stratum corneum. The stratum corneum consists of several layers of corneocytes that form the primary barrier to body water loss. Nevertheless a total of about 350 ml/24h of water is lost through skin surface by perspiratio insensibilis (unnoticed perspiration), not including secretion from sweat glands (Nislon 1977). Keratinocytes are the major cell type of viable layers of the epidermis and synthesize keratin which plays a major role in epidermis function. Keratin is characterized by its insolubility, its high sulfur content (cystein and cystin residues) and its conspicuous affinity for water. When corneocyte keratin is deprived of its water the tissue becomes brittle and breaks easily (Blank et al. 1953). The epidermis water content drops sharply beneath the stratum corneum. Although it is established that the stratum corneum is the primary barrier to transepidermal water transport in the epidermis (Scheuplein *et al.*, 1976), little is known on fluid transport within viable layers in the epidermis. The presence of aquaporin-3 (AQP3) was reported in rat skin (Frigeri et al. 1995). The purpose of this work was to investigate the presence of

Molecular Biology and Physiology of Water and Solute Transport
Edited by Hohmann and Nielsen, Kluwer Academic/Plenum Publishers, New York, 2000.

aquaporins in human skin and to study their role in water movements in the epidermis.

2. SKIN AQUAPORINS

In human skin homogenates, anti-AQP3 polyclonal anti-peptide antibodies strongly stained a diffuse, glycosylated band between 30 and 35 kDa and a single specie of ~ 50 kDa, probably representing AQP3 dimers (Fig. 1, lane 2). This pattern of staining is consistent with what was previously described for human AQP3 (Roudier *et al.*, 1998). PNGase-F treated human skin homogenate shows the staining of the unglycosylated form of the human AQP3 monomer of ~ 26 kDa (fig. 1, lane 1). No signal was observed when antibodies were incubated in the presence of the peptide used for immunization (not shown). No other aquaporin was detected in human epidermis (not shown).

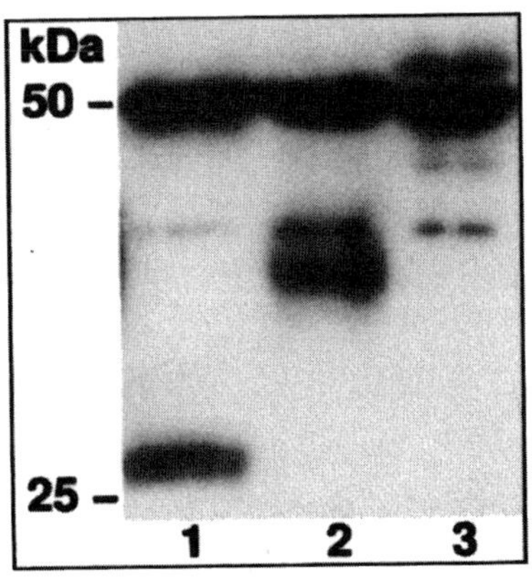

Figure 1. Western blotting of AQP3 in human skin and in reconstructed epidermis homogenates. Lane **1** PNGase-F-treated human skin; lane **2** human skin, Lane **3** reconstructed epidermis.

By indirect immunofluorescence, on 4 µm human skin cryosections, polyclonal anti-AQP3 antibodies strongly stained keratinocyte plasma membranes. The stratum corneum and the dermis were unstained (Fig. 2a). No staining was observed when antibodies were incubated in the presence of the AQP3 C-terminus peptide used for immunization (not shown). A similar AQP3 staining was observed in rat and mouse epidermis.

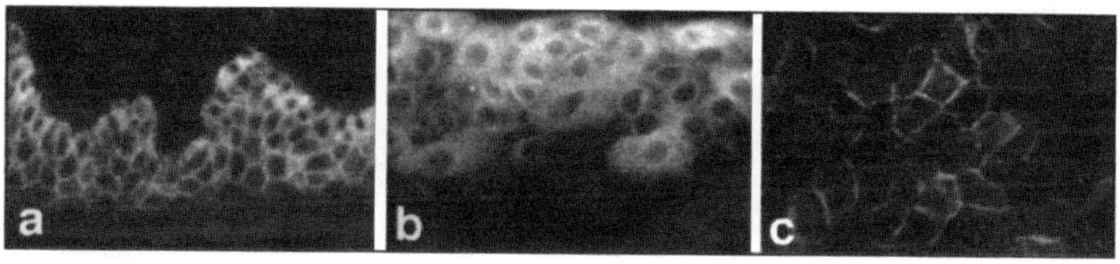

Figure 2. Immunofluorescence staining of AQP3 in human skin cryosection (**a**), keratinocytes grown in low calcium medium (**b**), and keratinocytes grown in 1,5 mM calcium (**c**).

3. AQP3 IN KERATINOCYTE DIFFERENCIATION

In human skin, interestingly, most cells on the basal layer of the epidermis showed cytoplasmic staining for AQP3 (fig. 2a). Calcium is known to play a major role in keratinocyte growth and differentiation (Dotto, 1998). In the presence of low [Ca^{2+}], keratinocyte cell cultures proliferate and increasing [Ca^{2+}] induce their differentiation. Accordingly, a calcium gradient is known to be present in the epidermis with low calcium concentration in suprabasal layers (Menon et al. 1985). In keratinocytes cultured in low [Ca^{2+}] (0.05mM), AQP3 showed a cytoplasmic localization (fig. 2b) whereas in the presence of 1.5 mM Ca^{2+}, AQP3 was localized at keratinocyte plasma membranes (fig. 2c). These observations therefore suggests that AQP3 localization in keratinocytes is closely associated to their state of differentiation.

4. FUNCTIONAL ROLE OF AQP3 IN THE EPIDERMIS

The functional role of aquaporins in kidney had been predicted even before they were identified. In contrast, the physiological role of aquaporins in most other mammalian tissues remains poorly understood. Little is known on fluid or water movements in the epidermis. The absence of AQP3 from the stratum corneum is consistent with its function of impermeable barrier, but what can be the role of AQP3 in keratinocytes and does it function as a water channel in this tissue ?

In order to address this question we have studied the water permeability of human skin and of a cell culture model of reconstructed epidermis.

Transepidermal water fluxes, in the presence of an osmotic gradient, were recorded. As could be expected, no water flux was measurable across intact isolated human skin (not shown). After stripping of the stratum corneum, however, osmotically (150mOsm gradient) induced transepidermal water fluxes could be recorded (Fig. 3, right). The transepidermal water flux was rapidly inhibited by 1mM $HgCl_2$ and this inhibition was partially reversed after the addition of 5mM beta-mercaptoethanol (Fig. 3, right).

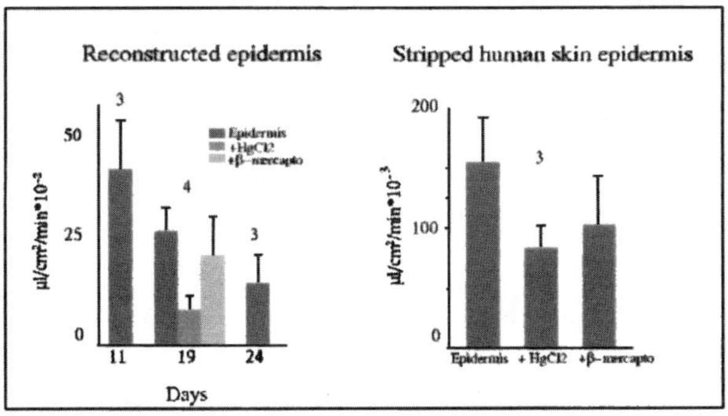

Figure 3. Transepidermal water fluxes, measured in the presence of a 150 mOsm gradient in reconstructed epidermis grown at the air/liquid interface and in stripped human epidermis show measurable water permeability.

Reconstructed epidermis were prepared according to the model of Asselineau et al. (1985). First, freshly isolated human keratinocytes were allowed to proliferate on filters in low Ca^{2+} culture medium until confluence. Then (day 0), filters were set at the air/liquid interface in medium containing 1.5 mM Ca^{2+} for up to 21 days, in order to build a pluristratified epithelium with a cornified layer as previously reported (Asselineau *et al.*, 1985). In Western blotting, reconstructed epidermis showed a strong signal for AQP3 dimers from day 0 to 21 days (Fig. 1, lane 3), despite monomeric AQP3 was not detected. From day 0 to day 7, cultures remained leaky and no osmotic water permeability could be measured. From day 7, osmotically induced water fluxes were recorded across reconstructed epidermis. Results showed a progressive decline in transepidermal water permeability over the period of 21 days (Fig. 3, left), consistent with the formation of a pluristratified barrier. Similarly to native human skin, water fluxes were inhibited by $HgCl_2$ in a reversible fashion.

5. CONCLUSION

Altogether our results suggest that AQP3 is abundant in human keratinocyte plasma membranes. Moreover, viable layers of human epidermis exhibit a significant, mercury sensitive, water permeability. This is consistent with a role for AQP3 in water transport within this tissue. However, the epidermis is a complex epithelium and the physiological significance of water transport within this tissue remains to be elucidated. The recent development of mice lacking various aquaporins, including AQP3, by Verkman and colleagues should provide useful tools to address this question.

ACKNOWLEDGMENTS

This work was supported in part by TMR FMRX-CT97-0128 from the European Community.

REFERENCES

Asselinau D Bernard B-A Bailly C and Darmon M 1985 Epidermal morphogenesis and induction of 67 kD keratin by culture at the liquid-air interface *Exp Cell Res* **159**:536-539

Blank I-H 1953 Further observation on factors which influence the water content of stratum corneum *J Invest Dermatol* **21**: 259-271

Dotto P 1998 The keratinocyte growth-differenciation swith *Front Biosci* **3**: 502-50

2.19 Mammalian physiology

IDENTIFICATION OF A NEW FORM OF AQP4 MRNA THAT IS DEVELOPMENTALLY EXPRESSED IN BRAIN

Brain specific AQP4 mRNA

[1,2]Sergey M. Zelenin, [1]Eli Gunnarson, [2]Tatyana Yu. Alikina, [2]Alexander A. Bondar, [1]Anita Aperia

[1]Department of Women and Children's Health, Karolinska Institutet, ALB/KS, Q2:09, 171 76 Stockholm, Sweden, and [2]Novosibirsk Institute of Bioorganic Chemistry, Russian Academy of Sciences, Russia. E-mail: serze@child.ks.se

1. INTRODUCTION

Aquaporin 4 (AQP4), a water channel that is expressed in the brain, has been suggested to play a role in the regulation of brain water homeostasis. There are at least two isoforms of the AQP4 protein, M1 and M23. The M1 form contains 22 more amino acids in the N-terminal than the M23 form. The M23 isoform has been reported to be encoded by 4 exon segments (1-4) (Ma *et al.,* 1996). The mRNA encoding the M1 protein has been reported to contain an additional nucleotide sequence corresponding to an exon segment (exon 0), upstream of exon 1 in the mouse AQP4 gene. There are however conflicting data (Ma *et al.,* 1996, Turtzo *et al.,* 1997) concerning the mRNA sequence of AQP4 and the number of distinct AQP4 mRNA isoforms. This has prompted us to re-examine the expression of AQP4 mRNA in mouse tissue.

The mouse AQP4 gene was cloned and partially sequenced. Three forms of AQP4 mRNA were identified, which we will name AQP4.M1 (encoding for M1), AQP4.M23X, a new form of AQP4 mRNA and AQP4.M23, the form previously described to encode for M23. In each case the AQP4 mRNA sequence was confirmed by comparison to sequence of genomic DNA.

We studied the age and tissue dependence of the expression of both forms of AQP4 mRNA AQP4.M23X and AQP4.M23 that presumably encode the M23 protein. The AQP4.M23X mRNA form was found to be expressed at relatively high levels in the adult brain and at low levels in the lung and kidney. In contrast, the AQP4.M23 mRNA form was detected at relatively low levels in the brain as compared to lung and kidney. The expression of the AQP4.M23X mRNA form in the brain was found to be age dependent.

2. MATERIALS AND METHODS

Cloning and structure analysis of AQP4 gene and AQP4 mRNA. A 1.6 kb PCR genomic DNA fragment was prepared using primers corresponding to exon 1 (sense) and 2 (anti-sense) of the AQP4 gene and used for screening of the mouse genomic library. The λDNA sample of the clone containing AQP4 gene was hydrolyzed by number restriction enzymes and genomic fragments covering total AQP4 gene structure were subcloned into plasmid pBlueScript (Stratagene). Using the "pBlueScript Exo/Mung DNA Sequencing System" (Stratagene) protocol we obtained gene fragments with predictable size of deletion. Sequencing was performed using Sequenase v.2.0 Kit (Amersham) and Termosequenase DNA sequencing Kit (Amersham).

Total RNA was extracted from adult mouse brain using Rneasy Total RNA Kit (QIAGEN). Reverse transcription and PCR were, with few modifications, carried out as described (Yasui *et al.*, 1996). PCR fragments were isolated by electrophoresis, extracted using QIAquick Gel Extraction Kit (QIAGEN) and used directly for sequencing.

Semiquantitative analysis of AQP4 mRNA expression. Total RNA was extracted from brain, kidney and lung tissues of 2-day-old and adult mice of the C57 Black strain by using Rneasy Total RNA Kit (QIAGEN). The adult mice were allowed free access to food and water. The infant mice were kept with their dams.

Reverse transcription and PCR were, with few modifications, carried out as described (Yasui *et al* 1996). All samples used for quantification were cycled in linear conditions at 28 and 30 cycles at least two times with three different RNA preparations. AQP4.M23X mRNA was quantified by semiquantitative RT PCR using the following primers AQ4.CF1 and antisense AQP4.460 or AQP4.078. AQP4.M23 mRNA was quantified by using the primers AQP4.001 and AQP4.460. β-Actin was amplified as internal control by using the β-Act.518 and β-Act.1202 primers. All primers

(Tab.1) were designed based on their reported sequences (Ma *et al.*, 1996, Sadano *et al.*, 1996) and the following criteria: GC content not more than 55%, and melting temperature was in range 56° - 60°C. All primers were analyzed against GeneBank sequence data for absent high homology to some alien nucleotide structure.

Table 1. Primers used for semiquantitative RT PCR.

Name	Structure of primer	Size of PCR fragment
AQ4.CF1	5'-TTATGGTTCACGGGTTTGGATG-3'	
AQP4.460	5'-TGGTGACTCCCAATCCTCCAAC-3'	749 bp
AQP4.078	5'- AAAGCCACCATGATGCTC-3'	367 bp
AQP4.001	5'-GGAAGGCTAGGTTGGTGACTTC-3'	
AQP4.460	5'-TGGTGACTCCCAATCCTCCAAC-3'	460 bp
β-Act.518	5'-TGGTCGTACCACAGGCATTGTG-3'	
β-Act.1202	5'-AAGGGTGTAAAACGCAGCTCAG-3'	726 bp

3. RESULTS AND DISCUSSION

Mouse AQP4 gene cloning, mapping and sequence analysis. Screening of a mouse genomic library with a 1.6 kb PCR fragment, corresponding to the intron between exon 1 and 2 of mouse AQP4 gene, yielded five independent clones. Restriction mapping revealed two clones that covered approximately a 27 kb genomic fragment and included 9 kb upstream of the 5' -end of exon 0 and 7 kb downstream of the 3' -end of exon 4. Comparison of genomic DNA sequence corresponding to the 5'-end flanking region of the AQP4 gene and AQP4 cDNA sequence data showed the presence of an exon, that was interspaced between exon 0 and exon 1 (Fig.1).

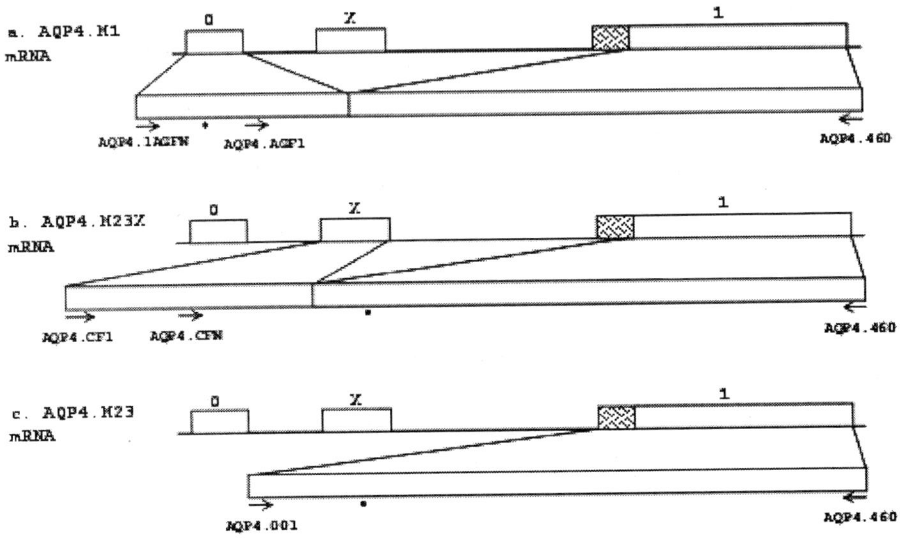

Figure 1. Scheme of the mouse AQP4 gene fragment, based on the nucleotide sequence of the genomic and cDNA fragments. **a - c.** Formation of alternative AQP4 mRNA forms. The squares indicate exon (0, X, 1) segments and corresponding mRNA structures. The dashed box represents part of exon 1 that is spliced out in AQP4.M1 mRNA and AQP4.M23X mRNA. Arrows show position of PCR primers. Position of ATG codon starting the M1 and M23 proteins shown as (♦) and (•), respectively.

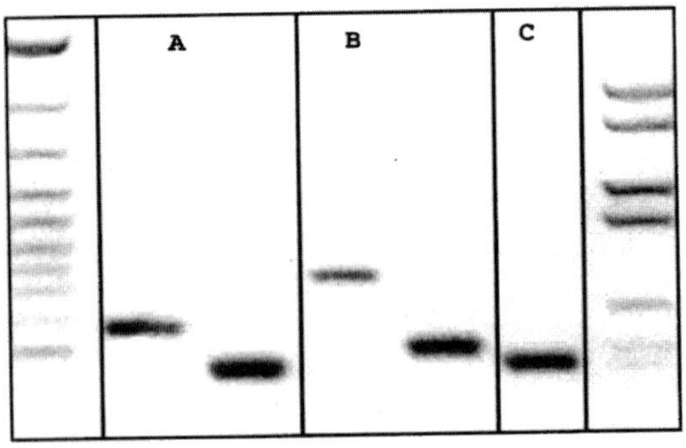

Figure 2. Detection of AQP4 mRNA isoforms by RT PCR. cDNA sample, corresponding mouse brain RNA, was used as a template for PCR amplifications using the identical anti-sense primer AQP4.460 and a number of sense primers. For details regarding positions of PCR primers please go to Figure 1. Panel A. PCR fragments, produced by using AQP4.1AGFN or AQP4.AGF1 oligonucleotides as sense primers, corresponding to exon 0. Panel B. PCR fragments, produced by using AQP4.CF1 or AQP4.CFN oligonucleotides as sense primers, corresponding to nucleotide sequence interspaced between exon 0 and exon 1. Panel C. PCR fragment, produced using AQP4.001 oligonucleotide as sense primer, corresponding to exon 1.

AQP4 mRNA structure analysis. By use of a set of primers, we identified 3 AQP4 mRNA, which will be named AQP4.M1 (encoding for M1), AQP4.M23X, a new form and AQP4.M23 mRNA (Fig. 2).

Several discrepancies were found with regard to previously reported forms of AQP4 mRNA. In this study, the sequence data of the AQP4 mRNA were confirmed by comparison with the mouse AQP4 gene sequence. We suggest that the differences in AQP4 mRNA structures between present and previous studies might be the consequences of artifacts during cDNA library construction.

The sequence of AQP4.M1 mRNA form (Fig. 3a) was similar to the sequence of the AQP4 M1 mRNA reported by Turtzo *et al.*, (1997) and MIWC2 reported by Ma *et al.*, (1996). However, in contrast to Turtzo *et al.*, (1997) and Ma *et al.*, (1996), we found additional nucleotides at the 5'-end of the exon 0. We did not confirm the presence of the first 32 nucleotides at the 5' end of the AQP4 M1 structure reported by Turtzo *et al.*, (1997). These 32 nucleotides are not present in structure of the genomic fragment.

A second form, AQP4.M23X mRNA (Fig. 3b) was similar to the AQP4 mRNA mMIWC3 described by Ma *et al.*, (1996). Comparison of genomic and cDNA structures did, however, reveal that AQP4.M23X mRNA includes a sequence that corresponds to an exon that is interspaced between exon 0 and exon 1 of the mouse AQP4 gene. Within this sequence we identified 206 additional nucleotides. Our analysis also indicated that alternative splicing of exon 1 had occurred. Analysis of the AQP4.M23X mRNA open reading frames indicated that the first optimal context for initiation of translation (Kozak 1996) was localized at the site for start of translation of the AQP4 M23 protein.

a

```
M1        - - - - - - - - - - - - - - - - - - - - - - - - - - - - - - -
MIWC2      - - - - - - - - - - - - - - - - - - - - - - - - - - - - - - -
AQP4.M1          5'-GCCACATGGTGCAGAATCTTTCCACCCCTACTCTCCAAAAACCCAATCAGACA
Genomic...cttactgattGCCACATGGTGCAGAATCTTTCCACCCCTACTCTCCAAAAACCCAATCAGACA

M1       - - - - - - - - - - -  GACCCAAGCTTAGATCTGGCTTTCAAAGGCGTCTGGACTCAAG
MIWC2    - - - - - - - - - - - - - - - - - - - - - - - - - - - - - - - -
AQP4.M1  AGTGCCCGTAATCTGACTCCCAGTGTACTGGAGCCCGGGGGCAGGCACTGAGCTGCACTCTGG
Genomic  AGTGCCCGTAATCTGACTCCCAGTGTACTGGAGCCCGGGGGCAGGCACTGAGCTGCACTCTGG

M1       CCAGGGAAGGCATGAGTGACGGAGCTGCAGCGAGGCGGTGGGG...
MIWC2   ._GAGGGAAGG ATGAGTGACGGAGCTGCAGCGAGGCGGTGGGG...
AQP4.M1  CCAGGGAAGGCATGAGTGACAGAGCTGCGGCAAGGCGGTGGGG...
Genomic  CCAGGGAAGGCATGAGTGACAGAGCTGCGGCAAGGCGGTGGGGgtaagtcct...
                    Met 1
```

Figure 3.

b

```
MIWC3          5'-AAAGTCATTTTAAAGTTTCTATTTATGGTTCACGGGTTTGGATGCTTTGTTT
AQP4.M23X                                5'- TTATGGTTCACGGGTTTGGATGCTTTGTTT
Genomic...  ggggggagctAAAGTCATTTTAAAGTTTCTATTTATGGTTCACGGGTTTGGATGCTTTGTTT

MIWC3       TTTTCTTTTTAAT_ _ _ _ _ _ _ _ _ _ _ _ _ _ _ _ _ _ _ _ _ _ _ _ _ _
AQP4.M23X   TTTTCTTTTTAATTTCTCTGATGAAAAAATGAGGTTCAAAGAGCAACTGAGTTTCTTGATTA
Genomic     TTTTCTTTTTAATTTCTCTGATGAAAAAATGAGGTTCAAAGAGCAACTGAGTTTCTTGATTA

MIWC3       _ _ _ _ _ _ _ _ _ _ _ _ _ _ _ _ _ _ _ _ _ _ _ _ _ _ _ _ _ _ _ _
AQP4.M23X   GGTCTCTGGAGAGGGGGACAGGTGTGTGAATGAGCAGGTGACAAAGGGTGCTGGAGGGAGGC
Genomic     GGTCTCTGGAGAGGGGGACAGGTGTGTGAATGAGCAGGTGACAAAGGGTGCTGGAGGGAGGC

MIWC3          _ _ _ _ _ _ _ _ _ _ _ _ _ _ _ _ _ _ _ _ _ _ _ _ _ _ _ _ _ _ _
AQP4.M23X   TGGCCTTTTCCAGGACCGCCTCCCCTGGGTGCTCTGCTATTGAGCAGGGACTGTTTCCTACC
Genomic     TGGCCTTTTCCAGGACCGCCTCCCCTGGGTGCTCTGCTATTGAGCAGGGACTGTTTCCTACC

MIWC3       _ _ _ _ _ _ _ _ _ _ _ _ _ _ _ _ _TTCTCTTTCCTCTCTCTGGGCCTCAGAAG
AQP4.M23X   CACCCTGCTCCCTGCACTTCCTGCTTGGTCCCCTTCTCTTTCCTCTCTCTGGGCCTCAGAAG
Genomic     CACCCTGCTCCCTGCACTTCCTGCTTGGTCCCCTTCTCTTTCCTCTCTCTGGGCCTCAGAAG

MIWC3       ACAGCACCTGTAATAGCACTTTGCCCCTCTGCCACCTGGCTACAACCCTGGACTGCTG...
AQP4.M23X   ACAGCACCTGTAATAGCACTTTGCCCCTCTGCCACCTGGCTACAACCCTGGACTGCTG...
Genomic     ACAGCACCTGTAATAGCACTTTGCCCCTCTGCCACCTGGCTACAACCCTGGACTGCTGgtg...
```

c

```
M1                    ...TAAGTGTGGACATTCCTGCAGTAGAGAGAGCATCATGGTGGCTTTCAAAGGA
MIWC2                 ...TAAGTGTGGACATTCCTGCAGTAGAGAGAGCATCATGGTGGCTTTCAAAGGA
AQP4.M1               ...TAAGTGTGGACATTCCTGCAGTAGAGAGAGCATCATGGTGGCTTTCAAAGGA
MIWC3                 ...TAAGTGTGGACATTCCTGCAGTAGAGAGAGCATCATGGTGGCTTTCAAAGGA
AQP4.M23X                ...TAAGTGTGGACATTCCTGCAGTAGAGAGAGCATCATGGTGGCTTTCAAAGGA
Genomic...  CTTCTTTCCAGTAAGTGTGGACATTCCTGCAGTAGAGAGAGCATCATGGTGGCTTTCAAAGGA
                                                            Met 23

M1          GTCTGGACTCAGGCTTTCTGGAAGGCAGTCTCAGCAGAATTTCTGGCCACGCTTATCTTTGTT
MIWC2       GTCTGGACTCAGGCTTTCTGGAAGGCAGTCTCAGCAGAATTTCTGGCCACGCTTATCTTTGTT
AQP4.M1     GTCTGGACTCAGGCTTTCTGGAAGGCAGTCTCAGCAGAATTTCTGGCCACACTTATCTTTGTT
MIWC3       GTCTGGACTCAGGCTTTCTGGAAGGCAGTCTCAGCAGAATTTCTGGCCACGCTTATCTTTGTT
AQP4.M23X   GTCTGGACTCAGGCTTTCTGGAAGGCAGTCTCAGCAGAATTTCTGGCCACACTTATCTTTGTT
Genomic     GTCTGGACTCAGGCTTTCTGGAAGGCAGTCTCAGCAGAATTTCTGGCCACACTTATCTTTGTT

M1          TTGCTCGGTGTGGGATCCACCATAAACTGGGGTGGCTCAGAAAACCCCTTACCTGTGGACATG
MIWC2       TTG_ _GGTGTGGGATCCACCATAAACTGGGGTGGCTCAGAAAACCCCTTACCTGTGGACATG
AQP4.M1     TTGCTCGGTGTGGGATCCACCATAAACTGGGGTGGCTCAGAAAACCCCTTACCTGTGGACATG
MIWC3       TTG_ _GGTGTGGGATCCACCATAAACTGGGGTGGCTCAGAAAACCCCTTACCTGTGGACATG
AQP4.M23X   TTGCTCGGTGTGGGATCCACCATAAACTGGGGTGGCTCAGAAAACCCCTTACCTGTGGACATG
Genomic     TTGCTCGGTGTGGGATCCACCATAAACTGGGGTGGCTCAGAAAACCCCTTACCTGTGGACATG

M1          GTCCTCATCTCCCTTTGCTTTGGACTCAGCATTGCTACCATGGTGCAGTGCTTTGGCCACATC
MIWC2       GTCCTCATCTCCCTTTGCTTTGGACTCAGCATTGCAACCATGGTGCAGTGCCTTGGCCACATC
AQP4.M1     GTCCTCATCTCCCTTTGCTTTGGACTCAGCATTGCTACCATGGTGCAGTGCTTTGGCCACATC
MIWC3       GTCCTCATCTCCCTTTGCTTTGGACTCAGCATTGCAACCATGGTGCAGTGCCTTGGCCACATC
AQP4.M23X   GTCCTCATCTCCCTTTGCTTTGGACTCAGCATTGCTACCATGGTGCAGTGCTTTGGCCACATC
Genomic     GTCCTCATCTCCCTTTGCTTTGGACTCAGCATTGCTACCATGGTGCAGTGCTTTGGCCACATC
```

Figure 3. (Continued)

M1	AGTGGTGGCCACATCAATCCCGCTGTGACTGTAGCCATGGTGTGCACACGAAAGATCAGCATC
MIWC2	AG**c**GGTGGCCACATCAATCC**A**GCTGTGACTGTAGCCATGGTGTGCACACGAAAGATCAGCATC
AQP4.M1	AGTGGTGGCCACATCAATCCCGCTGTGACTGTAGCCATGGTGTGCACACGAAAGATCAGCATC
MIWC3	AG**c**GGTGGCCACATCAATCC**A**GCTGTGACTGTAGCCATGGTGTGCACACGAAAGATCAGCATC
AQP4.M23X	AGTGGTGGCCACATCAATCCCGCTGTGACTGTAGCCATGGTGTGCACACGAAAGATCAGCATC
Genomic	AGTGGTGGCCACATCAATCCCGCTGTGACTGTAGCCATGGTGTGCACACGAAAGATCAGCATC
M1	GCTAAGTCCGTCTTCTACATCATTGCACAGTGCCTGGGGGCCATCATTGGAGCCGGCATCCTC
MIWC2	GCTAAGTCCGTCTTCTACATCATTGCACAGTGCCTGGGGGCCATCATTGGAGCCGGCATCCTC
AQP4.M1	GCTAAGTCCGTCTTCTACATCATTGCACAGTGCCTGGGGGCCATCATTGGAGCCGGCATCCTC
MIWC3	GCTAAGTCCGTCTTCTACATCATTGCACAGTGCCTGGGGGCCATCATTGGAGCCGGCATCCTC
AQP4.M23X	GCTAAGTCCGTCTTCTACATCATTGCACAGTGCCTGGGGGCCATCATTGGAGCCGGCATCCTC
Genomic	GCTAAGTCCGTCTTCTACATCATTGCACAGTGCCTGGGGGCCATCATTGGAGCCGGCATCCTC
M1	TACCTGGTCACACCTCCCAGTGTGGTTGGAGGATTGGGAGTCACCACG...
MIWC2	TACCTGGTCACACCTCCCAGTGTGGTTGGAGGATTGGGAGTCACCACG...
AQP4.M1	TACCTGGTCACACCTCCCAGTGTGGTTGGAGGATTGGGAGTCACCACG...
MIWC3	TACCTGGTCACACCTCCCAGTGTGGTTGGAGGATTGGGAGTCACCACG...
AQP4.M23X	TACCTGGTCACACCTCCCAGTGTGGTTGGAGGATTGGGAGTCACCACG...
Genomic	TACCTGGTCACACCTCCCAGTGTGGTTGGAGGATTGGGAGTCACCACGgtgagaactctt...

Figure 3. Nucleotide structures of the mouse AQP4 gene and related AQP4 mRNA fragments in relation to previously published sequence for mouse AQP4 mRNA. M1 reported by Turtzo *et al.*, (1997) and MIWC2, MIWC3 by Ma *et al.*, (1996). Nucleotides of the exons and related mRNA fragments are presented by capital letters and flanking AQP4 genomic structure by lowercase letters. Absent nucleotides in M1 and MIWC2 structure and deletion in MIWC3 cDNA are marked as "_". Dissimilar nucleotides in M1 and MIWC2 structures are presented in bold. Positions of ATG codon starting AQP4 protein are boxed. a. Nucleotides of the exon 0 and related AQP4.M1, MIWC2 and M1 mRNA fragments. b. Nucleotide structure of the mouse AQP4 gene region, including the exon, that is interspaced between exon 0 and exon 1 and related AQP4.M23X mRNA and MIWC3 nucleotide sequences. c. Nucleotides of the exon 1 fragment (spliced form) and related AQP4 cDNA fragments.

Tissue distribution of the AQP4.M23X and AQP4.M23 mRNA. The two AQP4 mRNA that presumably encode for the same AQP4 M23 protein, were analyzed with regard to tissue and developmental expression using semiquantitative RT-PCR.

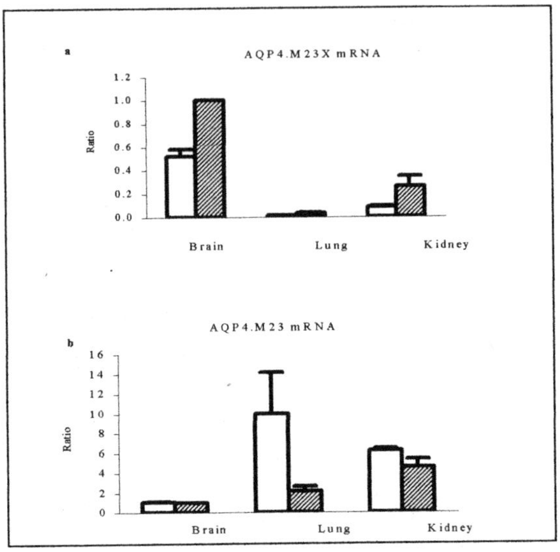

Figure 4. Relative ratios of AQP4.M23X and AQP4.M23 mRNA expressions in 2-day-old (open bars) and adult (hatched bars) mouse brain, lung and kidney. The relative expression of the AQP4.M23X mRNA transcript (related to the ß-actin transcript) in the adult brain was arbitrarily assigned a value of unity for comparison with the other tissues and ages. Bars represent mean and SEM (standard error of the mean) is indicated. a. Ratios based on calculations from semiquantitative RT-PCRs of the AQP4.M23X mRNA abundance. b. Ratios based on calculations from semiquantitative RT-PCRs of the AQP4.M23 mRNA abundance.

The AQP4.M23X mRNA was more abundantly expressed in brain than in the peripheral tissues examined, and might represent a brain specific form of AQP4 mRNA of the M23 protein (Fig. 4a). AQP4.M23 mRNA was less abundantly expressed in the brain than in lung and kidney (Fig. 4b). As a consequence the regulation of AQP4 M23 protein expression may be different in brain and in peripheral tissue. The biological significance of this needs to be taken into account in future studies.

The AQP4.M23X mRNA had a significantly lower expression in infant mouse brain than in the adult brain (Fig. 4a). In the lung the relative expression of the AQP4.M23 mRNA was significantly higher in the infant than the adult (Fig. 4b).

ACKNOWLEDGMENTS

The authors thank Lill-Britt Svensson for technical assistance. This work was supported in part by the Swedish Medical Research Council (03644),

Märta and Gunnar V. Philipson Foundation, INTAS grant no. INTAS-97-11404 and by the RFBR grant no.98-04-49369.

REFERENCES

Frigeri A Gropper MA Umenishi F Kawashima M Brown D Verkman AS 1995 Localisation of MIWC and GLIP water channel homologs in neuromuscular epithelial and glandular tissues *J Cell Sci* **108**:2993-3002

Kozak M 1996 Interpreting cDNA sequences: some insights from studies on translation. *Mammalian Genome* **7**: 563-574

Ma T, Yang B, Verkman AS 1996 Gene structure, cDNA cloning and expression of a mouse mercurial-insensitive water channel. *Genomics* **33**: 382-388

Menon G K Grayson S and Elias PM 1985 Ionic calcium reservoirs in mammalian epidermis ultrastructural localization by ion-capture cytochemistry *J Invest Dermatol* **84**: 508-512

Nilson GE 1977 On the measurement of evaporative water loss Methods and principal application Linkopping studies in Science and Technology *Dissertations* No **11**

Roudier N Verbavatz J-M Maurel C Ripoche P and Tacnet F 1998 Evidence for presence of aquaporin-3 in Human red blood cells *J Biol Chem* **273**: 8407-8412

Sadano H, Taniguchi S, Kakunaga T, Baba T 1988 cDNA cloning and sequence of a new type of actin in mouse B16 melanoma. *J Biol Chem* **263**: 15868-15871

Scheuplein RJ 1976 Percutaneous absorption after twenty-five years *J Invest Dermatol* **67**:31-38

Turtzo LC, Lee MD, Lu M, Smith BL, Copeland NG, Gilbert DJ, Jenkins NA, Agre P 1997 Cloning and chromosomal localization of mouse aquaporin 4: exclusion of a candidate mutant phenotype, ataxia. *Genomics* **41**: 267-270

Yasui M, Marples D, Belusa R, Eklof AC, Celsi G, Nielsen S, Aperia A 1996 Development of urinary concentrating capacity: role of aquaporin-2. *Am J Physiol* **271**: F461-468

2.20 Mammalian physiology

MODELS FOR ISOTONIC TRANSPORT ACROSS APICAL MEMBRANES OF EPITHELIAL CELLS

Studies on coexpression of SGLT1 and AQP1 in Xenopus oocytes

[1]Thomas Zeuthen, [1]Anne-Kristine Meinild, [2]Donald D. F. Loo, [2]Ernest M. Wright and [1]Dan A. Klaerke

[1]The Panum Institute, Blegdamsvej 3C, DK-2200 Copenhagen N, Denmark and [2]Department of Physiology, University of California School of Medicine, Los Angeles, CA 90095-1751, USA.
E-mail: T.Zeuthen@mfi.ku.dk

1. INTRODUCTION

The mechanisms and routes by which water is transported across epithelia are debated. It is generally held, and probably true, that the water transport is coupled in a secondary fashion to the transport of ions but the underlying mechanism is not understood at all. So far, most models for transepithelial water transport have been based upon osmotic mechanisms, in which water transport by the cellular route arises from transmembrane osmotic differences. Such models, however, have been unable to explain a number of well-established epithelial properties, for a review see Zeuthen (1992). With the finding that certain membrane proteins function as molecular water pumps, it has been possible to suggest a more general model for transepithelial water transport which would explain, for example, uphill and isotonic transport of water (Zeuthen, 1996 and 2000).

The possibility of a direct coupling between solute and water transport in membrane proteins was first suggested from studies of intracellular ion concentrations under various osmotic conditions (Zeuthen 1981, 1982, 1983). So far, all cotransporters of the symport type have been found to function as molecular water pumps, in which water is cotransported along with their non-aqueous substrates (Table 1). The coupling between water and the non-aqueous subtrates takes place by a mechanism within the protein which involve conformational changes of the protein. The coupling ratio, ie the number of water molecules cotransported per turnover of the protein, is a molecular property, independent of external parameter such as osmotic gradients, concentrations, and clamp potentials. It follows that the ratio between the number of water molecules and the number of non-aqueous substrates are fixed. Energy can be transferred within the protein from the downhill fluxes of the non-aqueous substrates to the uphill fluxes of water.

Table 1. Number of water molecules (n_w) cotransported per turnover in cotransport proteins

cotransporter	n_W	Method	Reference
K^+/Cl^-	500	*Microelectrodes*	Zeuthen 1994
H^+/lactate	500	*Microelectrodes*	Zeuthen *et al.*, 1996
Na^+/alanine	n.d.	*Microelectrodes*	Zeuthen 1995
Na^+/glucose			
rabbit	390	*expression*	Loo *et al.*, 1996, Zeuthen *et al.*, 1997
human	210	*expression*	Meinild *et al.*, 1998
Na^+/dicarboxylate	175	*expression*	Meinild *et al.*, 2000
Na^+/glutamate	424	*expression*	Boxenbaum *et al.*, 2000
Na^+/I^-	n.d.	*expression*	Loo *et al.*, 1996
Na^+/Cl^-/GABA	n.d.	*expression*	Loo *et al.*, 1996
H^+/aminoacid	n.d.	*expression*	Loo *et al.*, 1996

Expression was performed in Xenopus oocytes

2. TRANSPORT ACROSS APICAL MEMBRANES

The present investigation focuses upon the transport mechanisms in the apical membrane of typical leaky epithela such as the small intestine and the kidney proximal tubulus. The apical membrane incorporates several cotransporters, among them the Na^+- dependent glucose transporter SGLT1 which couples the secondary active influx of glucose to the inwardly driven Na^+ flux. The small intestine cell contains no known aquaporins, the osmotic water permeability of the kidney cell is determined by AQP1.

As a model of the apical cell membrane, water transport was studied by overexpressing the transport proteins in *Xenopus laevis* oocytes. The goal was to define the role of the cotransporter in generating water transport in the two situations: When only SGLT1 was expressed and when SGLT1 and AQP1 were coexpressed. Since the SGLT1 is electrogenic, transport rates can be monitored as a clamp current, while the water transport, J_{H2O}, can be recorded optically via the change in oocyte volume. We have previously established that the human SGLT1 cotransports 210 water molecules per 2 Na^+ and 1 glucose molecule and that SGLT1 is à water channel (Loo *et al.*, 1996, Zeuthen *et al.*, 1997, Meinild *et al.*, 1998, Loo *et al.*, 1999). Since the cotransport is hypertonic this will in turn result in additional osmotic water flow into the cell.

When glucose was added to oocytes expressing only SGLT1 under voltage clamped conditions, cotransport of water revealed itself as an initial rate of volume increase of about 20 pl sec^{-1}, for clamp currents of 1000 nA. After about 10 min the total water transport attained a steady state value of 2.2 ± 0.2 times the initial rate (paired experiments in 6 oocytes). At this point the transport is isotonic to the intracellular solution which could be estimated to be about 7 mosm l^{-1} hyperosmolar relative to the external solution. Removal of glucose reduced the rate of swelling by 42 ± 6%, the amount being identical to the rate of cotransport of water observed initially. Addition of Phlorizin (100 μM) reduced the rate of swelling by 75 ± 5% and the passive water permeability by 39 ± 3%. Phlorizin abolishes both the cotransport component of water transport and the water permeability of the SGLT1. We conclude that cotransport of water provided 42% of the total water flow, while osmotic flow was divided with about 23 % via the SGLT1 and 35 % via the membrane (Fig. 1A).

In small intestine the expression level of SGLT1 is about 10 times higher than in the oocyte. This would suggest that the osmotic component through the SGLT1 dominates in the epithelial cell and that the osmotic gradient required for isosmotic transport is about twice the one determined for the oocyte.

Coexpression of AQP1 with the SGLT1 increased the water permeability about 20-fold above that obtained with SGLT1 alone. Activation of the SGLT1 induced a rate of water flow similar to the one achieved after 10 min in the oocytes expressing only SGLT1. Given the rate of cotransport and the water permeability we conclude that about one third of the water transport is cotransported while two thirds are osmotic and mediated by a small osmotic gradient of about 0.5 mosm l^{-1} via the AQP1 (Fig. 1B). This result was confirmed by experiments in which Na^+- currents were mediated via gramicidin channels inserted in oocytes co-expressing SGLT1 and AQP1 but with the cotransport modality of the SGLT1 inactivated by the absence of sugar.

3. PHYSIOLOGICAL RELEVANCE

Our data suggest that SGLT1 has three roles in the generation of isotonic transport across the apical membrane. It cotransports water directly, it supplies a passive pathway for osmotic water transport, and it generates an

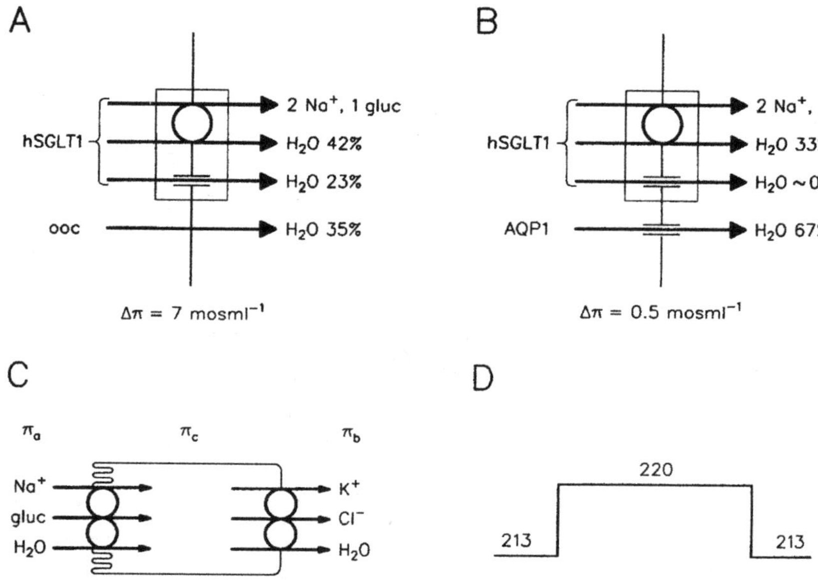

Figure 1. Steady state values of the cotransport and osmotic components of water transport across (A) a *Xenopus* membrane expressing hSGLT1, and (B) one co-expressing hSGLT1 and AQP1. In steady states, in the presence of sugar and under clamped conditions, both systems translocate a solution isotonic relative to the inside solution. The water transport has two components: One third that is cotransported via a mechanism within the SGLT1 and two thirds that are osmotically driven by the osmotic gradient set up by the SGLT1. The SGLT1 has a relatively low water permeability and membranes expressing only SGLT1 build up a relatively large osmotic gradient ($\Delta\pi = 7$ mOsm) in order to drive the osmotic flux through the SGLT1 itself and through the native oocyte membrane. AQP1 imparts relatively large osmotic water permeability to the membrane. Consequently, in the case of coexpression the osmotic component of water transport is achieved by a small osmotic gradient ($\Delta\pi = 0.5$ mOsm). C and D. In an epithelial model for isotonic transport, the cell is hyperosmolar relative to its surroundings, in the given example 220 versus 213 mOsm. The entry of water across the apical membrane will be energized at least in part by Na^+ coupled transporters as described. The capacity of the K^+/Cl^- cotransporter for uphill water transport will ensure the exit of water across the basolateral aspect of the cell. For clarity, other essential transporters, i.e. aquaporins, ATPases etc are not drawn.

osmotic driving force that can be employed by other pathways, as for example AQP1. The two situations investigated may serve as models for isotonic transport across apical (brush border) membranes of leaky epithelia. For membranes with relatively low passive water permeability (i.e. small intestine) it requires relatively large osmotic gradients to render the transported solution isotonic. For membranes with relative high water permeability (ie kidney proximal tubulus epithelium) it only requires small gradients.

In transgenic mice lacking AQP1 the rate of water transport in the proximal tubulus was reduced by a factor two and the water permeability was reduced by almost 80 % compared to wild type mice, Schnermann *et al.*, 1998, Verkmann 1999). A comparison with the oocyte models suggests that the mode of water transport in the transgenic mice resembles the case where only SGLT1 is expressed. Accordingly, the water permeability of the apical membrane in these animals should be determined by cotransporters and the lipid component. Our model would predict that the hyperosmolarity of the solution transported across the apical membrane should be larger than in the wild type animals.

In a model for transepithelial transport, cotransport of water participates as a new building block working together with the other transporters, channels, ATPases etc. In particular, the transport processes in the basolateral membrane must be taken into account as these participate in the build up of the cellular osmolarity and therefore of the osmotic gradient across the apical membrane. In the oocyte model the current clamp microelectrode takes the place of the basolateral membrane. The electrode current does not contribute significantly to the net influx of solute as it is carried by equal and opposite fluxes of K^+ and Cl^-. During transepithelial isotonic transport the basolateral bathing solution is isotonic to the apical bathing solution, and there is evidence of no hyperosmolarity in the lateral spaces between the cells (Zeuthen 1983, Ikonomow *et al.*, 1985). Accordingly, the epithelial cell is hyperosmolar to its surroundings. Water enter downhill across the apical membrane and leave the cell uphill across the serosal membrane, against the osmotic gradient, Fig. 1C and D. The ubiquitous K^+/Cl^- cotransporter, located at the basolateral membrane, plays a key role in this process. The capacity of this transporter for uphill water transport will ensure the exit of water from the cell, for a recent review see Zeuthen (2000).

ACKNOWLEDGMENTS

The Danish Research Council, the Novo-Nordisk Foundation, the National Institutes of Health (US), and the National Science Foundation (US) are thanked for financial support. The technical assistance of Tove Soland, Birthe Lynderup and Svend Christoffersen is gratefully acknowledged. Nanna MacAulay (nee Boxenbaum) is thanked for a critical reading of the manuscript.

REFERENCES

Boxenbaum N Gether U Klaerke D A and Zeuthen T 2000 Water transport in the Na^+/glutamate cotransporter *Proc Exp Biol* San Diego Abstract

Ikonomov O Simon M and Frömter E 1985 Electrophysiological studies on lateral intercellular spaces of necturus gallbladder epithelium *Pflugers Arch* **403**: 301-307

Loo DDF Zeuthen T Chandy G and Wright EM 1996 Cotransport of water by the Na^+/glucose cotransporter *Proc Nat Acad SciUSA* **93**: 13367-13370

Loo DF Hirayama BA Meinild A-K Chandy G Zeuthen T and Wright E 1999 Passive water and ion transport by cotransporters *J Physiol* **518**: 195-202

Meinild A-K Loo D D F Pajor A Zeuthen T and Wright E M 2000 Water transport by the renal Na^+/dicarboxylate cotransporter *Am J Physiol* in press

Meinild A-K Klaerke DA Loo DDF Wright EM and Zeuthen T 1998 The human Na^+/Glucose cotransporter is a molecular water pump *J Physiol* **5081**: 15-21

Schnermann J Chou C-L Ma T Traynor T Knepper MA and Verkman AS 1998 Defective proximal tubular fluid reabsorbtion in transgenic aquaporin-1 null mice *Proc Nat Acad Sci USA* **95** 9660-9664

Verkmann AS 1999 Lessons on Renal Physiology from Transgenic Mice Lacking Aquaporin Water Channels *J Am Soc Nephro* **10**: 1126-1135

Zeuthen T 2000 Molecular water pumps *Rev Physiol Biochem Pharm* Springer Verlag pp 97-151 in press

Zeuthen T 1981 Isotonic transport and intracellular osmolarity in the necturus gall-bladder epithelium In *Water transport across epithelia* (Ussing HH Bindslev N Lassen NA and Sten-Knudsen O eds) Munksgaard Copenhagen pp313-331

Zeuthen T 1982 Relations between intracellular ion activities and extracellular osmolarity in necturus gallbladder epithelium *J Membr Biol* **66**: 109-121

Zeuthen T 1983 Ion activities in the lateral intercellular spaces of gallbladder epithelium transporting at low external osmolarities *Jmemb rBiol* **76**: 113-122

Zeuthen T 1992 From contractile vacuole to leaky epithelia Coupling between salt and water fluxes in biological membranes *Biochim Biophys Acta* **1113**: 229-258

Zeuthen T 1994 Cotransport of K^+ Cl^- and H_2O by membrane proteins from choroid plexus epithelium of Necturus maculosus *J Physiol* **478**: 203-219
Zeuthen T 1995 Molecular mechanisms for passive and active transport of water *Int Rev Cytol* **160**: 99-161
Zeuthen T 1996 *Molecular Mechanisms of Water Transport* Springer Berlin & RG Landes Company pp 170
Zeuthen T Hamann S and La Cour M 1996 Cotransport of H^+ lactate and H_2O by membrane proteins in retinal pigment epithelium of bullfrog *J Physiol* **4971**: 3-17
Zeuthen T Meinild A-K Klaerke DA Loo DDF Wright EM Belhage B and Litman T 1997 Water transport by the Na^+/glucose cotransporter under isotonic conditions *Biol Cell* **89**: 307-312

3

PATHOPHYSIOLOGY

3.1 Pathophysiology

MISSENSE MUTATIONS IN THE MIP GENE, ENCODING THE MAJOR INTRINSIC PROTEIN OF THE LENS (AQUAPORIN-0), UNDERLIE CATARACTS IN HUMANS

[1,2]Peter J Francis, [1]Vanita Berry, [1]Shalesh Kaushal, [2,3]Anthony T Moore, [1]Shomi Bhattacharya

[1]Department of Molecular Genetics, Institute of Ophthalmology, 11-43 Bath Street, London. EC1V 9EL. UK. [2]Moorfields Eye Hospital, City Road, London. EC1V 2PD. UK. [3]Addenbrooke's Hospital, Hills Road, Cambridge. UK.. E-mail: p.j.francis@hgmp.mrc.ac.uk

1. INTRODUCTION

Cataract is the term used to describe opacification of the crystalline lens of the eye. It is the commonest cause of adult blindness world-wide and the most frequent treatable cause of visual impairment in childhood (Lambert & Drack 1996). Childhood cataract may be inherited, usually as an autosomal dominant trait, where it is both genetically heterogeneous and phenotypically diverse (Ionides *et al.*, 1999). Independent chromosomal loci have been mapped to 1p36, 1q21-25, 2q33-35, 13q, 16q22.1, 17p, 17q11-12, 17q24, 21q and 22q. Underlying mutations are now known in the crystallin genes at the loci on 2q, 21q and 22q, the connexin genes on 1q and 13q and the developmental regulator gene *PITX3* on 10q (Francis *et al.*, 1999).

2. THE MAJOR INTRINSIC PROTEIN OF THE LENS, MIP (AQUAPORIN-0)

Encoded by a gene cytogenetically localised to chromosome 12q14, the major intrinsic protein of the lens is a 28kD molecule, that constitutes a high

Molecular Biology and Physiology of Water and Solute Transport
Edited by Hohmann and Nielsen, Kluwer Academic/Plenum Publishers, New York, 2000.

proportion of the membrane protein of the differentiated lens fibre. MIP is now known to be a member of the aquaporin family of water channel proteins (Lee *et al.*, 1997).

Recently, two mouse models of cataract, the Cat^{Fr} (Fraser) and Cat^{lop} (lens opacity) have been shown to result from mutations in *Mip*, the murine homologue of the MIP gene. The Cat^{Fr} mouse has a transposon-induced splice site mutation that results in the translation of an additional 55 amino acids at the carboxy-terminus of the Mip protein. The *lop* mouse has a dominant missense mutation which leads to mistargetting of the protein and accumulation in the endoplasmic reticulum (Sheils and Bassnett 1996). For these reasons, MIP is considered a strong candidate for human cataractogenesis.

3. EXPERIMENTAL PROTOCOL

Linkage analysis: the patient databases at Moorfield's Eye Hospital, London, UK, were used to identify several English kindreds with autosomal dominant congenital cataract. All family members underwent a full ophthalmic assessment and DNA extracted from blood or buccal samples obtained from each individual. Linkage analysis (positional-candidate approach) was then performed using PCR-based microsatellite marker genotyping.

Sequence analysis: using gene specific primers, the entire coding region of the *MIP* gene was amplified from genomic DNA obtained from each member of families mapping to the *MIP* gene locus. Co-segregation of the disease with mutations was confirmed by restriction fragment length analysis.

4. RESULTS

After excluding a number of candidate loci for cataract, genotyping revealed two families (A and B) genetically linked to the *MIP* gene locus. Maximal lod scores were obtained in both families at marker D12S1676 (family A: $Z = 3.91$ at $\theta=0$; family B: $Z = 2.11$ at $\theta=0$). Haplotype analysis indicated that the cataract locus resides in the genetic interval D12S375-D12S1064.

Sequence analysis of the *MIP* gene identified in family A, a missense mutations in exon 2. The C→G transition results in the substitution of threonine for arginine at codon 138 (T138R). The sequence alteration results in the gain of an *AocI* restriction site, enabling co-segregation of the disease with the mutation to be confirmed by restriction fragment length analysis.

In family B, all affected individuals were found to have an A→G base change resulting in a glutamic acid-to-glycine substitution at codon 134 (E134G). The mutation results in the loss of a *BglII* restriction site. Co-segregation of the mutation with affected individuals was confirmed by restriction digest.

The cataract observed in family A, was variable or "polymorphic", a phenotype not previously recognised as an inherited phenotype. Discrete progressive congenital bilateral punctate lens opacities limited to mid-and peripheral lamellae were characteristic though some individuals had in addition asymmetric polar or cortical cataracts. Penetrance was complete. In contrast, family B had non-progressive congenital lamellar and sutural cataract. The condition was once again fully penetrant, with little variation in expressivity.

Both mutations occur within transmembrane helix 4 of the MIP protein and based on sequence homology result in the substitution of residues thought to participate in the water channel (Heymann *et al.,* 1998). The two mutations may therefore disrupt water transport by the protein (dominant-negative effect) or lead to misfolding with resultant mistargetting of the molecule as is seen in the *lop* mouse. In either instance, intracellular water homeostasis is likely to be disturbed possibly resulting in precipitation of the lens crystallin arrays observed as cataract.

5. CONCLUSIONS

Our data constitutes the first evidence implicating the major intrinsic protein of the lens, MIP (or aquaporin-0, AQP0), in human cataract. These studies provide the first confirmation that *MIP* plays a vital role in the maintenance of transparency of the human lens and further evidence of the important physiological role of aquaporins.

ACKNOWLEDGMENTS

We thank the families for their cooperation in this study. This work is supported by Wellcome Trust Grant 053416/Z/98.

REFERENCES

Francis P Berry V Moore A and Bhattacharya S 1999 Lens biology development and human cataractogenesis *Trends Genet* **15**: 191-196

Heymann J Agre P and Engel A 1998 Progress on the structure and function of aquaporin 1 *J Struc Biol* **121**: 191-206

Ionides A Francis P Berry V Bhattacharya S Moore A 1999 Clinical and geneticheterogeneity in autosomal dominant congenital cataract *Br J Ophthalmol* **83**: 802-808

Lambert S and Drack A 1996 Infantile Cataracts *Surv Ophthalmol* **40**: 427-458

Lee M King L and Agre P 1997 The aquaporin family of water channel proteins in clinical medicine *Medicine (Baltimore)* **76**: 141-156

Shiels A and Bassnett S 1996 Mutations in the founder of the *MIP* gene family underlie cataract development in the mouse *Nat Genet* **12**: 212-215

3.2 Pathophysiology

URINARY EXCRETION OF AQUAPORIN-2 WATER CHANNEL IS DOMINANT IN PATHOLOGICAL STATE OF ARGININE VASOPRESSIN-INDUCED IMPAIRED WATER EXCRETION

San-e Ishikawa, Takako Saito, Akinori Fukagawa, Tomoatsu Nakamura, Ikuyo Kusaka, Minori Higashiyama, Shoichiro Nagasaka and Toshikazu Saito
Division of Endocrinology and Metabolism, Department of Medicine, Jichi Medical School, 3311-1 Yakushiji Minamikawachi, Tochigi 329-0498 Japan. E-mail: saneiskw@jichi.ac.jp

1. INTRODUCTION

There is impaired water excretion in patients with syndrome of inappropriate secretion of antidiuretic hormone (SIADH), congestive heart failure, liver cirrhosis with ascites and adrenal insufficiency (Bartter and Schwartz 1967, Bichet *et al.*, 1982, Schrier 1988a, b). In these clinical settings there is hyponatremia to a various extent. Non-suppressible release of arginine vasopressin (AVP) is found despite hypoosmolality, which should suppress plasma AVP to undetectable levels (Bartter and Schwartz 1967, Bichet *et al.*, 1982, Schrier 1988a, b, Ishikawa *et al.*, 1982, 1986).

In response to AVP concentrated urine is produced by water reabsorption across renal collecting duct (Ishikawa 1993, Knepper and Rector 1995). Aquaporin-2 (AQP-2) is an AVP-regulated water channel of collecting duct cells (Fushimi *et al.*, 1993). Exogenous and endogenous AVP stimulates the expression of AQP-2 mRNA in kidney (Fushimi *et al.*, 1993, Matsumura *et al.*, 1997). Approximately 3% of AQP-2 in the collecting duct is excreted into urine, and urinary excretion of AQP-2 has a positive correlation with plasma AVP levels in normal subjects (Kanno *et al.*, 1995, Rai *et al.*, 1997, Saito *et al.*, 1997). There is evidence that urinary excretion of AQP-2 is exaggerated in patients with impaired water excretion (Saito *et al.,* 1998). In the present study we further determined whether augmented hydroosmotic effect of AVP contributes to the impairment in water excretion by measurement of urinary excretion of AQP-2 in the hyponatremic patients with impaired water excretion.

2. SUBJECTS AND METHODS

2.1 Subjects

Eighteen patients with impaired water excretion and 12 normal subjects were studied. The patient group had 10 males and 8 females, aged 48-77 years-old. Serum sodium (Na) levels were 126.1±2.8 mEq/l on admission. They included 5 patients with SIADH, 10 hypopituitarism, and 3 salt-wasting syndrome. The control subjects were 7 males and 5 females, aged 19-74 years-old. An acute oral water load was carried out in these two groups of subjects. After an overnight fast, the study was started at 8:00 h. The subjects were allowed free access to water before the start of the study. After urination, blood collection was made. Thereafter, water (20 ml/kg BW) was ingested for 30 min. Thirty-min urine collections were made, and blood samples were collected at 1 h intervals for 4 h. Urine samples were subjected to measurement of urine volume, urinary osmolality (Uosm) and urinary excretion of creatinine and AQP-2, and blood samples to the measure of plasma osmolality (Posm) and plasma AVP levels. Plasma AVP levels were measured by RIA using AVP RIA kits (Mitsubishi Chemistry, Tokyo) (Ishikawa *et al.,* 1986). Urinary AQP-2 was determined as described in our previous report (Saito *et al.,* 1998). The present study was approved by the Ethical Committee of Jichi Medical School Hospital for the human study. We obtained informed consent from all the subjects who joined the present protocol.

2.2 Statistical analysis

Serum Na, Posm, percent excretion of water load, Uosm, plasma AVP, and urinary excretion of AQP-2 were expressed as the means ±SEM. The values were analyzed by Student's t test to compare the difference. Simple linear regression analysis was performed to calculate correlations. The statistical package of StatView for Macintosh, version 4.5, was employed for the present analysis. A P value less than 0.05 was considered significant.

3. RESULTS

Table 1 shows the results of an acute oral water load. Percent excretion of the water load was only 28.2±3.9% in the patient group, which was significantly less than that of 77.3±6.2% in the control. The minimal Uosm was 122.9±16.4 mOsm/kg H2O, which was found in 90-120 min in the control subjects. Uosm was not sufficiently diluted in the patient group, as the minimal Uosm remained as high as 437.3±54.8 mOsm/kg H2O. There was no difference in the minimal plasma AVP levels between the two groups of subjects, but the minimal plasma AVP levels were relatively high as compared to the respective Posm in the patients. Urinary excretion of AQP-2 was 363.1±21.7 fmol/mg creatinine in the patients, which was significantly higher than that in the control. Also, the minimal urinary excretion of AQP-2 remained high in the patient group.

Table 1. Changes in Posm, Uosm, plasma AVP levels and urinary excretion of AQP-2 (UAQP-2) in the acute water load in patients with impaired water excretion and control subjects

Group	N	Posm (mOsm/kg H_2O)		Uosm (mOsm/kg H_2O)		Plasma AVP (pg/ml)		UAQP-2 (fmol/mg creatinine)	
		basal	minimal	basal	minimal	basal	minimal	basal	minimal
Control	12	288.9 ± 3.4	282.0 ± 5.4	837.9 ± 100.8	122.9 ± 16.4	2.5 ± 0.5	1.2 ± 0.2	191.9 ± 33.6	132.6 ± 30.7
Patient	18	279.8 ± 2.6*	269.8 ± 2.7**	591.7 ± 68.2	437.3 ± 54.8***	1.5 ± 0.2*	1.3 ± 0.2	363.1 ± 21.7*	235.3 ± 42.5*

*P<0.05, **P<0.01 and ***P<0.001 vs. the control group.

The relationship between plasma AVP levels and urinary excretion of AQP-2 during the acute water load is shown in Fig.1. Urinary excretion of AQP-2 had a positive correlation with plasma AVP levels in the control subjects (r=0.31, P<0.05). In contrast, the urinary excretion of AQP-2 was apparently high compared with the respective plasma AVP levels in the patient group, and thus the correlation between plasma AVP levels and urinary excretion of AQP-2 disappeared.

4. DISCUSSION

We demonstrated that the impaired water excretion was closely linked with the non-suppressible levels of plasma AVP despite hypoosmolality in the patient group. The plasma AVP levels themselves remained normal values, and they were comparable levels to those in the control subjects. In contrast, the basal and minimal urinary excretion of AQP-2 remained high in the patients with impaired water excretion, which were significantly greater than those in the control subjects. For evaluation of the antidiuretic effect, plasma AVP levels have to be compared with the respective low Posm, whereas urinary excretion of AQP-2 can be shown as an absolute value. Approximately 3% of AQP-2 in collecting duct cells is excreted into urine (Rai *et al.,* 1997), and the enhanced urinary excretion of AQP-2 should be closely linked with the upregulation of AQP-2 mRNA expression in the kidney. These upregulation has been clarified in the experimental models of SIADH, congestive heart failure, and liver cirrhosis with ascites (Fujita *et al.,* 1995, Asahina *et al.,* 1995, Nielsen *et al.,* 1997, Xu *et al.,* 1997).

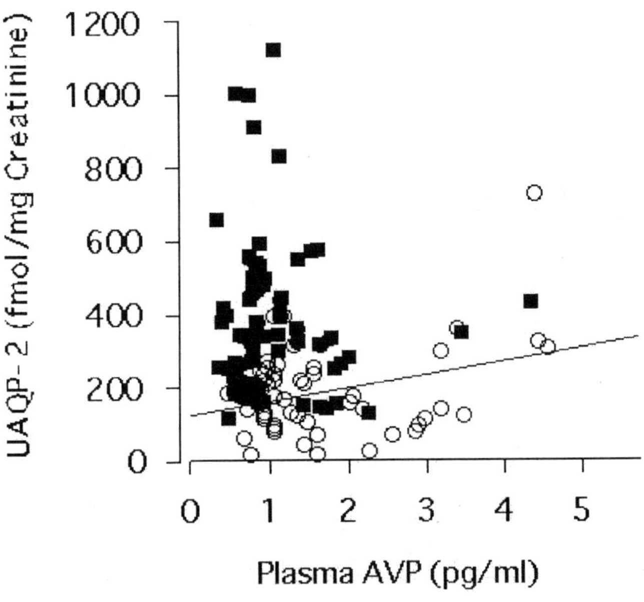

Figure 1. The relationship between plasma AVP levels and urinary excretion of AQP-2 in an acute water load in the control subjects (○) and the patients with impaired water excretion (■). A positive correlation is obtained in the control subjects (r=0.31, n=54, P<0.05).

As shown in Figure 1 and described in the previous report (Saito *et al.,* 1997), in the normal subjects there is a positive correlation between urinary excretion of AQP-2 and plasma AVP levels. Urinary excretion of AQP-2 in the patients was distinct from the controls. The urinary excretion of AQP-2 was more manifest than that expected from the plasma AVP levels in the patients with impaired water excretion, and thus the correlation disappeared.

Augmented hydroosmotic action of AVP in kidney is evident despite the non-suppressible, but normal value of plasma AVP in the pathological state of impaired water excretion under hypoosmotic state. There may be an unknown response mechanism in kidney that detects the non-suppressible levels of plasma AVP under hypoosmotic condition. These results indicate that hydroosmotic action of AVP is exaggerated more than that expected from plasma AVP levels in the pathological state of impaired water excretion dependent on non-suppressible release of AVP.

REFERENCES

Asahina Y Izumi N Enomoto N Sasaki S Fushimi K Marumo F and Sato C 1995 Increased gene expression of water channel in cirrhotic rat kidneys *Hepatology* **21**: 169-173

Bartter FC and Schwartz W B 1967 The syndrome of inappropriate secretion of antidiuretic hormone *Am J Med* **42**: 790-806

Bichet D G Szatalowicz V Chaimovitz C and Schrier RW 1982 Role of vasopressin in abnormal water excretion in cirrhotic patients *Ann Intern Med* **96**: 413-417

Fujita N Ishikawa S Sasaki S Fushimi K Marumo F and Saito T 1995 Role of water channel AQP-CD in water retention in SIADH and cirrhotic rats *Am J Physiol* **269**: F926-F931

Fushimi K Uchida S Hara Y Hirata Y Marumo F and Sasaki S 1993 Cloning and expression of apical membrane water channel of rat kidney collecting tubule *Nature* **361**: 549-552

Ishikawa S 1993 Cellular action of arginine vasopressin in the kidney *Endocr J* **40**: 373-386

Ishikawa S Saito T Okada K Tsutsui K and Kuzuya T 1986 Effect of vasopressin antagonist on water excretion in inferior vena cava constriction *Kidney Int* **30**: 49-55

Ishikawa S and Schrier RW 1982 Effect of arginine vasopressin antagonist on renal water excretion in glucocorticoid and mineralocorticoid deficient rats *Kidney Int* **22**: 587-593

Kanno K Sasaki S Hirata Y Ishikawa S Fushimi K Nakanishi S Bichet DG and Marumo F 1995 Urinary excretion of aquaporin-2 in patients with diabetes insipidus *N Engl J Med* **332**: 1540-1545

Knepper MA and Rector Jr FC 1995 Urinary concentration and dilution IN: The Kidney (eds) Brenner BM Rector Jr FC Saunders Philadelphia pp 532-570

Matsumura Y Uchida S Furuno M Marumo F and Sasaki S 1997 Transcriptional regulation of aquaporin-2 water channel gene by cAMP *J Am Soc Nephrol* **8**: 861-867

Nielsen S Terris J Anderson D Ecelbarger C Frokler J Jonassen T Marples D Knepper M A and Peterson J S 1997 Congestive heart failure in rats is associated with increased expression and targeting of aquaporin-2 water channel in collecting duct *Proc Natl Acad Sci USA* **94**: 5450-5455

Rai T Sekine K Kanno K Hata K Miura M Mizushima A Marumo F and Sasaki S 1997 Urinary excretion of aquaporin-2 water channel protein in human and rats *J Am Soc Nephrol* **8**: 1357-1362

Saito T Ishikawa S Ando F Okada N Nakamura T Kusaka I Higashiyama M Nagasaka S and Saito T 1998 Exaggerated urinary excretion of aquaporin-2 in the pathological state of impaired water excretion dependent upon arginine vasopressin *J Clin Endocrinol Metab* **83**: 4034-4040

Saito T Ishikawa S Sasaki S Nakamura T Rokkaku K Kawakami A Honda K Marumo F and Saito T 1997 Urinary excretion of aquaporin-2 in the diagnosis of central diabetes insipidus *J Clin Endocrinol Metab* **82**: 1823-1827

Schrier RW 1988a Pathogenesis of sodium and water retention in high-output and low-output cardiac failure nephrotic syndrome cirrhosis and pregnancy *N Engl J Med* **319**: 1065-1072

Schrier RW 1988b Pathogenesis of sodium and water retention in high-output and low-output cardiac failure nephrotic syndrome cirrhosis and pregnancy *N Engl J Med* **319**: 1127-1134

Schrier RW Berl T and Anderson RJ 1979 Osmotic and nonosmotic control of vasopressin release *Am J Physiol* **236**: F321-F332

Xu D Martin PY Ohara M St John J Pattison T Meng X Morris K Kim JK and Schrier RW 1997 Upregulation of aquaporin-2 water channel expression in chronic heart failure rat *J Clin Invest* **99**: 1500-1505

3.3 Pathophysiology

DYSREGULATION OF RENAL AQUAPORINS AND SODIUM TRANSPORTERS IN EXPERIMENTAL CHRONIC AND ACUTE RENAL FAILURE IN RAT

[1,2]Tae-Hwan Kwon, [3]Jørgen Frøkiær, and [1]Søren Nielsen
[1]Department of Cell Biology, Institute of Anatomy, University of Aarhus, DK-8000 Aarhus C, Denmark, sn@ana.au.dk; [2]Department of Physiology, School of Medicine, Dongguk University, Kyungju 780-714, Korea; and [3]Department of Clinical Physiology, Aarhus University Hospital and Institute of Experimental Clinical Research, DK-8200 Aarhus N, Denmark

1. INTRODUCTION

Renal failure, both acute and chronic, is associated with polyuria, an urinary concentrating defect, and altered renal sodium handling, although in both cases there are a wide range of glomerular and tubular abnormalities that contribute to the overall renal dysfunction. Recently, we examined the molecular mechanisms underlying altered water and sodium handling in kidneys subjected to the reduction of functioning renal mass (experimental chronic renal failure model induced by 5/6 nephrectomy) or to the ischemic injury (ischemia-induced acute renal failure model).

2. REDUCED EXPRESSION OF AQP2 AND AQP3 IN KIDNEYS OF RATS WITH CHRONIC RENAL FAILURE

Patients with advanced chronic renal failure (CRF) have urine which remains hypotonic to plasma despite the administration of supramaximal doses of vasopressin (Tannen *et al.*, 1969). This vasopressin-resistant hyposthenuria specifically implies abnormalities in collecting duct water reabsorption in CRF patients. Consistently, Fine et al., observed that isolated and perfused cortical collecting ducts dissected from remnant kidneys of severely uremic rabbits exhibited a significantly decreased water flux and adenylate cyclase activity in the response to vasopressin (Fine *et al.*, 1978). Importantly, they demonstrated that the 8-bromo-cAMP failed to induce a normal hydrosmotic response in cortical collecting duct from remnant kidneys. Several micropuncture and microcatheterization studies have also indicated that the impaired urinary concentrating ability may, at least partly, be caused by impairment of vasopressin stimulated water reabsorption in the collecting duct in CRF (Buerkert *et al.*, 1979; Wilson *et al.*, 1979). As an extension of these observations, Teitelbaum and McGuinness demonstrated that RT-PCR of total RNA from the inner medulla of CRF rat kidneys revealed virtual absence of V_2 receptor mRNA (Teitelbaum *et al.*, 1995). Thus, these studies provide firm evidences for significant defects in the collecting duct water permeability. Consistent with these observations, we have shown both decreased collecting duct water channel AQP2 (Fig 1) and AQP3 expression and a vasopressin-resistant downregulation of AQP2 in a 5/6 nephrectomy-induced CRF rat model (Kwon *et al.*, 1998). Immunocytochemistry and immunoelectron microscopy confirmed a marked reduction in AQP2 and AQP3 expression in the principal cells. This suggests that reduced AQP2 and AQP3 expression levels may be one factor involved in impaired collecting duct water permeability and in reduced vasopressin responsiveness in CRF.

3. ALTERED EXPRESSION OF SODIUM TRANSPORTERS IN KIDNEYS FROM RATS WITH CHRONIC RENAL FAILURE

As the population of surviving nephrons is reduced in chronic renal failure (CRF), the remaining tissue undergoes hypertrophy and the function is dramatically altered. It is well documented that when total renal mass is reduced, single nephron glomerular filtration rate (SNGFR) in the remaining nephrons markedly increases (Hayslett 1979). The elevated filtration rate leads to an increased reabsorptive burden on the residual nephrons. This together with the compensatory renal hypertrophy is associated with marked alterations in tubular reabsorptive capacities of sodium and water (Buerkert *et al.*, 1979; Fine *et al.*, 1978; Hayslett *et al.*, 1968). To address the mechanisms involved in the altered sodium handling, we determined the total kidney levels and the densities per nephron of the major renal sodium transporters in rats with experimental CRF (Kwon *et al.*, 1999). Two weeks after 5/6 nephrectomy (reducing the total number of nephrons to approximately 24 ± 8%), the rats were azotemic and displayed increased urinary Na^+ excretion. Semiquantitative immunoblotting revealed significant reduction in the total kidney levels of the proximal tubule sodium transporters type-3 Na/H exchanger (NHE-3, 48% of control), type-II Na/Pi cotransporter(NaPi-II,13%), and Na-K-ATPase (30%). However, the densities per nephron of NHE-3, NaPi-II, and Na-K-ATPase were not significantly altered in remnant kidneys despite the extensive hypertrophy of remaining nephrons. Immunocytochemistry confirmed the reduction in NHE-3 and Na-K-ATPase immunolabeling in the proximal tubule. In contrast, there was no significant reduction in the total kidney levels of the TAL and distal convoluted tubule NaCl transporters BSC-1 (type-1 bumetanide sensitive Na-K-2Cl cotransporter; NKCC2) and TSC (thiazide-sensitive Na-Cl cotransporter). This correspond to a 3.6- and 2.5-fold increase in densities per nephron, respectively and this was confirmed by immunocytochemistry.

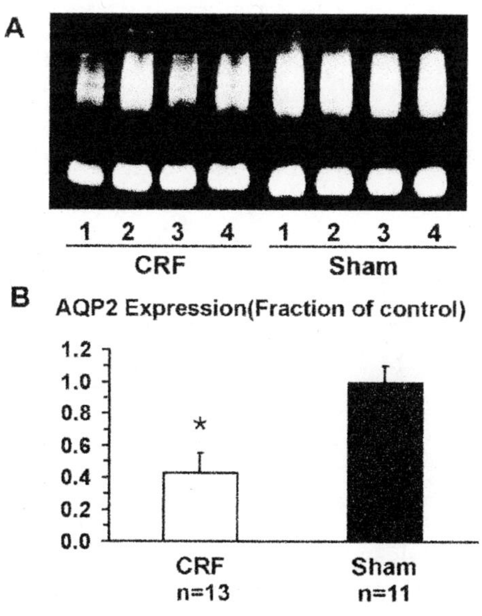

Figure 1. Immunoblot of membrane fractions of total kidneys from CRF and sham operated rats. A) The immunoblot was reacted with affinity purified anti-AQP2 and revealed 29 kDa and 35-50 kDa AQP2 bands, representing non-glycosylated and glycosylated forms of AQP2 (densities were 66 ± 10% in CRF rats and 100 ± 5% in sham operated rats, respectively, $P<0.05$). B) Densitometric analysis of all samples from CRF and sham operated rats (corrected according to loading and loading fraction of total kidney mass) revealed a marked decrease in AQP2 expression from 100 ± 11% in sham operated controls (n=11) to 43 ± 12% in CRF rats (n=13). $*P<0.05$.

Therefore, it appears that changes in sodium excretion are modulated by proximal nephron segments in response to the reduction of renal mass. We have demonstrated a significant decrease in total kidney NHE-3, NaPi-II, and Na-K-ATPase levels associated with reduced immunolabeling of NHE-3 and Na-K-ATPase in the proximal tubule in rats with CRF. Moreover, estimated densities of NHE-3 and Na-K-ATPase per nephron in remnant kidneys revealed a modest increase, but not significant. The hypertrophy, manifested by an increase in tubular diameter and length of the proximal tubule in remnant kidneys, appears to offer an explanation for this modest increase in densities per nephron of NHE-3 and Na-K-ATPase. This may provide an explanation for the previously observed increase in absolute sodium reabsorption in the proximal tubule from CRF animals (Fine *et al.*, 1978; Trizna *et al.*, 1981). However, at the same time, the inability to increase the densities per nephron of these NaCl transporters proportionately to the increased SNGFR and tubular fluid flow rate may provide an explanation for 1) the previously observed reduced fractional reabsorption of sodium in proximal tubule, 2) the glomerulotubular imbalance, and 3) the increased urinary excretion of sodium in CRF. In contrast, there are compensatory increases in BSC-1 and TSC densities and maintained Na-K-ATPase expression in TAL and distal convoluted tubule. This indicates a compensatory increase which may be partly due to increased blood vasopressin (Kim *et al.*, 1999) and aldosterone levels (Kim *et al.*, 1998) and increased delivery of NaCl to the distal nephron in CRF.

4. REDUCED ABUNDANCE OF AQUAPORINS IN RATS WITH BILATERAL ISCHEMIA-INDUCED ACUTE RENAL FAILURE: PREVENTION BY ALPHA-MSH

Ischemia-induced experimental acute renal failure (ARF) in rats is characterized by structural alterations in renal tubule, in association with an impairment in urinary concentration. The proximal tubule (S3 segment) and thick ascending limb are known to be main sites of ischemic injury (Venkatachalam *et al.*, 1978; Brezis *et al.*, 1984).

Several studies have shown defects in collecting duct water reabsorption, proximal tubule water reabsorption as well as defects in solute handling in post-ischemic kidneys (Johnston *et al.*, 1984; Hanley *et al.*, 1980). In an isolated tubule and microperfusion study of the rabbit kidney, it was observed that water reabsorption in the proximal tubule and cortical collecting duct was significantly reduced following ischemia (Hanley *et al.*, 1980). Moreover, it has been demonstrated that there are no differences in either basal or vasopressin-induced cAMP levels in outer or inner medulla in rats with ARF compared to sham operated rats (Anderson *et al.*, 1982), supporting the view that there are defects in collecting duct water reabsorption in postischemic kidneys. It is also well known that kidneys subjected to injury by ischemia are unable to establish or maintain a high medullary solute content (Beck *et al.*, 1995). Moreover, a decreased ability of the thick ascending limb to lower perfusate chloride ion concentration was observed as well (Hanley *et al.*, 1980). These studies suggest that there are defects in both countercurrent multiplication and collecting duct water permeability in response to ischemic damage.

Consistent with these findings, it was recently demonstrated that the levels of AQP2 (Fig 2) and AQP3 in the collecting duct as well as AQP1 in proximal tubule are significantly reduced in response to renal ischemia (Fernandez-Llama *et al.*, 1999; Kwon *et al.*, 1999). The decreased levels of aquaporins were associated with impaired urinary concentration in rats with both oliguric or nonoliguric ARF. Moreover, both the reduced expression of AQPs and reduced urinary concentration was significantly prevented by co-treatment with alpha-melanocyte-stimulating hormone (Kwon *et al.*, 1999). The results together suggest that decreased levels of aquaporins in both the proximal tubule and collecting duct in post-ischemic kidneys may play a significant role in the impairment of urinary concentration encountered in oliguric, maintenance and polyuric phases of experimental ischemia-induced ARF.

5. DECREASED ABUNDANCE OF MAJOR NA+ TRANSPORTERS IN KIDNEYS OF RATS WITH ISCHEMIA-INDUCED ACUTE RENAL FAILURE

Ischemia-induced acute renal failure (ARF) is known to be associated with significant impairment of tubular sodium reabsorption. We examined whether temporary bilateral renal ischemia (30, 40, or 60 min) and reperfusion (1-5 days) affect the abundance of several renal sodium transporters and urinary sodium excretion in rats (Kwon *et al.*, In Press). Bilateral renal ischemia for 30, 40, or 60 min produced mild, moderate, and severe ARF.

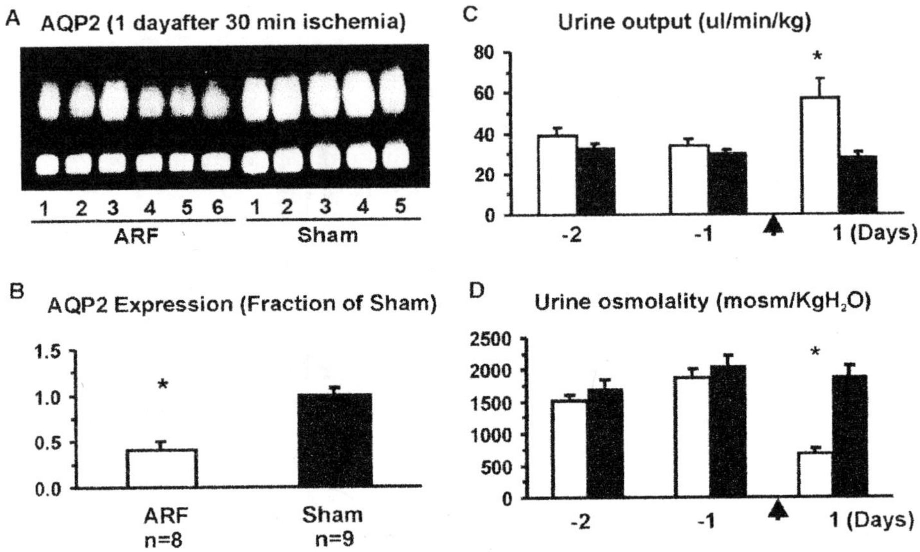

Figure 2. Semiquantitative immunoblotting of membrane fractions of whole kidneys, changes in urine output and urine osmolality from ARF rats (one day after 30 min bilateral renal ischemia: 30/1d) and sham operated rats. A) The immunoblot is reacted with anti-AQP2. B) Densitometric analysis of all samples from ARF (30/1d) and sham operated rats reveals a marked decrease in AQP2 expression from 100 ± 7% in sham operated controls to 40 ± 11% in ARF rats. C) Time course of the changes in urine output and D) urine osmolality in ARF rats (30/1d, n=8, open bars) and sham operated rats (n=9, solid bars). Urine output is significantly increased and urine osmolality is significantly decreased in ARF rats (30/1d). In contrast there are no changes in urine output and urine osmolality after sham operation in control rats. *$P < 0.05$.

In rats with mild ARF (30 min), semiquantitative immunoblotting revealed that the abundance of the two proximal tubular.sodium transporters NHE-3 and NaPi-II was significantly decreased to 28 ± 6% and 14 ± 6% of control levels, respectively at day 1. Moreover, whole kidney Na-K-ATPase levels were also significantly decreased to 51 ± 11% of sham levels. In contrast, there was no significant decrease in the BSC-1 and TSC levels at day 1. Consistent with this, fractional urinary excretion of sodium (FENa) was significantly increased in mild ARF (30min), and rates of urinary sodium excretion (UNa x V) was unchanged, despite a marked reduction in GFR.

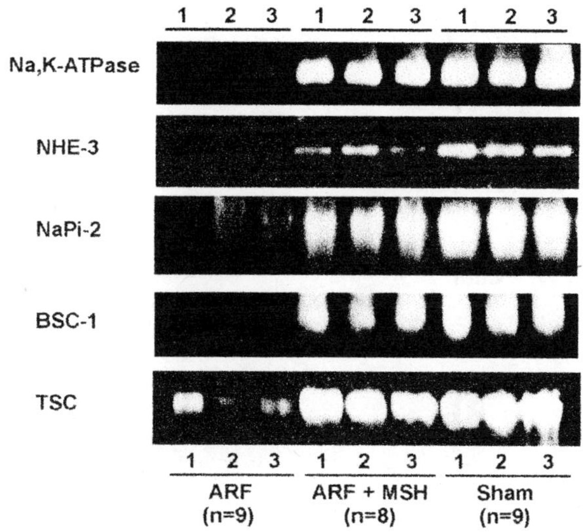

Figure 3. Effect of alpha-MSH treatment on the expression of several sodium transporters (Na,K-ATPase, NHE-3, NaPi-2, BSC-1 and TSC) in rats with ARF (2 days after 40 min bilateral renal ischemia). Semiquantitative immunoblotting demonstrates that alpha-MSH treatment during reperfusion in rats with ARF significantly reduce the decline in sodium transporter expressions following renal ischemia, compared to ARF rats which were not treated with alpha-MSH.

The sodium transporter levels were normalized within 5 days in rats with mild ischemic injury (30 min), and associated with the normalized renal sodium handling. Severe ischemic injury (60 min) resulted in a marked decrease in the abundance of sodium transporters (Na-K-ATPase, NHE-3, NaPi-II, BSC-1 and TSC) both at day1 and day 5 after ischemia. Consistent with this, FENa was significantly increased at at day 1 (12.5 ± 2.5%) and day 5 (3 ± 0.8%). Immunocytochemistry confirmed the decline in abundance of sodium transporters following renal ischemia. Intravenous alpha-MSH treatment, which is known to reduce the ischemia-induced injury, partially prevented the ischemia-induced downregulation of renal sodium transporters and reduced the high FENa to control levels (Fig 3 and Fig 4).

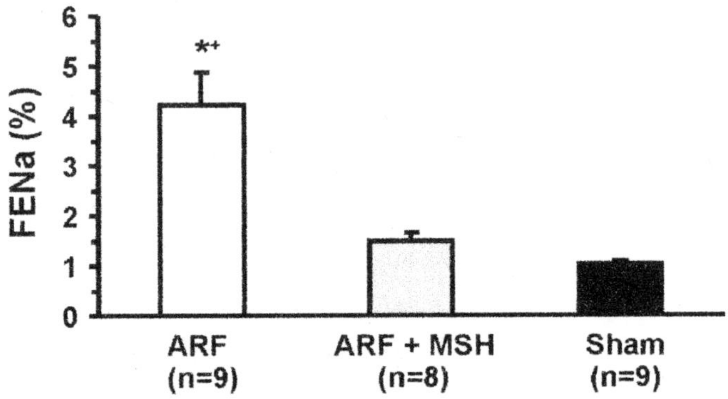

Figure 4. Effect of alpha-MSH treatment on FENa levels in rats with ARF (2 days after 40 min bilateral renal ischemia) and sham operated control rats. Rats with ARF have significantly increased FENa, two days after renal ischemia (4.2 ± 0.6%). In contrast, rats with ARF that were treated with alpha-MSH shows a markedly reduced FENa (1.5 ± 0.6%), compared to rats with ARF that were not treated with alpha-MSH. The FENa in alpha-MSH treated rats with ARF is similar to that of sham operated rats (1 ± 0.1%). $^{*}P < 0.05$, ARF rats are compared with sham operated rats. $^{+}P < 0.05$, ARF rats are compared with alpha-MSH treated ARF rats.

Therefore, we conclude that reduced levels of sodium transporters along the nephron may play a critical role in the impairment of tubular reabsorption of filtered sodium in post-ischemic kidney. Thus, this contributes to increased sodium excretion during the recovery phase of ischemia-induced ARF, despite a marked improvement in GFR.

6. CONCLUSION

We demonstrated that altered expression of renal AQPs and sodium transporters are associated with deranged urinary concentration and urinary sodium excretion in experimental CRF (5/6 nephrectomy) and ARF (ischemia-induced) models that have altered renal water and sodium handling. The marked reduction of renal AQPs and major renal sodium transporters may be significantly responsible for the reduced urinary concentration and increased urinary sodium excretion in experimental renal failure models.

ACKNOWLEDGMENTS

Support for this study was provided by the Commission of the European Union (EU-TMR Program and EU-Biotech Program), the Karen Elise Jensen Foundation, Danish Medical Research Council, and the Dongguk University.

REFERENCES

Anderson R J Gordon JA Kim J Peterson LM and Gross PA 1982 Renal concentration defect following nonoliguric acute renal failure in the rat *Kidney Int* **21**: 583-591

Beck F X Ohno A Dorge A and Thurau K 1995 Ischemia-induced changes in cell element composition and osmolyte contents of outer medulla *Kidney Int* **48**: 449-457

Brezis M Rosen S Silva P and Epstein FH 1984 Selective vulnerability of the medullary thick ascending limb to anoxia in the isolated perfused rat kidney *J Clin Invest* **73**: 182-190

Buerkert J Martin D Prasad J Chambless S and Klahr S 1979 Response of deep nephrons and the terminal collecting duct to a reduction in renal mass *Am J Physiol* **236**: F454-64

Fernandez-Llama P Andrews P Turner R Saggi S Di Mari J Kwon T-H NielsenS Safirstein R and Knepper MA 1999 Role of collecting duct aquaporins in polyuria of post-ischemic acute renal failure in rats *J Am Soc Nephrol* **10**: 1658-1668

Fine L G Schlondorff D Trizna W Gilbert RM and Bricker NS 1978 Functional profile of the isolated uremic nephron Impaired water permeability and adenylate cyclase responsiveness of the cortical collecting tubule to vasopressin *J Clin Invest* **61**: 1519-1527

Fine L G Trizna W Bourgoignie JJ and Bricker NS 1978 Functional profile of the isolated uremic nephron Role of compensatory hypertrophy in the control of fluid reabsorption by the proximal straight tubule *J Clin Invest* **61**: 1508-1518

Hanley M J 1980 Isolated nephron segments in a rabbit model of ischemic acute renal failure *Am J Physiol* 239: F17-F23

Hayslett J P 1979 Functional adaptation to reduction in renal mass *Physiol Rev* **59**: 137-164

Hayslett J P Kashgarian M and Epstein FH 1968 Functional correlates of compensatory renal hypertrophy *J Clin Invest* **47**: 774-799

Johnston P A Rennke H and Levinsky NG 1984 Recovery of proximal tubular function from ischemic injury *Am J Physiol* **246**: F159-F166

Kim G -H Ecelbarger CA Mitchell C Packer RK Wade JB and Knepper MA 1999 Vasopressin increases Na-K-2Cl cotransporter expression in thick ascending limb of Henle's loop *Am J Physiol* **276**: F96-F103

Kim G -H Masilamani S Turner R Mitchell C Wade JB and Knepper MA 1998 The thiazide-sensitive Na-Cl cotransporter is an aldosterone-induced protein *Proc Natl Acad Sci U S A* **95**: 14552-14557

Kwon T -H Frokiaer J Fernandez Llama P Knepper MA and Nielsen S 1999 Reduced abundance of aquaporins in rats with bilateral ischemia-induced acute renal failure: prevention by alpha-MSH *Am J Physiol* **277**: F413-F427

Kwon T -H Frokiaer J Fernandez-Llama P Maunsbach AB Knepper MA and Nielsen S 1999 Altered expression of Na transporters NHE-3 NaPi-II Na-K-ATPase BSC-1 and TSC in CRF rat kidneys *Am J Physiol* **277**: F257-F270

Kwon T -H Frokiaer J Han JS Knepper MA and Nielsen S 2000 Decreased abundance of Na transporters Na-K-ATPase NHE-3 NaPi-II BSC-1 and TSC in rats with ARF *Am J Physiol* In Press

Kwon T -H Frokiaer J Knepper MA and Nielsen S 1998 Reduced AQP1 -2 and -3 levels in kidneys of rats with CRF induced by surgical reduction in renal mass *Am J Physiol* **275**: F724-F741

Tannen R L Regal EM Dunn MJ and Schrier RW 1969 Vasopressin-resistant hyposthenuria in advanced chronic renal disease *N Engl J Med* **280**: 1135-1141

Teitelbaum I and McGuinness S 1995 Vasopressin resistance in chronic renal failure Evidence for the role of decreased V2 receptor mRNA *J Clin Invest* **96**: 378-385

Trizna W Yanagawa N Bar Khayim Y Houston B and Fine LG 1981 Functional profile of the isolated uremic nephron Evidence of proximal tubular "memory" in experimental renal disease *J Clin Invest* **68**: 760-767

Venkatachalam M A Bernard DB Donohoe JF and Levinsky NG 1978 Ischemic damage and repair in the rat proximal tubule: differences among the S1 S2 and S3 segments *Kidney Int* **14**: 31-49

WilsonD R and Sonnenberg H 1979 Medullary collecting duct function in the remnant kidney before and after volume expansion *Kidney Int* **15**: 487-501

3.4 Pathophysiology

DYSREGULATION OF AQP2 IN BILATERAL AND UNILATERAL URETERAL OBSTRUCTION

[1]Chunling Li, [2]Weidong Wang, [2]Søren Nielsen and [1]Jørgen Frøkiær
1Institute of Experimental Clinical Research, and 2Department of Cell Biology, Institute of Anatomy, University of Aarhus, DK-8000 Aarhus, Denmark, skejjf@aau.dk,

1. INTRODUCTION

Urinary tract obstruction is a complex disease which results in profound changes in renal hemodynamics and tubule function ultimately resulting in an impaired urinary concentrating capacity.

During the obstruction glomerular filtration rate (GFR) falls following the onset of complete ureteral obstruction (Harris *et al.*, 1981). Renal blood flow (RBF) starts to decline at approximately 3 to 5 hours after onset of obstruction, reaching about 30 to 50 percent of control values by 24 hours (Blackshear *et al.*, 1979, Provoost *et al.*, 1981) in bilateral ureteral obstruction (BUO). Similar RBF changes are seen in response to a 24 hours unilateral ureteral obstructions (UUO). Immediately following the release of a 24-hours period of BUO, RBF falls due to an increased afferent arteriolar resistance, and GFR remains low. This may be due to increased tubuloglomerular feedback as a result of the increased distal delivery of electrolytes. After longer periods (24 hours) of UUO, GFR reaches 80% of normal by 24 hours after release (Wright 1982). Associated with the hemodynamic abnormalities renal tubule function is disturbed in response to urinary tract obstruction. Major defects appear to be located in distal segments of the nephron. Thus, a common characteristic of obstructive nephropathy is a decreased ability to concentrate the urine caused by a reduced medullary solute concentration and impaired responsiveness to antidiuretic hormone (ADH). In addition, alterations in sodium reabsorption and reabsorption or secretion of other solutes such as potassium, magnesium, calcium, and hydrogen have also been described in obstructive uropathy (Saulo Klahr 1997).

2. DYSREGULATION OF AQP2 IN BUO

The vasopressin regulated water channel aquaporin2 (AQP2) is predominantly located in the apical domains of the collecting duct principal cells (Fushimi *et al.*, 1993, Nielsen *et al.*, 1993). Multiple studies have shown that AQP2 is the predominantly vasopressin-sensitive and regulated both by short-term (Kuwahara *et al.*, 1995, Nielsen *et al.*, 1995, Sabolic *et al.*, 1995, Yamamoto *et al.*, 1995) and long-term mechanisms. AQP2 expression levels are reduced in well-known models of nephrogenic diabetes insipidus (NDI) (lithium-induced NDI) associated with severe polyuria suggesting that AQP2 is important for maintaining a normal urinary concentrating capacity (Deen *et al.*, 1994). In order to examine whether altered AQP2 expression was associated with the well characterized urinary concentrating defect observed in rats after release of BUO AQP2 expression levels were examined in a rat model in which both ureters were reversibly obstructed. This study demonstrated that AQP2 expression in the kidney inner medulla during a period of 24 h of BUO was downregulated to 26±8% of control levels (Fig.1). Release of obstruction after 24 hours of BUO was associated with an immediate onset of a predominant osmotic-dependent polyuria. Forty-eight hours after release of BUO, the reduction in AQP2 expression persisted (19±8%, $p<0.001$), concurrent with a marked non-osmotic POD. Immunofluorescence and immunoelectron microscopy confirmed the reduced levels of AQP2 in collecting duct principal cells. Seven days after release, the renal excretion of water and electrolytes had almost normalized. However, downregulation of AQP2 was only partly reversed (49±14% of control levels). In parallel rats subjected to a 24 hours period of thirst 7 days after release of BUO showed a markedly impairment in urine osmolality suggesting that the persistent downregulation of AQP2 in response to BUO play an important role for the impaired urinary concentrating and slow recovery from POD (Frokiaer *et al.*, 1996).

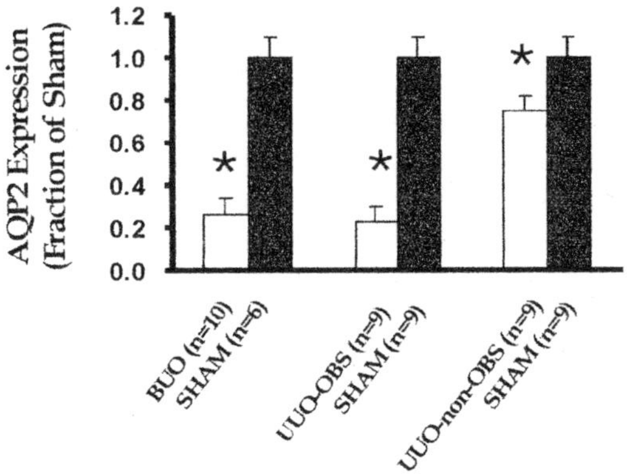

Figure 1. Densitometry of AQP2 levels in kidneys with 24 h BUO, obstructed kidneys with 24 h UUO and non-obstructed kidneys with 24 h UUO compared with levels in kidneys from sham-operated animals. alues represent combined densites of both AQP2 bands. 24 h BUO resulted in a dramatic decrease in AQP2 expression level (26±8% of sham level, n=10). Values are 23±7% in obstructed kidneys with 24 h UUO (n=9) compared with kidneys from sham levels (100±10%, n=9). In non-obstructed kidneys with 24 h UUO (n=9) values are 75±7% compared with kidneys from sham levels (100±10%, n=9). *$p<0.05$.

3. DYSREGULATION OF AQP2 IN UUO

In contrast to BUO, where there are extensive changes in systemic solute balance (Frokiaer *et al.*, 1996), the absolute excretion of sodium and water in UUO does not differ significantly from that in normal animals, since the non-obstructed kidney compensates for the lack of function in the obstructed kidney (Yarger 1991). Following release of obstruction in UUO, glomerular filtration rate is reduced and the fractional excretion of both sodium and water is substantially increased, and consequently the ability of the postobstructed kidney to concentrate urine is markedly impaired even in the presence of exogenous vasopressin (Wilson 1980). However, these changes are compensated for by the contralateral kidney.

In rats with UUO the relative role of intrarenal and systemic factors can be evaluated. In rats with 24 hours of UUO semiquantitative immunoblotting revealed a marked decrease in AQP2 in the obstructed kidney to 23±7% of sham levels. This downregulation persisted 24 h after release of UUO. Further more, there was a significant but less extensive downregulation of AQP2 in the nonobstructed kidneys to 75±7% of sham levels (Fig.1).

Consistent with impairment of collecting duct water reabsorption, free water clearance was greatly elevated in the obstructed kidneys (-2±1 μl/min/kg, determined immediately after release) and only moderately elevated in nonobstructed kidneys compared with sham-operated controls. Also AQP2 mRNA levels were reduced in obstructed kidneys. These experiments revealed that changes in AQP2 expression are reciprocal to the changes in free water clearance demonstrating a functional association between these two parameters. The much lower AQP2 protein and mRNA levels in obstructed vs. nonobstructed kidneys are consistent with intrarenal factors playing a major role for downregulation of AQP2 (Frøkiær *et al.*, 1996).

4. CONCLUSION

Downregulation of renal AQP2 in rats with BUO and UUO is associated with impairment of urinary concentration and POD. These findings suggests that intrarenal factors may play a major role for downregulation of AQP2.

ACKNOWLEDGEMENTS

The work cited from our laboratory was supported partly by grants from EU-Commission (FMRX-CT97-0124 and BIO4-CT98-0024), Karen Elise Jensen Foundation and the Danish Research Council.

REFERENCES

Blackshear JL Spielman WS Knox FG Romero JC 1979 Dissociation of renin release and renal vasodilation by prostaglandin synthesis inhibitors *Am J Physiol* **237**: F20-24

Deen PMT Verdijk MAJ Knoers NVAM Wierenga B Monnens LAH Van Os CH Van Oost BA 1994 Requirement of human renal water channel aquaporin-2 for vasopressin-dependent concentration of urine *Science* **264**: 92-95

Frokiaer J Marples D Knepper MA Nielsen S 1996 Bilateral ureteral obstruction downregulates expression of vasopressin- sensitive AQP-2 water channel in rat kidney *Am J Physiol* **270**: F657-F668

Frøkiær J Marples D Mønster B Nielsen UB Knepper MA Nielsen S 1997 Downregulation of aquaporin-2 parallels changes in renal water excretion in unilateral obstruction of the ureter *Am J Physiol* **273**: F213-223

Fushimi K Uchida S Hara Y Hirata Y Marumo F Sasaki S 1993 Cloning and expression of apical membrane water channel of rat kidney collecting tubule *Nature* **361**: 549-552

Harris RH Gill JM 1981 Changes in glomerular filtration rate during complete ureteral obstruction in rats *Kidney Int* **19**: 603-608

Kuwahara M Fushimi K Terada Y Bai L Marumo F Sasaki S 1995 cAMP-dependent phosphorylation stimulates water permeability of aquaporin-collecting duct water channel protein expressed in Xenopus oocytes *J Biol Chem* **270**: 10384-10387

Nielsen S Chou C-L Marples D Christensen EI Kishore BK Knepper MA 1995 Vasopressin increases water permeability of kidney collecting duct by inducing translocation of aquaporin-CD water channels to plasma membrane *Proc Natl Acad Sci USA* **92**: 1013-1017

Nielsen S DiGiovanni SR Christensen EI Knepper MA Harris HW 1993 Cellular and subcellular immunolocalization of vasopressin-regulated water channel in rat kidney *Proc Natl Acad Sci USA* **90**: 11663-11667

Provoost AP Molenaar JC 1981 Renal function during and after a temporary complete unilateral ureter obstruction in rats *Invest Urol* **18**: 242-46

Sabolic I Katsura T Verbavatz JM Brown D 1995 The AQP2 water channel: effect of vasopressin treatment microtubule disruption and distribution in neonatal rats (published erratum appears in J Membr Biol 1995 May;145(1):107-8) *J Membr Biol* **143**: 165-175

Saulo Klahr 1997 Obstructive Nephropathy: Pathophysiology and Management In: Robert WSchrier (ed) Renal and Electrolyte Disorders vol 1 558-559

Wilson DR 1980 Pathophysiology of obstructive nephropathy *Kidney Int* **18**: 281-292

Wright FS 1982 Effects of urinary tract obstruction on glomerular filtration rate and renal blood flow *Sem Nephrol* **2**: 5-16

Yamamoto T Sasaki S Fushimi K Ishibashi K Yaoita E Kawasaki K Marumo F Kihara I 1995 Vasopressin increases AQP-CD water channel in apical membrane of collecting duct cells in Brattleboro rats *Am J Physiol* **268**: C1546-C1551

Yarger WE 1991 Urinary Tract Obstruction In: Brenner BM Rector FC (eds) The Kidney vol 2 4 ed Philadelphia: Saunder; 1768-1808

3.5 Pathophysiology

UPREGULATION OF AQUAPORIN-2 GENE EXPRESSION IN PATHOLOGICAL STATE OF IMPAIRED WATER EXCRETION IN GLUCOCORTICOID-DEFICIENT RATS

Takako Saito, San-e Ishikawa, Minori Higashiyama, Tomoatsu Nakamura, Ikuyo Kusaka, Shoichiro Nagasaka, [1]Sei Sasaki and Toshikazu Saito
Division of Endocrinology and Metabolism Department of Medicine, Jichi Medical School, Tochigi and [1]Department of Medicine, Tokyo Medical and Dental University, Tokyo, Japan.
E-mail: saneiskw@jichi.ac.jp

1. INTRODUCTION

Impairment in water excretion occurs in patients with hypopituitarism and adrenal insufficiency (Slessor 1951, Amhed *et al.,* 1967, Ishikawa *et al.,* 1991), and in experimental models of glucocorticoid- and mineralocorticoid-deficiency (Green *et al.,* 1970, Linas *et al.,* 1980, Ishikawa and Schrier 1982, Robinson *et al.,* 1983). There is evidence that arginine vasopressin (AVP) plays a crucial role in impaired water excretion in these pathological states (Amhed *et al.,* 1967, Ishikawa *et al.,* 1991, Linas *et al.,* 1980, Ishikawa and Schrier 1982). We demonstrated the removal of water retention by an AVP V_2 receptor antagonist in glucocorticoid-deficient rats (Ishikawa and Schrier 1982).

AVP produces the water permeability of collecting duct, by mediating adenosine 3',5'-cyclic monophosphate (cAMP) (Ishikawa 1993, Knepper and Brenner 1995). Aquaporin-2 (AQP-2) is an AVP-regulated water channel of collecting duct cells (Fushimi *et al.,* 1993). The expression of AQP-2 mRNA is increased by exogenous and endogenous AVP in various pathophysiological conditions (Fushimi *et al.,* 1993, Nielsen *et al.,* 1993), and it is abolished by the presence of an AVP V_2 receptor antagonist

Molecular Biology and Physiology of Water and Solute Transport
Edited by Hohmann and Nielsen, Kluwer Academic/Plenum Publishers, New York, 2000.

(Hayashi *et al.,* 1994, Fujita *et al.,* 1995). Recent studies have demonstrated the upregulation of AQP-2 gene expression in kidney in pathological states of impaired water excretion in syndrome of inappropriate secretion of antidiuretic hormone (SIADH), congestive heart failure, and liver cirrhosis with ascites (Fujita *et al.,* 1995, Asahina *et al.,* 1995, Nielsen *et al.,* 1997, Xu *et al.,* 1997).

We determined whether AQP-2 is involved in pathogenesis of impaired water excretion in glucocorticoid-deficient rats. In addition, what alteration in AVP V_2 receptor in glucocorticoid deficiency was examined.

2. METHODS

2.1 Animals

Male Sprague-Dawley rats weighing 250-300 g were used. Under light ether anesthesia, the animals were adrenalectomized through bilateral retroperitoneal incisions, and then implanted osmotic minipumps (Alza model 2ML4, Palo Alto, CA) subcutaneously along the back. A group of glucocorticoid-deficient rats were administered aldosterone at a rate of 42 ng/100g/h by osmotic pumps throughout the experiments. Control group was administered both aldosterone (42 ng/100g/h) and dexamethasone (58 ng/100g/h). After the recovery, they were placed in cages with free access to water and rat chows. The two groups of rats were subjected to the studies on day 7-14. One ml blood was collected to measure serum Na and plasma osmolality (Posm), and thereafter the animals were sacrificed by decapitation. Blood from trunk was collected to measure plasma AVP levels (Ishikawa *et al.,* 1986). Both kidneys were removed to determine AQP-2 mRNA and AQP-2 protein expressions, AVP V_2 receptor mRNA expression and AVP receptor binding (Fushimi *et al.,* 1993, Fujita *et al.,* 1995, Jard *et al.,* 1987).

2.2 An acute water load

Acute oral water load was studied 2 hr after removal of food and water. The rats were again 30 ml/kg of deionized water by gastric tube under light ether anesthesia. The rats were then placed in individual metabolic cages and allowed to awaken. Spontaneously voided urine samples were collected at 20-min intervals for 3 hrs. Urine volume and urinary osmolality (Uosm) were measured.

A few days later the acute oral water loads were performed again. At the present occasion, the rats were killed by decapitation one hr after the water load. Trunkal blood and both kidneys were obtained to determine plasma AVP levels and expressions of AQP-2 mRNA and protein. The same experiments were carried out in the glucocorticoid-deficient rats and the control rats receiving 30 mg/kg of the nonpeptide AVP V_2 receptor antagonist, OPC-31260 (Yamamura *et al.,* 1992).

2.3 Statistical analysis

Values of percent excretion of water load, Uosm, serum Na, Posm, plasma AVP levels, ratio of the density of AQP-2 mRNA expression to that of β-actin mRNA, and density of AQP-2 in immunoreactive bands were analyzed by Student's *t* test. A P value of less than 0.05 was considered significant.

3. RESULTS

Serum Na levels were 139.3±0.3 mEq/l in the glucocorticoid-deficient rats, which were less than that of 146.5±0.8 mEq/l in the control (n=6, P<0.05). Figure 1 shows the expression of AQP-2 mRNA in the kidney of

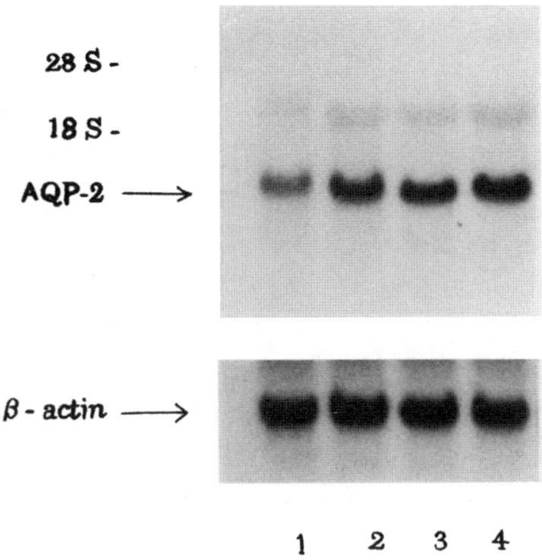

Figure 1. Expression of AQP-2 mRNA in the kidney of the glucocorticoid-deficient and the control rats under ad libitum water drinking. There are two transcripts at 1.5 kb and 4.4 kb. Lanes 1-2: control rats, lanes 3-4: glucocorticoid deficient (GD) rats.

the glucocorticoid-deficient and the control rats. A transcript at 1.5 kb expresses AQP-2 mRNA. Densitometric analysis showed that the expression of AQP-2 mRNA increased by 164.2±3.7% in the glucocorticoid-deficient rats, as compared to the control (n=6, P<0.01). Similar results were obtained with AQP-2 protein in western blot analysis (Fig. 2). Immunoblots showed the bands at 29 and 36-45 kDa. High-molecular-weight-mass band of 36-45 kDa was the glycosylated form of AQP-2 protein. The glucocorticoid-deficient rats increased AQP-2 protein in renal medulla by 137.2±5.8% compared with the control (n=6, P<0.05).

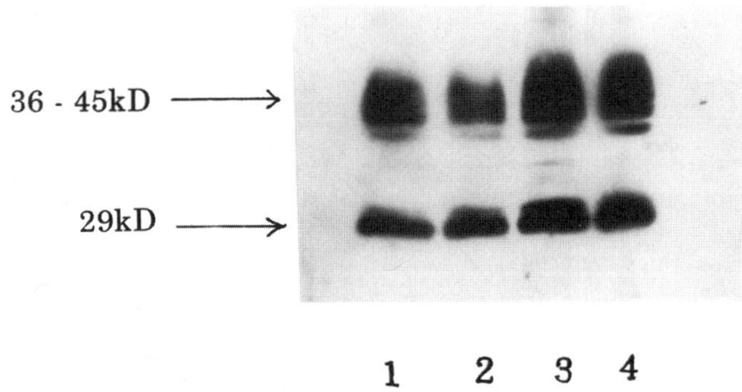

Figure 2. Western blot analysis of AQP-2 protein in the kidney of the glucocorticoid-deficient and the control rats under ad libitum water drinking. Lanes 1-2: control rats, lanes 3-4: glucocorticoid-deficient (GD) rats.

The results of an acute oral water load test is depicted in Table 1. Percent excretion of the water load was 38.7±11.7% in the glucocorticoid-deficient rats, a value less than that of 139.3±10.1% in the control (P<0.01). The minimal Uosm was 379.4±50.8 mOsm/kg H_2O in the glucocorticoid-deficient rats, that was greater than the control (P<0.01). The administration of 30 mg/kg OPC-31260 totally restored the impairment in water excretion in the glucocorticoid-deficient rats. Plasma AVP levels were 2.6±0.8 pg/ml in the glucocorticoid-deficient rats, and was decreased to only 1.4±0.1 pg/ml 60 min after the water load. In the control rats this maneuver decreased plasma AVP levels from 2.9±0.5 to 1.2±0.1 pg/ml.

The expression of AQP-2 mRNA one hr after the water load in the glucocorticoid-deficient rats is shown in Fig. 3. Its expression was still increased in the glucocorticoid-deficient rats, that was 141.5±17.3% vs. the control rats (n=6, P<0.05). When the study was carried out in the rats receiving OPC-31260, an increase in the expression of AQP-2 mRNA totally disappeared in the glucocorticoid-deficient rats (89.3±8.8%). Similar results were obtained with AQP-2 protein expression in the kidney (Fig. 4).

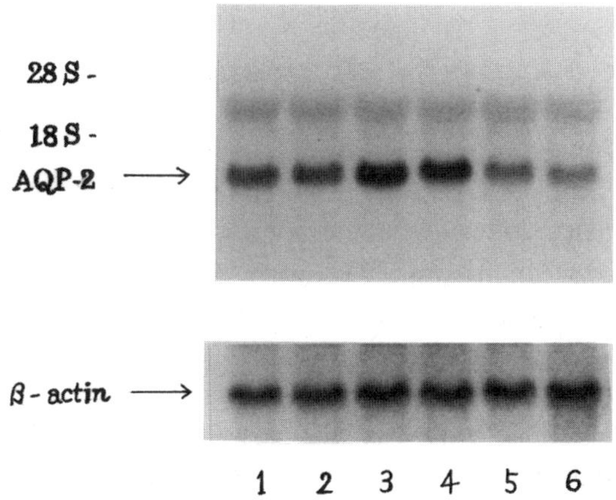

Figure 3. AQP-2 mRNA expression in the kidney of the glucocorticoid-deficient rats and the control rats one hr after the acute water load (30 ml/kg BW). Lanes 1-2: control rats, lanes 3-4: glucocorticoid-deficient (GD) rats receiving the vehicle for OPC-31260, lanes 5-6: GD rats receiving 30 mg/kg OPC-31260.

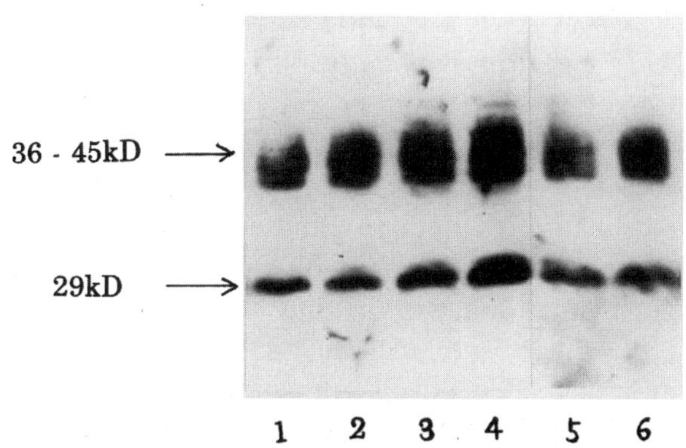

Figure 4. AQP-2 protein expression in the kidney of the glucocorticoid-deficient rats and the control rats one hr after the acute water load. Lanes 1-2: control rats, lanes 3-4: glucocorticoid-deficient (GD) rats receiving the vehicle for OPC-31260, lanes 5-6: GD rats receiving 30 mg/kg OPC-31260.

Membrane fractions of kidney medulla was studied to determine whether glucocorticoid deficiency affects [^{3}H]AVP receptor binding. There was no difference in Kd and Bmax between the glucocorticoid-deficient and the control rats (Kd: 1.51×10^{-10} vs. 1.75×10^{-10} M, Bmax: 8.45 vs. 7.29 pmol/mg protein). The expression of AVP V_2 receptor mRNA was analyzed by semiquantitative multiprimer RT-PCR. There was no difference in the expression of AVP V_2 receptor mRNA between the two groups of rats.

4. DISCUSSION

We confirmed impaired water excretion in glucocorticoid deficiency (Green *et al.,* 1970, Linas *et al.,* 1980, Ishikawa and Schrier 1982, Robinson *et al.,* 1983). The percent excretion of the water load was reduced to 38.7% and the minimal Uosm remained as high as 379.4 mOsm/kg H_2O. The plasma AVP levels were high compared with the Posm, but they were not sufficiently suppressed by the acute water load in the glucocorticoid-deficient rats. As indicated the impairment in water exretion, the expression of AQP-2 mRNA and protein in kidney was significantly increased in the glucocorticoid-deficient rats as compared to the control. The enhanced expression of AQP-2 mRNA and protein totally disappeared in the glucocorticoid-deficient rats receiving the AVP V_2 receptor antagonist, in association with the normalization of renal water excretion. These findings strongly indicate that the augmentation in AQP-2 mRNA expression is dependent on plasma AVP. In addition, urinary excretion of AQP-2 is exaggerated in the patients with hypopituitarism, particularly in hypofunction of pituitary-adrenocortical axis (Saito *et al.,* 1998). As approximately 3% of AQP-2 in collecting duct cells is excreted into urine (Rai *et al.,* 1997), and urinary excretion of AQP-2 has a positive correlation with plasma AVP levels (Saito *et al.,* 1997), urinary AQP-2 excretion participates in the understanding of AVP dependency. Upregulation of AQP-2 mRNA expression in kidney was evident in the glucocorticoid-deficient rats, which could be tightly linked with the exaggerated urinary excretion of AQP-2 in the pituitary-adrenocortical dysfunction.

What mechanisms underlie in the upregulation of AQP-2 mRNA expression in the glucocorticoid-deficient rats? Enhanced antidiuresis has been understanding to depend on non-suppressible levels of plasma AVP despite hypoosmolality, though plasma AVP levels are not so high, in most clinical settings of water retention (Ishikawa and Schrier 1982, Ishikawa *et al.,* 1986, Saito *et al.,* 1998). As a result, the exaggerated action of AVP is found. [^{3}H]AVP receptor binding study revealed that there was no difference in Bmax and Kd between the glucocorticoid-deficient and the control rats. Also, AVP V_2 receptor mRNA expression was almost equal in these two groups. Therefore, at the receptor site the initiation of intracellular signaling is kept in the normal range under chronically relative AVP excess state. In the present study, it is clear that the expression of AQP-2 mRNA was upregulated in the glucocorticoid-deficient rats. Chronically non-suppressible levels of plasma AVP does not always downregulate the V_2 receptor function, and could rather maintain its long-term stimulation on V_2 receptor-mediated signaling. In fact, in the patients with hypopituitarism the urinary excretion of AQP-2 is more manifest than that expected from plasma

AVP levels (Saito *et al.,* 1998). However, there is no evidence that other factor except for AVP is involved in regulation of AQP-2 in collecting duct cells. These findings can not rule out the unknown mechanism for enhancing post-receptor signaling of AVP.

REFERENCES

Amhed AB George BC Gonzalez-Auvert T and Dingman JF 1967 Increased plasma arginine vasopressin in clinical adrenocortical insufficiency and its inhibition of glucocorticoids *J Clin Invest* **46**: 111-123

Asahina Y Izumi N Enomoto N Sasaki S Fushimi K Marumo F and Sato C 1995 Increased gene expression of water channel in cirrhotic rat kidneys *Hepatology* **21**: 169-173

Fujita N Ishikawa S Sasaki S Fujisawa G Fushimi K Marumo F and Saito T 1995 Role of water channel AQP-CD in water retention in SIADH and cirrhotic rats *Am J Physiol* **269**: F926-F931

Fushimi K Uchida S Hara Y Hirata Y Marumo F and Sasaki S 1993 Cloning and expression of apical membrane water channel of rat kidney collecting duct tubule *Nature* **361**: 549-552

Green HH Harrington AR and Valtin H 1970 On the role of antidiuretic hormone in the inhibition of acute water diuresis in adrenal insufficiency and the effects of gluco- and mineralocorticoids in recovering the inhibition *J Clin Invest* **49**: 1724-1736

Hayashi M Sasaki S Tsuganezawa H Monkawa T Kitajima W Konishi K Fushimi K Marumo F and Saruta T 1994 Expression and distribution of aquaporin of collecting duct are regulated by vasopressin V_2 receptor in rat kidney *J Clin Invest* **94**: 1778-1783

Ishikawa S 1993 Cellular action of arginine vasopressin in the kidney *Endocr J* **40**: 373-386

Ishikawa S Fujisawa G Tsuboi Y Okada K Kuzuya T and Saito T 1991 Role of antidiuretic hormone in hyponatremia in patients with isolated adrenocorticotropic hormone deficiency *Endocrinol Japon* **38**: 325-330

Ishikawa S Saito T Okada K Tsutsui K and Kuzuya T 1986 Effect of vasopressin antagonist on water excretion in inferior vena cava constriction *Kidney Int* **30**: 49-55

Ishikawa S and Schrier RW 1982 Effect of arginine vasopressin antagonist on renal water excretion in glucocorticoid and mineralocorticoid deficient rats *Kidney Int* **22**: 587-593

Jard S Lombard C Marie J and Devillies G 1987 Vasopressin receptors from cultured mesangial cells resemble V1a type *Am J Physiol* **253**: F41-F49

Knepper MA and Rector Jr FC 1995 Urinary concentration and dilution In *The Kidney* (BM Brenner and Jr FC Rector eds) Saunders Philadelphia pp 532-570

Linas SL Berl T Robertson GL Aisenbrey GA Schrier RW and Anderson RJ 1980 Role of vasopressin in the impaired water excretion of glucocorticoid deficiency *Kidney Int* **10**: 58-67

Nielsen S DiGiovanni SR Christensen EI Knepper MA and Harris HW 1993 Cellular and subcellular immunolocalization of vasopressin-regulated water channel in rat kidney *Proc Natl Acad Sci USA* **90**: 11663-11667

Nielsen S Terris J Anderson D Ecelbarger C Frokler J Jonassen T Marples D Knepper MA and Peterson JS 1997 Congestive heart failure in rats is associated with increased expression and targeting of aquaporin-2 water channel in collecting duct *Proc Natl Acad Sci USA* **94**: 5450-5455

Rai T Sekine K Kanno K Hata K Miura M Mizushima A Marumo F and Sasaki S 1997 Urinary excretion of aquaporin-2 water channel protein in human and rat *J Am Soc Nephrol* **8**: 1357-1362

Robinson AG Seif SM Verbalis JG Brownstein MJ 1983 Quantitation of changes in the content of neurohypophyseal peptides in hypothalamic nuclei after adrenalectomy *Neuroendocrinology* **36**: 347-350

Saito T Ishikawa S Ando F Okada N Nakamura T Kusaka I Higashiyama M Nagasaka S and Saito T 1998 Exaggerated urinary excretion of aquaporin-2 in the pathological state of impaired water excretion dependent upon arginine vasopressin *J Clin Endocrinol Metab* **83**: 4034-4040

Saito T Ishikawa S Sasaki S Nakamura T Rokkaku K Kawakami A Honda K Marumo F and Saito T 1997 Urinary excretion of aquaporin-2 in the diagnosis of central diabetes insipidus *J Clin Endocrinol Metab* 82: 1823-1827

Slessor A 1951 Studies concerning the mechanism of water retention in Addison disease and in hypopituitarism *J Clin Endocrinol* **11**: 700-723

Xu DL Martin PY Ohara M St John J Pattison T Meng X Morris K Kim JK and Schrier RW 1997 Upregulation of aquaporin-2 water channel expression in chronic heart failure rat *J Clin Invest* **99**: 1500-1505

Yamamura Y Ogawa H Yamashita H Chihara T Miyamoto H Nakamura S Onogawa T Yamashita T Hosokawa T Mori T Tominaga M and Yabuuchi Y 1992 Characterization of a novel aquaretic agent OPC-31260 as an orally effective nonpeptide vasopressin V_2 receptor antagonist *Br J Pharmacol* **105**: 787-791

3.6 Pathophysiology

THE ROLE OF AQUAPORINS IN CEREBRAL VOLUME REGULATION AND EDEMA FORMATION

[1,2]Zsolt Vajda, [1]Dominique Promeneur, [2]Tamás Dóczi, [3]Endre Sulyok, [1]Jørgen Frøkiær, [4]Ole P. Ottersen and [1]Søren Nielsen
[1]Department of Cell Biology, University of Aarhus, Aarhus, Denmark. [2]Department of Neurosurgery, Medical University of Pécs, Pécs, Hungary; [3]County Children's Hospital, Pécs, Hungary; [4]Institute of Basic Medical Sciences, University of Oslo, Oslo, Norway. E-mail skejjf@aau.dk

1. INTRODUCTION

Regulation of tissue water content and brain volume is of critical importance for the normal functioning of the central nervous system (CNS), which, surrounded by the rigid cranium, is highly sensitive to any increase in the intracranial pressure. Alterations in cerebral water homeostasis and distribution may lead to neuronal and glial swelling due to accumulation of intracellular water. This condition, known as cytotoxic brain edema, is associated with many intracranial neuropathological states, including traumatic and ischemic brain injury, neoplasms, and metabolic diseases like systemic hyponatremia (Klatzo 1994). Failure of the compensatory mechanisms may result in fatal outcome due to decreased cerebral blood flow and mechanical tissue damages, whereas perturbed water homeostasis at the cellular level may affect ion homeostasis and neuronal excitability. Although numerous investigations have been performed to elucidate the underlying molecular basis and pathophysiology of brain edema, little is known about the regulation of water transport across the blood-brain barrier and between extra- and intracellular compartments of the brain parenchyma.

Molecular Biology and Physiology of Water and Solute Transport
Edited by Hohmann and Nielsen, Kluwer Academic/Plenum Publishers, New York, 2000.

2. AQUAPORINS IN THE CENTRAL NERVOUS SYSTEM

The discovery and characterization of the aquaporin (AQP) family of membrane water channels by Agre and colleagues (Agre *et al.,* 1995, King *et al.,* 1996) provided molecular insight into fundamental processes of water transport across plasma membranes. Two AQPs are expressed abundantly in the mammalian brain: AQP1 in the apical plasma membranes of the cells of choroid plexus of the ventricles, where it has been suggested to participate in the secretion of cerebrospinal fluid (Nielsen *et al.,* 1993) and AQP4 in plasma membranes of ependymal cells and astrocytes. Immunogold electron microscopy revealed abundant AQP4 labeling of astrocytes and of subpopulations of ependymal cells with a highly polarized distribution in glial membranes in direct contact with capillaries and astrocyte end-feet forming the glia limitans (Nielsen *et al.,* 1997), where it has been shown, by freeze-fracture studies, to be present in square arrays (Rash *et al.,* 1998). This localization pattern suggests a role for AQP4 in water transport between different water compartments of the CNS. Strong AQP4 expression has also been described in astrocytes and ependymocytes in osmosensory areas, including the supraoptic nucleus and subfornical organ (Jung *et al.,* 1994, Hasegawa *et al.,* 1994, Nielsen *et al.,* 1997) suggesting a role also in osmosensory mechanisms (Wells 1998). In a recent study, using in situ hybridization, Venero et al.,. reported the presence of AQP4 mRNA in GFAP negative neurons and in choroid plexus cells (Venero *et al.,* 1999).

Several studies have provided additional insight of the role of AQP4 in the central nervous system. It has been demonstrated that AQP4 is also abundant in Müller glial cells in a highly polarized distribution which closely matches the compartmentation of potassium conductance and the inward rectifying K+ channel Kir4.1 (Nagelhus *et al.,* 1999). This supports the view that AQP4 may serve a role in activity dependent water and K+ redistribution, by short-circuiting osmotic gradients and by directing water flux to specific extracellular compartments. Recent studies on the postnatal development of AQP4 mRNA and protein expression in rat brain revealed that the ontogeny of AQP4 co-incides with the maturation of the processes underlying water and K^+ homeostasis, pointing to the critical role of AQP4 in brain water homeostasis (Wen *et al.,* 1999).

Two recent studies have revealed that AQP4 undergoes regulation, and the protein kinase C (PKC) second messenger system has been suggested to participate in this. Han et al.,. have recently demonstrated, using transfected *Xenopus oocytes*, that the PKC activator phorbol esters stimulate the phosphorylation of AQP4 protein and decrease its water conductance (Han *et al.*, 1998). Nakahama et al., have reported a marked decrease in the expression of AQP4 mRNA levels in response to phorbol ester treatment in cultured rat astrocytes (Nakahama *et al.*, 1999).

3. THE ROLE OF AQP4 IN BRAIN EDEMA

The brain has been shown to behave like an imperfect osmometer by activating volume regulatory mechanisms in response to osmotic perturbations (Gullans *et al.*, 1993). Volume regulation is mediated by cellular accumulation or loss of osmotically active inorganic ions and small organic solutes, followed by osmotically driven water shifts between extra- and intracellular compartments (McManus *et al.*, 1995). Systemic hyponatremia is often accompanied by the development of severe brain edema. The reduced intravascular osmolarity may promote a transport of water across the capillary endothelium and the glial cell membrane, resulting in cell swelling. Since AQP4 is highly concentrated in specific glial and ependymal cell membrane domains associated with blood-brain and liquor-brain interfaces it can be hypothesized that a protective downregulation of AQP4 under hypoosmotic stress would protect against glial water uptake and development of acute brain edema. Theoretically this may be achieved by reducing the conductance of AQP4 by changes at the molecular level or by altering the density or distribution of AQP4 in the plasma membrane.

In agreement with this hypothesis, it has been shown recently, that AQP4-deficient knock-out mice had much better survival and neurological outcome in brain edema models produced by acute water intoxication or middle cerebral artery ligation, than their wild-type counterparts (Manley *et al.*, 2000). One hour following the combined intraperitoneal administration of distilled water (~20% of body weight) and 8-deamino-arginine vasopressin (dDAVP) the mortality rate dramatically decreased from 76% in wild-type controls to 8% in the knock-out mice. Quantification of the cross-sectional area of astrocytic foot processes adjacent to brain capillaries revealed a significantly reduced swelling of these processes in AQP4 knock-out mice.

Similar findings of reduced brain swelling were seen in AQP4 knock-out mice following focal cerebral ischemia induced by permanent middle cerebral artery (MCA) occlusion. The extent of the edema as measured by the enlargement of the affected hemisphere was significantly reduced in the mice lacking AQP4. These observations strongly highlight the role of AQP4 in mediating the influx of water into the cells of the CNS.

In our recent study we have examined the possibility of downregulation and /or subcellular redistribution of AQP4 protein in the brain in response to systemic hypoosmotic challenge (Vajda *et al.*, 2000). Rats were rendered acutely or chronically hyponatremic by repeated combined administration of hypotonic dextrose solution i.p. and dDAVP s.c. Immunoblots of brain and cerebellar samples incubated with the antibody against the intracellular C-terminal of the AQP4 protein revealed increased immunoreactivity of AQP4 in brain and cerebellum in response to systemic hypoosmolarity and this elevation proved to be consistent in the acute (4 h) and the chronic (48 h) stage as well. In contrast, AQP4 mRNA levels remained unchanged after 4 h of hypoosmolarity, suggesting that the observed increase in AQP4 immunoreactivity is not a consequence of increased transcription. If the AQP4 protein level is unchanged, then the increased immunosignal in tissue sections and immunoblots must reflect an increased labeling efficiency, i.e., an increased molecular ratio between bound antibodies and AQP4. An increased labeling efficiency would occur if the C-terminus of AQP4 were modified so as to increase its affinity to the specific antibodies. The rapid onset of the increased AQP4 immunosignal further supports the idea that it is caused by conformational changes rather than changes in protein expression.

An other important aspect of the study was to test if there were any signs of subcellular redistribution of AQP4 in response to hyponatremia. In the kidney it has been demonstrated that vasopressin regulates the osmotic water permeability of collecting duct principal cells by inducing a translocation of AQP2 from intracellular vesicles to the apical plasma membrane (Nielsen *et al.*, 1995). The cellular and subcellular distribution as well as the structural organization of AQP4 in the brain has been recently described in electron microscopical studies (Nielsen *et al.*, 1997, Rash *et al.*, 1998). These studies have revealed a highly polarized distribution of the protein with marked expression in astroglial membrane domains facing capillaries and pia mater, i.e., at sites where AQP4 is concentrated in square arrays. In contrast, AQP4 is present at low level in the glial processes facing the neuropil and is absent from intracellular vesicles (Nielsen *et al.*, 1997). This intuitively suggests that AQP4 distribution between plasma membrane and intracellular vesicles normally is not regulated. In agreement, neither light microscopic immunocytochemistry nor high-resolution immunogold electron microscopy provided any evidence of a subcellular redistribution of AQP4 protein in response to systemic hyponatremia. However it cannot be excluded that AQP4 redistribution takes place within the plasma membrane of the glial process facing capillaries. This cannot be resolved by conventional electron microscopy but requires freeze-fracture immunogold labeling. Interestingly, freeze-fracture studies have previously demonstrated that square arrays, now known to incorporate AQP4 (Rash *et al.*, 1998), are dissociated in response to several experimental neuropathological conditions (for references see Rash *et al.*, 1998). Thus it remains possible that the oligomeric organization of AQP4 may be perturbed in response to hyponatremia and that this may have an adaptive effect.

4. CONCLUSION

The discovery of the aquaporin membrane water channel family provided new insight into the molecular mechanisms of cellular water metabolism. Two members of the aquaporin family have been described to be present in the central nervous system so far. AQP1 is localized in the apical membranes of the choroid plexus cells and has been suggested to participate in the secretion of the cerebrospinal fluid. AQP4 is predominantly present in membrane domains associated with blood-brain and blood-CSF interfaces, which predicts its involvement in maintaining the water homeostasis of the central nervous system. Recent studies have demonstrated the role of AQP4 in the pathophysiology of cellular brain edema, making it a possible target for anti-edematous drug design.

ACKNOWLEDGMENTS

The work cited from our laboratory was supported partly by grants from EU-Commission (TMR and Biotech programs), Karen Elise Jensen Foundation and the Danish Research Council.

REFERENCES

Agre P Brown D and Nielsen S 1995 Aquaporin water channels: unanswered questions and unresolved controversies *Curr Opin Cell Biol* 1995 **7:** 472-483

Gullans SR and Verbalis JG 1993 Control of brain volume during hyperosmolar and hypoosmolar conditions *Annu Rev Med* **44:** 289-301

Han Z Wax MB and Patil RV 1998 Regulation of aquaporin-4 water channels by phorbol ester-dependent protein phosphorylation *J Biol Chem* **273:** 6001-6004

Hasegawa H Ma T Skach W Matthay MA and Verkman AS 1994 Molecular cloning of a mercurial-insensitive water channel expressed in selected water-transporting tissues *J Biol Chem* **269:** 5497-500

Jung JS Bhat RV Preston GM Guggino WB Baraban JM and Agre P 1994 Molecular characterization of an aquaporin cDNA from brain: candidate osmoreceptor and regulator of water balance *Proc Natl Acad Sci USA* **91:** 13052-13056

King LS and Agre P 1996 Pathophysiology of the aquaporin water channels *Annu Rev Physiol* **58:** 619-648

Klatzo I 1994 Evolution of brain edema concepts *Acta Neurochir Suppl (Wien)* **60:** 3-6

Manley GT Fujimura M Ma T Noshita N Filiz F Bollen AW Chan P and Verkman AS 2000 Aquaporin-4 deletion in mice reduces brain edema after acute water intoxication and ischemic stroke *Nat Med* **6:** 159-163

McManus ML Churchwell KB and Strange K 1995 Regulation of cell volume in health and disease *N Engl J Med* **333:** 1260-1266

Nagelhus EA Horio Y Inanobe A Fujita A Haug FM Nielsen S Kurachi Y and Ottersen OP 1999 Immunogold evidence suggests that coupling of K+ siphoning and water transport in rat retinal Muller cells is mediated by a coenrichment of Kir41 and AQP4 in specific membrane domains *Glia* **26:** 47-54

Nakahama K Nagano M Fujioka A Shinoda K and Sasaki H 1999 Effect of TPA on aquaporin 4 mRNA expression in cultured rat astrocytes *Glia* **25:** 240-246

Nielsen S Smith BL Christensen EI and Agre P 1993 Distribution of the aquaporin CHIP in secretory and resorptive epithelia and capillary endothelia *Proc Natl Acad Sci USA* **90:** 7275-7279

Nielsen S Chou CL Marples D Christensen EI Kishore BK and Knepper MA 1995 Vasopressin increases water permeability of kidney collecting duct by inducing translocation of aquaporin-CD water channels to plasma membrane *Proc Natl Acad Sci USA* **92:** 1013-1017

Nielsen S Nagelhus EA Amiry-Moghaddam M Bourque C Agre P and Ottersen OP 1997 Specialized membrane domains for water transport in glial cells: high-resolution immunogold cytochemistry of aquaporin-4 in rat brain *J Neurosci* **17:** 171-180

Rash JE Yasumura T Hudson CS Agre P and Nielsen S 1998 Direct immunogold labeling of aquaporin-4 in square arrays of astrocyte and ependymocyte plasma membranes in rat brain and spinal cord *Proc Natl Acad Sci USA* **95:** 11981-11986

Vajda Z Promeneur D Doczi T Sulyok E Frokiaer J Ottersen OP and Nielsen S 2000 Increased aquaporin-4 immunoreactivity in rat brain in response to systemic hyponatremia *Biochem Biophys Res Commun* **270:** 495-503

Venero JL Vizuete ML Ilundain AA Machado A Echevarria M and Cano J 1999 Detailed localization of aquaporin-4 messenger RNA in the CNS: preferential expression in periventricular organs *Neuroscience* **94:** 239-250

Wells T 1998 Vesicular osmometers vasopression secretion and aquaporin-4: a new mechanism for osmoreception? *Mol Cell Endocrinol* **136:** 103-107

Wen H Nagelhus EA Amiry-Moghaddam M Agre P Ottersen OP and Nielsen S 1999 Ontogeny of water transport in rat brain: postnatal expression of the aquaporin-4 water channel *Eur J Neurosci* **11:** 935-945

3.7 Pathophysiology

DYSREGULATION OF AQUAPORINS AND SODIUM TRANSPORTERS IN EXPERIMENTAL NEPHROGENIC DIABETES INSIPIDUS IN RAT

[1]Weidong Wang, [2]Tae-Hwan Kwon, [3]Chunling Li, [4]David Marples [3]Jørgen Frøkiær, and [1]Søren Nielsen

[1]Department of Cell Biology, Institute of Anatomy, University of Aarhus, DK-8000 Aarhus, Denmark; [2]Department of Physiology, School of Medicine, Dongguk University, Kyungju, 780-714, Korea; [3]Institute of Experimental Clinical Research, University of Aarhus, DK-8000 Aarhus C, Denmark; [4]School of Biomedical Sciences, University of Leeds, United Kingdom. E-mail: skejjf@aau.dk

1. INTRODUCTION

The renal tubule is composed of a series of distinct segments which function together to permit the regulated excretion of water and solutes. The aquaporins (AQPs) are a family of membrane proteins that function as water channels. AQP1 is highly abundant in the proximal tubule and descending thin limb and several studies have emphasized its important role in the constitutive water reabsorption in these segments and its role for urinary concentration (Chou *et al.*, 1999, Agre *et al.*, 1998, Nielsen *et al.*, 1995). In the kidney collecting duct at least three aquaporins are known to be expressed and they may participate in the vasopressin-regulated water reabsorption. AQP2 is the apical water channel of collecting duct principal cells and is the chief target for regulation of collecting duct water permeability by vasopressin (Nielsen *et al.*, 1993, Nielsen *et al.*, 1995). Water transport across the basolateral plasma membrane of collecting duct principal cells is thought to be mediated by aquaporin-3 (AQP3) (Ecelbarger, *et al.*,. 1995) and aquaporin-4 (AQP4). A series of studies have demonstrated that altered expression and apical targeting of AQP2 play a significant role in water balance disorders (Nielsen, *et al.*, 1999).

The proximal tubule is the site of reabsorption of approximately two thirds of the NaCl that enters the tubular fluid by glomerular filtration. An important fraction of the sodium reabsorption in the proximal tubule occurs by the apical Na^+/H^+ exchanger, resulting in a net process of sodium and bicarbonate reabsorption. Several lines of evidence including immunolocalization of type-3 Na^+/H^+ exchanger (NHE-3) to the apical plasma membrane domains in the proximal tubule, have made it clear that NHE-3 is the isoform responsible for apical membrane Na^+/H^+ exchanger activity (Amemiya *et al.*, 1995). Furthermore, in renal proximal tubules, transport of phosphate (Pi) through the proximal apical membrane is largely performed by the type II Na-Pi cotransporter (NaPi-2) (Biber *et al.*, 1996). The Na,K-ATPase in the basolateral membrane of proximal tubule cells (Kashgarian *et al.*, 1985) is essential for the efficient proximal sodium reabsorption. Concentration of the urine requires establishment and maintenance of a hypertonic medullary interstitium. The loop of Henle generates a high osmolality in renal medulla via the countercurrent multiplier, which is dependent on the NaCl reabsorption by the thick ascending limb (TAL). Furthermore, urinary dilution is also dependent on NaCl reabsorption in the thick ascending limb and distal convoluted tubule. The apically expressed Na-K-2Cl cotransporter (rat type 1 bumetanide-sensitive cotransporter BSC-1 or NKCC2 (Xu *et al.*, 1994, Kaplan *et al.*, 1996, Ecelbarger *et al.*, 1996), NHE-3, and the Na,K-ATPase are key components responsible for sodium reabsorption by the thick ascending limb. The Na-Cl cotransporter (thiazide-sensitive Na-Cl cotransporter: TSC) is expressed in the apical plasma membrane of distal convoluted tubule of the kidney (Obermuller *et al.*, 1995) and is also responsible for sodium and chloride reabsorption in this segment (Kim *et al.*, 1998).

Dysregulation of aquaporins and sodium transporters have been shown to be involved in multiple hereditary and acquired water balance disorders including nephrogenic diabetes insipidus (NDI). Most cases of acquired NDI are due to metabolic disorders (hypokalemia, hypercalcemia), drugs (e.g., lithium), acute and chronic renal failure, or postobstructive diuresis. In principle, aquaporin and sodium transporter expression might be decreased (and be a causative factor in the poluria) or increased (in an attempt to compensate for some other defect in the concentrating mechanism) in NDI.

2. LITHIUM-INDUCED NDI

The most common drug-induced form of NDI is due to lithium treatment. In the western world, about one American in 1000 is on lithium treatment

(Peet *et al.*, 1993), and as many as 50% of these patients will have a significant urinary concentrating defect (Boton *et al.*, 1987). In a rat model of this condition, plasma lithium levels in the therapeutic range are associated with a profound polyuria. Previously, it had been assumed that this reflected an impaired ability of vasopressin to stimulate adenylate cyclase (Christensen *et al.*, 1985). Marples et al., used a rat model of lithium-induced NDI, in which lithium was administered orally for several weeks inducing very severe polyuria and NDI in rats (Kim *et al.*, 1998). Direct studies demonstrated that a smaller fraction of the AQP2 was found in the plasma membrane in lithium treated animals than in controls, despite their polyuria, and presumed dehydration, but the most striking finding was that there was a 95% decrease in total AQP2 levels in the cells (Marples *et al.*, 1995). Clearly such a decrease in AQP2 would have a profound effect on the ability of the cells to mount an antidiuretic response, and may explain at least partly the polyuria seen in these animals.

In addition to downregulation of AQP2, it may be possible to speculate that 1) lithium-induced NDI may be associated with significant alterations in the expression of collecting duct water channels AQP3, AQP4 and of proximal nephron water channel AQP1; 2) lithium-induced NDI may be associated with marked alterations in the expression of major renal sodium transporters. The altered expression of sodium transporters may be, at least partly, responsible for the urinary concentrating defects, or may reflect the secondary changes made by the trials for the preservation of water and sodium against marked polyuria and dehydration; and 3) changes in expression of sodium transporters may influence the transcellular lithium transport and absorption, and hence the intracellular accumulation of lithium.

Recently, Kwon et al., have demonstrated that the expression of collecting duct water channels AQP2 and AQP3 was severely reduced following 4 weeks of lithium treatment and this was associated with an impairment of urinary concentration (Kwon *et al.*, 2000 in press). In contrast to AQP2 and AQP3, no major reduction in AQP4 expression was observed by immunoelectron microscopy and the expression of proximal nephron water channel AQP1 was not decreased but maintained at the control levels in the lithium-induced polyuric condition. Moreover, Li-treated rats had a marked increase in urinary sodium excretion rates. However, the expression of several major Na^+ transporters in the proximal tubule, loop of Henle, and distal convoluted tubule were not reduced by the chronic Li therapy, except NaPi-abundance levels (52 ± 5%, $P<0.05$). Therefore, downregulation of AQP2 and AQP3 appears to play a crucial role for the development of Li-induced polyuria. In contrast, the increased or unchanged expression of

NHE-3, BSC-1, Na,K-ATPase and TSC indicates that these Na^+ transporters do not participate in the development of Li-induced polyuria.

3. HYPERCALCEMIA

It is well established that oral administration of dihydrotachysterol causes hypercalcemia in rats. This is associated with a significant increase in total plasma calcium levels, as well as marked functional changes in urine output and urine osmolality. It has been shown that AQP2 expression levels were significantly reduced in this hypercalcemia-induced acquired NDI (Earm *et al.,* 1998, Sands *et al.,*. 1998). Earm et al., and Sands et al., also demonstrated that intracellular trafficking of the AQP2 from intracellular vesicles to the apical plasma membranes was markedly reduced in chronically hypercalcemic rats (Sands *et al.,* 1998, Earm *et al.,* 1998).

It is also known that chronic hypercalcemia is associated with a significant reduction in medullary solute content, particularly the NaCl contents (Levi *et al.,* 1983, Berl 1987). Therefore, it may be speculated that reduced expression of major renal sodium transporters may play a critical role in the impairment of tubular sodium and water reabsorption as well as in the impairment of the urinary concentration in rats with hypercalcemia. Thus, in addition to the changes in expression of renal aquaporins, changes in expression of renal sodium transporters may play a fundamental role in disturbances in extracellular fluid volume regulation seen in hypercalcemia.

Recent studies (Wang *et al.,* 2000 abstract) has shown a significant decrease in AQP2 and phosphorylated-AQP2 (p-AQP2) abundance in inner medulla of hypercalcemic rats to 52 % and 36% of control levels, respectively. Inner medullary AQP3 abundance was also decreased to 45% of control levels and in addition whole kidney AQP1 abundance was decreased to 60%. Moreover, semiquantitative immunoblottings revealed significant reductions in the abundance of two sodium transporters abundance: NaPi-II in proximal tubules (25%) and BSC-1 in thick ascending tubules (36%) of hypercalcemic rats. In contrast, there was a significant increase in the abundance of the distal convoluted tubule NaCl transporter TSC (272%).

Therefore, hypercalcemia causes substantial decrease of AQP2 and AQP3 in inner medullary collecting ducts as well as a reduced expression in p-AQP2 that is phosphorylated in the PKA phosphorylation consensus site (serine 256) of AQP2, in association with the progressive development of polyuria and decreased urinary osmolality. This strongly suggests that dysregulation of aquaporins play important roles in the development of polyuria associated with hypercalcemia. Also, increased urinary sodium excretion in hypercalcemia may be a result of reduced expression of several renal sodium transporters including NaPi-2 and BSC-1. The decreased expression of sodium transporter BSC-1 in thick ascending limb indicated potential defects in the establishment of driving forces for the countermultiplication. In contrast, the increased TSC expression in distal convoluted tubule showed a obvious compensation in response to the marked polyuria and increased urinary sodium excretion. The decreased NaPi-II expression may reduce sodium reabsorption and augment phosphate excretion in the proximal tubule

4. HYPOKALEMIA

Similar to the lithium treatment, hypokalemia causes a decrease in the urinary concentrating ability. Marples et al., demonstrated that hypokalemia results in a significant decrease in AQP2 expression in both cortex and inner medulla of rat kidney, in parallel with the development of polyuria. It was also showen that AQP2 expression is not affected by the loop diuretic furosemide, which ablates medullary hypertonicity, hence an effect of interstitial osmolality or tubule flow on AQP2 expression was excluded (Marples *et al.,* 1996).

5. POSTOBSTRUCTIVE POLYURIA

This will be covered in a parallel review. Briefly, it is now well established that unilateral or bilateral ureteral obstruction and release of obstruction is associated with severe dysregulation of renal aquaporins and sodium transporters in parallel with a marked decrease in urinary concentrating ability (Frokiaer *et al.,* 1997, Nielsen *et al.,* 1999).

6. SUMMARY

Reduced or altered aquaporins and sodium transporters expression in acquired NDI (e.g., lithium treatment, hypercalcemia, hypokalemia and ureteral obstruction) strongly suggest that aquaporins and sodium transporters may play critical roles in the development of polyuria.

ACKNOWLEDGMENTS

The work cited from our laboratory was supported partly by grants from EU-Commission (TMR and Biotech programs), Karen Elise Jensen Foundation, Danish Research Council.

REFERENCES

Agre P Bonhivers M and Borgnia MJ 1998 The aquaporins blueprints for cellular plumbing systems *J Biol Chem* **273**: 14659-14662

Amemiya M Loffing J Lotscher M Kaissling B Alpern RJ and Moe OW 1995 Expression of NHE-3 in the apical membrane of rat renal proximal tubule and thick ascending limb *Kidney Int* **48**: 1206-1215

Berl T 1987 The cAMP system in vasopressin-sensitive nephron segments of the vitamin D-treated rat *Kidney Int* **31**: 1065-1071

Biber J Custer M Magagnin S Hayes G Werner A Lotscher M Kaissling B and Murer H 1996 Renal Na/Pi-cotransporters *Kidney Int* **49**: 981-985

Boton R Gaviria M and Batlle DC 1987 Prevalence pathogenesis and treatment of renal dysfunction associated with chronic lithium therapy *Am J Kidney Dis* **10**: 329-345

Chou C L Knepper MA A Hoek N Brown D Yang B Ma T and Verkman AS 1999 Reduced water permeability and altered ultrastructure in thin descending limb of Henle in aquaporin-1 null mice *J Clin Invest* **103**: 491-496

Christensen S Kusano E Yusufi AN Murayama N and Dousa TP 1985 Pathogenesis of nephrogenic diabetes insipidus due to chronic administration of lithium in rats *J Clin Invest* **75**: 1869-1879

Earm JH Christensen BM Frokiaer J Marples D Han JS Knepper MA and Nielsen S 1998 Decreased aquaporin-2 expression and apical plasma membrane delivery in kidney collecting ducts of polyuric hypercalcemic rats *J Am Soc Nephrol* **9**: 2181-2193

Ecelbarger CA Terris T Frindt G Echevarria M Marples D Nielsen S and Knepper MA 1995 Aquaporin-3 water channel localization and regulation in rat kidney *Am J Physiol* **269**: F663-F672

Ecelbarger CA Terris J Hoyer JR Nielsen S Wade JB and Knepper MA 1996 Localization and regulation of the rat renal Na(+)-K(+)-2Cl- cotransporter BSC-1 *Am J Physiol* **271**: F619-F628

Frokiaer J Christensen BM Marples D Djurhuus JC Jensen UB Knepper MA and Nielsen S 1997 Downregulation of aquaporin-2 parallels changes in renal water excretion in unilateral ureteral obstruction *Am J Physiol* **273**: F213-F223

Kaplan MR Plotkin MD Lee WS Xu ZC Lytton J and Hebert SC 1996 Apical localization of the Na-K-Cl cotransporter rBSC1 on rat thick ascending limbs *Kidney Int* **49**: 40-47

Kashgarian M Biemesderfer D Caplan M and Forbush B III 1985 Monoclonal antibody to NaK-ATPase: immunocytochemical localization along nephron segments *Kidney Int* **28**: 899-913

Kim GH Masilamani S Turner R Mitchell C Wade JB and Knepper MA 1998 The thiazide-sensitive Na-Cl cotransporter is an aldosterone-induced protein *Proc Natl Acad Sci USA* **95**: 14552-14557

Kwon T-H Laursen UH Marples D Maunsbach AB Knepper MA Frokiaer J Nielsen S 2000 Altered expression of renal aquaporins and Na+ transporters in rats with lithium-induced nephrogenic diabetes insipidus *Am J Physiol*

Levi M Peterson L and Berl T 1983 Mechanism of concentrating defect in hypercalcemia Role of polydipsia and prostaglandins *Kidney Int* **23**: 489-497

Marples D Christensen S Christensen EI Ottosen PD and Nielsen S 1995 Lithium-induced downregulation of aquaporin-2 water channel expression in rat kidney medulla *J Clin Invest* **95**: 1838-1845

Marples D Frokiaer J Dorup J Knepper MA and Nielsen S 1996 Hypokalemia-induced downregulation of aquaporin-2 water channel expression in rat kidney medulla and cortex *J Clin Invest* **97**: 1960-1968

Nielsen S Chou CL Marples D Christensen EI Kishore BK and Knepper MA 1995 Vasopressin increases water permeability of kidney collecting duct by inducing translocation of aquaporin-CD water channels to plasma membrane *Proc Natl Acad Sci USA* **92**: 1013-1017

Nielsen S DiGiovanni SR Christensen EI Knepper MA and Harris HW 1993 Cellular and subcellular immunolocalization of vasopressin-regulated water channel in rat kidney *Proc Natl Acad Sci USA* **90**: 11663-11667

Nielsen S Kwon TH Christensen BM Promeneur D Frokiaer J and Marples D 1999 Physiology and pathophysiology of renal aquaporins *J Am Soc Nephrol* **10**: 647-663

Nielsen S Pallone T Smith BL Christensen EI Agre P and Maunsbach AB 1995 Aquaporin-1 water channels in short and long loop descending thin limbs and in descending vasa recta in rat kidney *Am J Physiol* **268**: F1023-F1037

Obermuller N Bernstein P Velazquez H Reilly R Moser D Ellison DH and Bachmann S 1995 Expression of the thiazide-sensitive Na-Cl cotransporter in rat and human kidney *Am J Physiol* **269**: F900-F910

Peet M and Pratt JP 1993 Lithium Current status in psychiatric disorders *Drugs* **46**: 7-17

Sands JM Flores FX Kato A Baum MA Brown EM Ward DT Hebert SC and Harris HW 1998 Vasopressin-elicited water and urea permeabilities are altered in IMCD in hypercalcemic rats *Am J Physiol* **274**: F978-F985

Wang W Kwon TH Li C Knepper MA Frøkiær J and Nielsen S 2000 Altered expression of renal aquaporins and Na-transporters in the rats with vitamin D-induced chronic hypercalcemia Abstract in MIP2000 Goteborg Sweden

Xu JC Lytle C Zhu TT Payne JA Benz E Jr and Forbush B III 1994 Molecular cloning and functional expression of the bumetanide-sensitive Na-K-Cl cotransporter *Proc Natl Acad SciUSA* **91**: 2201-2205

4

MOLECULAR BIOLOGY AND PHYSIOLOGY OF WATER AND SOLUTE TRANSPORT IN PLANTS

4.1 Plant physiology

MODULATION OF AQUAPORIN GENE EXPRESSION IN *ARABIDOPSIS* LEADS TO ALTERED MEMBRANE WATER PERMEABILITY

[1]François Barrieu, [2]Raphael Morillon and [2]Maarten J. Chrispeels
[1]Laboratoire de Biologie Cellulaire et Moléculaire du Développement des Plantes, Université de Bordeaux I, Avenue des Facultés, 33405, Talence, France [2]Department of Biology, University of California San Diego, La Jolla, CA 92093-0116 USA. E-mail: mchrispeels@ucsd.edu

1. INTRODUCTION

A single plant contains at least thirty different MIP genes, most of which appear to be expressed under normal growth conditions. The expression of some may be linked to specific physical stresses such as water deficit. The proteins are found in the tonoplast (vacuolar membrane) and the plasma membrane. It is not known whether all these genes encode aquaporins (or aqua-glyceroporins) because the activities of many have not yet been tested. On a phylogenetic tree we can distinguish the tonoplast aquaporins (TIPs) from the plasma membrane aquaporins (PIPs) and among the latter there are different sub-families. The activity of putative aquaporins has been tested almost exclusively in heterologous systems, primarily in *Xenopus* oocytes. *In planta* function must be inferred from such tests. TIPs appear to always be very active when expressed in oocytes, and among the PIPs, the PIP2 family is active, but the PIP1 family is largely inactive. This inactivity is not caused by the failure of these proteins to be transported to the plasma membrane (see Chaumont *et al.*, this volume) but could be due to the inability of oocytes to activate PIPs. Such findings point up the need to assay aquaporin activity *in planta*. Because plants have so many aquaporin genes, single knockouts are probably not a particularly effective way to understand function. However, using an antisense construct, it is possible to downregulate an entire family of sequence related genes. At this time, only the study by Kaldenhoff *et al.* (1998) in which the PIP1 family of *Arabidopsis* was down regulated by antisense expression addresses the issue of *in planta* function of aquaporins. We therefore decided to create transgenic plants that overexpress or underexpress plasma membrane aquaporins of the very active PIP2 family.

2. METHODS

The entire coding region of the RD28 cDNA (Yamaguchi-Shinozaki *et al.*, 1992), member of the PIP2 family, was amplified using the polymerase chain reaction procedure and cloned in an antisense or sense orientation between the doubly-enhanced cauliflower mosaic virus 35S promoter and the nopaline synthase gene termination sequences of the binary plasmid pJIT60. *In planta* transformation of *A. thaliana* was carried out with *Agrobacterium tumefaciens*. T1 seeds were collected and sown on sterile media containing 50 µg/mL kanamycin to select the transformants. Surviving T1 plantlets were transferred to soil to set seeds (T2). The segregation frequency of the T2 generation with regard to kanamycin resistance was determined on selective media. T2 plants were assumed to be homozygous for the transgene if all of 50 to 100 progeny seedlings were kanamycin resistant. Seeds (T3) of transgenic lines segregating for kanamycin resistance in a Mendelian ratio of 3:1 typical for a single integration locus were collected and again sown on selective media.

For each generation, we also checked the abundance of RD28 and related aquaporins by immunoblotting of leaf extracts. The serum used for these experiments is likely to recognize not only RD28 but also the other members of the *Arabidopsis* PIP2 family.

Protoplasts were isolated from leaves of the T4 plants and the osmotic water permeability (P_{os}) of the protoplasts was determined as described by Ramahaleo *et al.* (1999). The technique is illustrated in figure 1A. To change the medium surrounding a protoplast that is attached to a pipette, one moves the vessel (not the protoplast) to the right, so that the protoplast finds itself in a new medium. The position of the pipette, the protoplast and the microscope remain the same, and a video recorder is used to image the swelling (or shrinking) of the protoplast. This allows one to calculate the P_{os} from the swelling rate. The record obtained for two protoplasts with vastly differing swelling rates is shown in figure 1B.

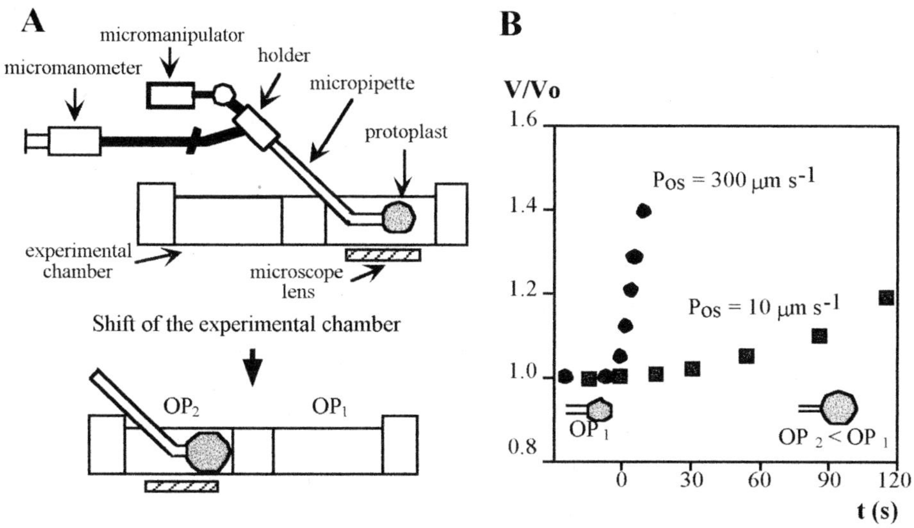

Figure 1. A) Experimental system. The movement of the experimental chamber to the right allows one to expose a protoplast that is in the OP_1 solution (400 mmol kg^{-1}) to the OP_2 solution (200 mmol kg^{-1}). B) Curves of volume variation for two different protoplasts. The osmotic water permeability of each protoplast is indicated on the graph.

Table 1. Osmotic water permeability of leaf protoplasts from WT, antisense and overexpressing plants of the RD28 aquaporin. For one row, each number represent the number of protoplast having this range of water permeability (first column).

P_{os} Range ($\mu m\ s^{-1}$)	Antisense	WT	Overexpressors
0.625 – 1.25	3	0	0
1.25 – 2.5	12	0	0
2.5 – 5	9	1	0
5 – 10	6	7	3
10 – 20	0	8	10
20 – 40	0	2	6
40 – 80	0	3	4
80 – 160	0	4	1
160 – 320	0	2	1
320 – 640	0	3	3
640 – 1280	0	0	2

3. RESULTS

Surprisingly, the P_{os} of individual protoplasts obtained from the leaves of WT plants spans a 100-fold range: from 5 to 500 μm s^{-1}, with most of the values around 10 to 20 μm s^{-1} (Table 1). This is in agreement with similar experiments with other plants (Ramahaleo *et al.*, 1999). The RD28 overexpressing plants showed a similar distribution of P_{os} values. On the other hand, in the antisense plants, the range is very different with most of the P_{os} values between 1.25 and 5 and a complete absence of values over 10.

A semi-quantitative estimation of the RD28 protein content of the plants shows that in the sense-expressing plants the band is 3 to 4 times more intense than the WT, and in the antisense-expressing plants, the band is about one quarter as intense as the WT (data not shown). Thus, there are substantial differences in the amount of PIP2 aquaporins in these plants.

The results allow the following tentative conclusion: overexpression of RD28 does not result in an increase in P_{os}. Water flow through plant cell membranes could be regulated in two different ways: first by changing the abundance of aquaporins and second, by altering the activity of the proteins. Our results with the sense expressing plants can then be explained by a possible downregulation of the other aquaporins located in the plasma membrane or by the negative regulation of the activity of the overexpressed RD28 proteins themselves. We know that the activity of some PIPs can be regulated *in vitro* by phosphorylation (Johansson *et al.*, 1998). Further experiments are now in progress to find out if the abundance of other PIPs is modified in the overexpressing plants or if the stability of the P_{os} is linked to the regulation of plasma membrane aquaporins activity *in planta*.

Down regulation of PIP2 expression, on the other hand, greatly reduces the P_{os} of all the highly permeable protoplasts. Whether this down regulation affects the water economy of the plants is not clear. It is quite surprising to observe that, when the antisense plants are grown in soil, they have no marked visible phenotype. These data suggest that plants can adapt to a reduced membrane water permeability, at least when grown in normal physiological conditions. Because the expression of some aquaporin genes can be linked to water deficit (Yamaguchi-Shinozaki *et al.*, 1992; Barrieu *et al.*, 1999), we are now planning additional experiments to test the ability of the antisense plants to react to water stress conditions.

REFERENCES

Barrieu F Thomas D Marty-Mazars D Chaumont F Charbonnier M and Marty F 1999 Desiccation and osmotic stress increase the abundance of mRNA of the tonoplast aquaporin BobTIP26 in cauliflower cells *Planta* **209**: 77-86

Johansson I Karlsson M Shukla VK Chrispeels MJ Larsson C and Kjellbom P 1998 Water transport activity of the plasma membrane aquaporin PM28A is regulated by phosphorylation *Plant Cell* **10**: 451-459

Kaldenhoff R Grote K Zhu JJ and Zimmermann U 1998 Significance of plasmalemma aquaporins for water-transport in *Arabidopsis thaliana Plant J* **14**: 121-128

Ramahaleo T Morillon R Alexandre J and Lassalles JP 1999 Osmotic water permeability of isolated protoplasts: modification during development *Plant Physiol* **119**: 885-896

Yamaguchi-Shinozaki K Koizumi M Urao S and Shinozaki K 1992 Molecular cloning and characterization of 9 cDNAs for genes that are responsive to desiccation in *Arabiopsis thaliana*: sequence analysis of one cDNA clone that encodes a putative transmembrane channel protein *Plant Cell Physiol* **33**: 217-224

4.2 Plant physiology

THE RESPONSE OF PLANTS TO SALINITY INVOLVES ROOT WATER CHANNELS

Micaela Carvajal, MariCarmen Martínez-Ballesta and Vicente Martínez
Dpto. Nutrición y Fisiología Vegetal. Centro de Edafología y Biología Aplicada del Segura, CSIC. P.O. Box 4195. 30080 Murcia. Spain. E-mail: mcarvaja@natura.cebas.csic.es

1. INTRODUCTION

Irrigation with saline water occurs in many semi-arid regions of the world, where it decreases the growth and yields of crop plants. There is general agreement that whole plant growth responses to salinity are multigenic and that a better knowledge of the underlying physiology is required, in order to understand why some plants are more resistant than others. It has been hypothesised that plant growth is initially inhibited by cellular responses to the osmotic effect of external salt, but in a later response, growth is inhibited by the toxic effect of excessive salt accumulation within the plant. In our experiments, we have centred our attention on the effects of NaCl on water channels and their role in the response of plants to salinity. Firstly, pepper plants were stressed with two isotonic solutions: NaCl and increased macronutrients of the Hoagland nutrient solution (HNS). Our results suggested that the putative reduction in water channel function of NaCl-treated plants did not seem to be due to the osmotic effect. In other experiments we showed that the ameliorative effect of Ca^{2+} on NaCl stressed melon plants could be related to water channel function. Also, osmotic water permeability (P_f) measurements were carried out on protoplasts treated with NaCl plus okadaic acid (an inhibitor of phosphatase activity). The results showed that the negative effect of NaCl on water channel activity was not due to a high ion concentration effect on channel pores or to the increase in osmotic pressure. We suggest that it was due to a direct action of NaCl on protein regulation. However, in other experiments with *Arabidopsis*, a reduction of mRNA expression of the aquaporin PIP1a after exposure to NaCl has been observed. Therefore, we hypothesise that there should be two different effects of NaCl on water channels, one at the protein level (gating regulation) and the other through gene expression.

2. EFFECTS OF SALINITY

Water with a high NaCl content occurs in many arid and semi-arid regions of the world, due to natural processes or the over-exploitation of aquifers. In coastal regions where horticulture is practised, there is a risk of sea-water incursions if abstraction of water, for irrigation, leads to tidal intrusion of saline water into aquifers. In the southeast of Spain, for example, of the nearly 6000 analyses of irrigation water made over the past five years, about 53% had an EC>3.5 dS m^{-1}. With that EC, there would be few moderately tolerant crops that could be grown and a considerable amount of water would need to be leached through the soil profile to reduce salinity to an acceptable level.

The general effect of salinity is to reduce the growth rate, resulting in smaller leaves, occasionally fewer leaves, and reduced height. Therefore, irrigation of crops with saline water produces an inhibition of growth and fruit yield (Greeway and Munns, 1980). The initial and primary effect of salinity, especially at low to moderate concentrations, is due to its osmotic effect (Jacoby, 1994). Depending upon the composition of the saline solution, ion toxicities or nutritional deficiencies may arise because of a predominance of a specific ion or competition between cations and anions (Berstein *et al.*, 1974). Calcium deficiency symptoms are common when the Na/Ca ratio is high in soil water (Shannon and Grieve, 1999).

2.1 Osmotic Adjustment

When plants are grown at high salinity, rapid rates of ion uptake could result in a build-up of high ion concentration in the cell walls of roots and leaves, particularly when cells reach flux equilibrium. This would have an adverse effect on the water relations of the individual cells (Greenway and Munns, 1980). The stresses imposed by salinity relate to ion composition and to ion concentration within the plants. When dissolved salt concentrations in soil solutions increase, water energy gradients decrease, making it more difficult for water and nutrients to move through root membranes and into the plant. The rate of water and solute uptake slows, but does not cease. With time, the solute-rich soil water increases ion concentration within the plant's aqueous transportation stream. This osmotic effect, encountered at the root membrane, applies at all the plant's internal membranes served by its conductive tissue (Volkmar *et al.*, 1998).

Osmotic adjustment involves the net accumulation of solutes in a cell in response to a fall in the water potential of the environment of the cell. As a consequence of this net accumulation, the osmotic potential of the cell is lowered, which in turn attracts water into the cell and tends to maintain turgor pressure. Therefore, osmotic adjustment is regarded as an important adaptation of plants to salinity. A variety of so-called compatible solutes have been identified and characterised as having roughly similar properties: a low polar charge, high solubility and large hydration shell (Paleg *et al.*, 1985). However, the production of sufficient osmotica is metabolically expensive, potentially limiting the plant production by consuming significant quantities of carbon that could otherwise be used for growth (Greenway and Munns, 1980). The alternative to producing organic osmotica is to accumulate a high concentration of ions from the external medium, for which the energetical cost is lower. Nevertheless, this causes another problem because such a high concentration of osmotic ions may interfere with normal biochemical activities within the cell (Volkmar *et al.*, 1998).

3. WATER CHANNELS

Water channels or aquaporins allow water to pass freely across cellular membranes, following osmotic or hydrostatic pressure gradients (Chrispeels and Maurel, 1994). These proteins provide an organism with the possibility to accelerate water movement across membranes, but diffusion will still occur in parallel. Furthermore, the ability to increase or decrease the water permeability of a cell seems to justify the enormous effort in expressing large amounts of these proteins (Schäffner, 1998). Therefore, the responses to stresses in plants have been related to aquaporins. For example, in glycophytes, expression of aquaporin PIPs (plasma membrane intrinsic proteins) genes was induced by desiccation (Fray *et al.*, 1994; Guerrero *et al.*, 1990). In rice, an initial decline of aquaporin expression after water stress, followed by up-regulation after one day has been observed (Malz and Sauter, 1999). For *Mesembrianthemum cristalinum* and *Nicotiana excelsior* expression of aquaporin genes was down regulated for the first 30 h after salt stress (Yamada *et al.*, 1995, 1997).

Although the hydraulic conductivity of tissues could be regulated by changing the level of specific aquaporins, regulation could also occur by changing the activity of the proteins. In our previous work (Carvajal *et al.*,

1999), pepper plants were stressed with two isotonic solutions: NaCl 30 mM and increased macronutrients (6x) of the Hoagland nutrient solution (HNS). Root hydraulic conductance, L_0, was always less in NaCl plants than in controls and HNS plants. In control and HNS plants, $HgCl_2$ treatment (50 μM) caused a sharp decline in L_0, to values similar to those of NaCl-stressed roots, but L_0 was restored by treating with 5 mM DTT. However, in NaCl roots, only a slight effect of Hg^{2+} and DTT was observed. The results suggest that the putative reduction in water-channel function of NaCl-treated plants did not seem to be due to the osmotic effect. Therefore, we suggested that salinity possibly affects water channel function in pepper plant roots by both gene expression and by biochemical changes in the water channel protein, i.e. phosphorylation, which has been reported to be a mechanism for molecular gating of aquaporin (Johannson *et al.*, 1996).

Light microscopy is a reliable method for determining individual cell membrane properties. It is based on the recording of variations cell dimensions with time, induced by an osmotic change in the external medium. In our work, we used light microscopy for measuring osmotic water permeability (P_f) in protoplasts treated with 90 mM NaCl for 1h. The calculated P_f declined strongly with NaCl, which is in agreement with the results obtained for the measurements of L_0. The evidence that most of the water crossed the plasma membrane through aquaporins in control protoplasts came from the observation of the decrease of P_f by around 85% after the Hg application (Carvajal *et al.*, 2000). Okadaic acid is an inhibitor of protein phosphatase, an enzyme that works in opposition to protein kinase. It has been reported that a protein phosphatase closed water channels and was activated when spinach plants grew under water deficiency (Johansson *et al.*, 1998). In our work (Martínez-Ballesta *et al.*, in press), application of okadaic acid to control protoplasts did not have any effect on P_f, probably due to the water channels in the protoplast plasma membrane remaining in their natural open state. When 90 mM NaCl was added, P_f decreased. This effect could be a consequence of an inhibition of the protein kinase or an activation of the phosphatase. NaCl addition to protoplasts treated with okadaic acid showed a lower effect on P_f than when NaCl was applied alone. Therefore, a higher proportion of water channels seemed to remain open. When NaCl was added first to the protoplast suspension and okadaic acid was applied 90 min later, P_f was again less affected than in the NaCl treatment. Therefore, the Pf of okadaic acid treated-protoplasts was independent of the time of NaCl application. It has been proposed that NaCl could affect water channels by blocking their pores (E. Steudle, pers. comm.) or by the increase in osmotic potential. However, these results indicate that the inhibition of water channel activity in NaCl treated plants or protoplasts was probably due to a direct action on the proteins that regulate aquaporin opening and closing. However, in other experiments with *Arabidopsis* (Carvajal *et al.*, unpublished results), a reduction of mRNA expression of the aquaporin PIP1a after 1 d of exposure to NaCl has been observed. Therefore, we hypothesise that there should be two different effects of NaCl on water channels, one at the protein level (gating regulation) and the other through gene expression.

3.1 Calcium and Water Channels

Calcium seems to play an important role in the water transport of plants growing under salt stress, as its concentration in the nutrient solution can determine the restoration of root hydraulic conductivity (Azaizeh and Steudle, 1991). Calcium ions are well known to have regulatory roles in metabolism, and sodium ions may compete with calcium ions for membrane-binding sites. It has been hypothesised that high calcium levels can protect the cell membrane from the adverse effects of salinity (Cramer *et al.*, 1986).

Calcium and apoplastic water potential have been reported to be involved in the phosphorylation of the water channel PM28A, present in the spinach plasma membrane (Johansson *et al.*, 1996). In vitro phosphorylation of this aquaporin, due to protein kinase, was strictly dependent on submicromolar (physiologically relevant) Ca^{2+} concentrations (Johansson *et al.*, 1996). These findings agree with those reporting that *in vitro* phosphorylation of TIP is Ca^{2+} dependent (Johnson and Chrispeels, 1992) and suggest a role for cytosolic Ca^{2+} in regulating cell turgor. Therefore, calcium seems to be directly involved in the opening and closing of water channels, which is in accordance with our results. We have shown that either L_0 or P_f of melon plant roots or membranes, was positively affected by this element in saline conditions. Also, by studying the percentages of inhibition due to Hg, we could estimate the amount of active Hg-sensitive water channels, which was lower in the NaCl treatment than in NaCl+Ca plants and protoplasts. Therefore, the reduction of water channel function or gene expression observed in plants under saline stress can be partially restored by calcium. This element may act by stabilising the plasma membrane and then activating the protein kinase (Johansson *et al.*, 1996) or by directly activating the protein kinase.

4. CONCLUDING REMARKS

The discovery of water channels has changed how we envisage the control of water transport in plants. Aquaporins belong to the family of MIP proteins from which, i.e. in *Arabidopsis*, eight have been identified as aquaporins. The presence of so many actively transcribed aquaporins suggests functional differences between the proteins themselves or in their regulation. Therefore, the gene multiplicity observed may reflect demands for the control of water permeability of membranes by differential gene expression and/or by specialised function.

Although the hydraulic conductivity of tissues could be regulated by changing the abundance of specific aquaporins, regulation should also occur by changing the activity of the proteins. Therefore, evidence points towards the need to control water channel activity under stress. While the flux of water is strictly dependent on osmotic differences between opposite sides of the membrane that contains such proteins, there is control over the open and closed states of the channel, possibly by protein phosphorylation. The equilibrium between open and closed conformations seems to be affected by growth regulator-mediated signals. Therefore, differences in the regulation of aquaporin function may reflect the tolerance to stresses in plants. In consequence, if aquaporins dominate trans-membrane water movement, and can be regulated like other channels, then this introduces the possibility of control of water movement under salt stress conditions. However, we hypothesise that there should be two different effects of NaCl on water channels, one at the protein level (gating regulation) and the other through gene expression.

REFERENCES

Azaizeh H and Steudle E 1991 Effects of salinity on water transport of excised maize (*Zea mays* L) roots *Plant Physiol* **97**: 1136-1145

Berstein L Francois LE and Clark RA 1974 Interactive effects of salinity and fertility on yields of grains and vegetables *Agron J* **66**: 412-421

Carvajal M Cerdá A and Martínez V 2000 Does calcium ameliorate the negative effect of NaCl on melon root water transport by regulating aquaporin activity? *New Phytol* **145**: 439-447

Carvajal M Martínez V and Alcaraz CF 1999 Physiological function of water-channels as affected by salinity in roots of paprika pepper *Physiol Plant* **105**: 95-101

Chrispeels MJ and Maurel C 1994 Aquaporins: the molecular basis of facilitated water movement through living plant cells? *Plant Physiol* **105**: 9-13

Cramer GR Läuchli A and Epstein E 1986 Effects of NaCl and $CaCl_2$ on ion activities in complex nutrient solutions and root growth of cotton *Plant Physiol* **81**: 792-797

Fray RG Wallace A Grierson D and Lycett GW 1994 Nucleotide sequence and expression of a ripening and water stress-related cDNA from tomato with homology to the MIP class of membrane channel proteins *Plant Mol Biol* **24**: 539-543

Guerrero FD Jones JT and Mullet JE 1990 Turgor-responsive gene transcription and RNA levels increase rapidly when pea shoots are wilted Sequence and expression of three inducible genes *Plant Mol Biol* **15**: 11-26

Greenway H and Munns R 1980 Mechanism of salt tolerance in nonhalophytes *Ann Rev Plant Physiol* **31**: 149-190

Jacoby B 1994 Mechanisms involved in salt tolerance by plants In *Handbook of Plant and Crop Stress* (M Pessarakli ed) pp 97-123

Johansson I Karlsson M Shukla VK Chrispeels MJ Larsson C and Kjellbom P 1998 Water transport activity of the plasma membrane aquaporin PM28A is regulated by phosphorylation *Plant Cell* **10**: 451-459

Johansson Y Larsson C Ek B and Kjellbom P 1996 The major integral proteins of spinach leaf plasma membranes are putative aquaporins and are phosphorylated in response to Ca^{2+} and water potential *Plant Cell* **8**: 1181-1191

Johnson KD and Chrispeels MJ 1992 Tonoplast-bound protein kinase phosphorylates tonoplast intrinsic protein *Plant Physiol* **100**: 1787-1795

Malz S and Sauter M 1999 Expression of two PIP genes in rapidly growing internodes of rice is not primarily controlled by meristem activity or cell expansion *Plant Mol Biol* **40**: 985-995

Martínez-Ballesta MC Martínez V and Carvajal M 2000 Regulation of water channel activity in whole roots and in protoplasts from roots of melon plants grown under saline conditions *Aust J Plant Physiol* (in press)

Paleg LG Stewart GR and Starr R 1985 The effect of compatible solutes on proteins *Plant Soil* **89**: 46-57

Schäffner AR 1998 Aquaporin function structure and expression: are there more surprises to surface in water relations? *Planta* **204**: 131-139

Shannon MC and Grieve CM 1999 Tolerance of vegetable crops to salinity *Sci Hortic* **78**: 5-38

Volkmar KM Hu Y and Steppuhn H 1998 Physiological responses of plants to salinity: A review *Can J Plant Sci* **78**: 19-27

Yamada S Katsuhara M Kelly WB Michalowski CB and Bohnert HJ 1995 A family of transcripts encoding water channel proteins: tissue-specific expression in the common ice plant *Plant Cell* **7**: 1129-1142

Yamada S Komori T Myers PN Kuwata S Kubo T and Imaseki H 1997 Expression of plasma membrane water channel genes under water stress in *Nicotiana excelsior Plant Cell Physiol* **38**: 1226-1231

4.3 Plant physiology

CHARACTERIZATION OF PLASMA MEMBRANE MIP PROTEINS IN MAIZE

[1]François Chaumont, [1]Valérie Van Wilder, [1]Karolina Fetter, [2]François Barrieu and [2]Maarten J. Chrispeels
[1]Unité de Biochimie Physiologique, Université catholique de Louvain, Croix du Sud 2-20, 1348 Louvain-la-Neuve, Belgium; [2] Dept of Biology, University of California, San Diego, La Jolla. E-mail: E-mail: chaumont@fysa.ucl.ac.be

1. INTRODUCTION

A large number of major intrinsic protein (MIP) homologs have been identified in plant species (Tyerman *et al.*, 1999; Kjellbom *et al.*, 1999). They are classified in two main groups according to their sequence identity with MIPs localized in either the plasma membrane (PIP, plasma membrane intrinsic protein) or the vacuolar membrane (TIP, tonoplast intrinsic protein). In addition, a number of sequences together with a MIP protein from soybean (Glycine max L.), NOD26, found in the symbiosome membrane surrounding nitrogen-fixing bacteria, form a third cluster. In maize, more than 40 MIP cDNAs have been identified (Chaumont, Barrieu, Chrispeels and Jung, unpublished data) and sequence comparisons enabled us to assign a putative subcellular membrane location (TIP, PIP and NOD26-like protein). Evidence that MIPs form water channels (aquaporins) is mostly derived from experiments with heterologous expression systems such as Xenopus oocytes. Expression of aquaporins in this heterologous system increases considerably the osmotic water permeability (P_f) value of the plasma membrane. Most plant MIP homologues that have been tested for water transport activity have been found to be water channel or aquaporins. Some plant MIPs transport glycerol in addition to water. However, because the primary sequences of MIPs do not allow us to deduce which solute is being transported, functional testing is always necessary.

Maize (*Zea mays*) is an important crop that has been extensively used to study plant-water relations. We are currently determining the function and regulation of maize plasma membrane MIP proteins.

2. PLASMA MEMBRANE MIP PROTEINS CLUSTER IN TWO SEQUENCE SUBGROUPS WITH DIFFERENTIAL AQUAPORIN ACTIVITY

We are characterizing three maize plasma membrane PIPs, ZmPIP1a, 1b and 2a (Chaumont *et al.*, 2000). A comparison of the amino acid sequences with other known plant PIPs reveals that two main subgroups are observed on a dendrogram. They correspond to the PIP1 and PIP2 subfamilies, previously identified in *Arabidopsis thaliana* (Schäffner 1998). The PIP1 subfamily proteins include ZmPIP1a and ZmPIP1b and are characterized by a longer amino-terminal extension and a shorter carboxy-terminal end in comparison to the PIP2 subfamily proteins including ZmPIP2a.

2.1 ZmPIP1a, 1b and 2a have a different aquaporin activity

The biological activity of ZmPIP1a, 1b and 2a has been determined with Xenopus oocyte swelling assays (Chaumont *et al.*, 2000). When transferred into a hypotonic medium, oocytes injected with the *ZmPIP2a* cRNAs rapidly increased in volume, indicating the presence of a facilitated water-transport pathway. The P_f of the oocyte membrane increased 8-fold over the control. Surprisingly, oocytes injected with *ZmPIP1a* and *ZmPIP1b* cRNAs swell very slowly and have a P_f similar to the water-injected oocyte. Control experiments using [^{35}S] amino acids and subcellular fractionations show that this lack of activity of maize PIP1 proteins is not caused by their failure to reach the plasma membrane of the oocytes. Additionally, ZmPIP1b also does not appear to facilitate the transport of any of the small solutes tried (glycerol, choline, ethanol, urea, amino acids). A review of the literature shows that most PIP1 proteins identified in other plants have no- or a very low aquaporin activity in oocytes. Arabidopsis PIP1 proteins are the only exception.

2.2 ZmPIP1b and 2a co-operate to increase the P_f

To determine whether non-functional ZmPIP1b interacts with functional ZmPIP2a, P_f measurements were carried out in oocytes simultaneously injected with cRNA encoding ZmPIP1b and ZmPIP2a. In these experiments, we injected a low but constant amount of *ZmPIP2a* cRNA (3.125 ng) together with increasing amounts of ZmPIP1b cRNA. Figure 1 shows that P_f of oocytes co-expressing ZmPIP1b and ZmPIP2a increased significantly in comparison with P_f of *ZmPIP2a* injected oocytes and this increase was proportional to the amount of injected *ZmPIP1a* cRNA. The finding of a positive cooperative effect suggests that injecting *ZmPIP1b* cRNAs increased the translation/membrane targeting of ZmPIP2a and/or that ZmPIP1b and ZmPIP2a monomers interact in the plasma membrane to increase the P_f. The use of biochemical techniques will help to investigate this hypothesis.

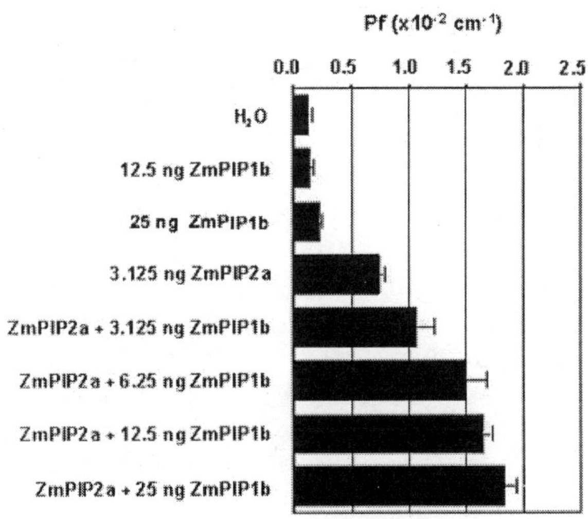

Figure 1. Osmotic water permeability of individual oocytes co-expressing ZmPIP1b and ZmPIP2a. P_f values of oocytes injected with water (H_2O) and cRNA encoding ZmPIP1b (12.5 and 25 ng), ZmPIP2a (3.125 ng) and ZmPIP2a (3.125 ng) together with ZmPIP2a (3.125, 6.250, 12.5 and 25 ng), derived from measurements of the volume variation. Data are expressed as the mean ± SE of data from 7 to 10 cells

3. SUBCELLULAR LOCATION OF ZMPIPs

The postulated plasma membrane localization of ZmPIPs relies entirely on amino acid sequence comparisons with proteins previously shown to be located in the plasma membrane and needs to be demonstrated. We fused *ZmPIP1b* and *ZmPIP2a* cDNAs to the reporter gene coding for the Green Fluorescence Protein (GFP). GFP alone and GFP linked to the tonoplast aquaporin ZmTIP1 (Chaumont *et al.*, 1998) were used as controls. These constructs were introduced into transgenic tobacco (Chaumont *et al.*, 2000) or expressed transiently in tobacco protoplasts (Figure 2). Cells were observed using a confocal microscope. Strong green fluorescent signals were observed in the plasma membrane surrounding cells expressing *ZmPIP1b::GFP* and *ZmPIP2a::GFP* (Figure 2B). A green fluorescence was also detected in the perinuclear region and transcellular strands extending from the nucleus to the plasma membrane suggesting that ZmPIP::GFP might be associated with organelles of the secretory pathway. In control cells expressing the soluble GFP, the fluorescence signal was observed in the peripheral cytoplasm and the nucleoplasm (Figure 2A). Finally, protoplasts expressing *ZmTIP1::GFP* appeared differently labeled with fluorescence signals associated with membranes that may correspond to the vacuolar membrane and vesicles (Figure 2C).

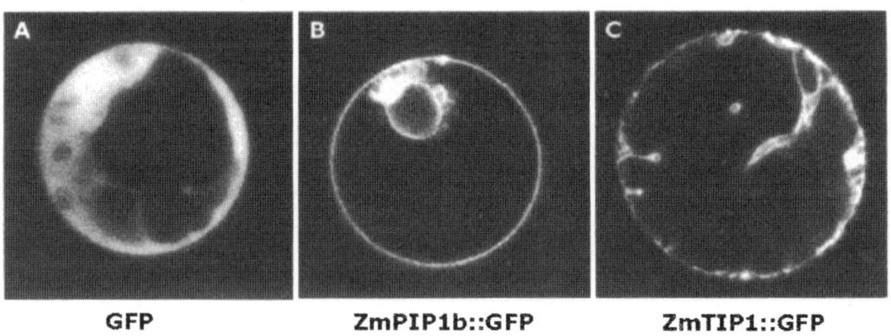

Figure 2. Localization of GFP (A), ZmPIP1b::GFP (B) and ZmTIP1::GFP (C) transiently expressed in tobacco protoplasts. Transverse sections of protoplasts were obtained using a confocal microscope. The white signals correspond to the green fluorescence labeling.

The complex patterns of ZmPIP::GFP and ZmPIP::TIP labeling could be due to the overproduction of the fusion proteins and the resulting saturation of the secretory pathway leading to a non-specific green fluorescence labeling. However, the immunocytochemistry and subcellular fractionation analysis of Arabidopsis and *Mesembryanthemum crystallinum* PIP proteins suggests a complex location and a possible association with different endomembranes (Tyerman *et al.,* 1999).

4. IN VITRO AND IN VIVO PHOSPHORYLATION OF MAIZE PIPs

Several studies indicate that some plant MIP proteins are in vitro and in vivo phosphorylated and that phosphorylation increases water channel activity in Xenopus oocytes (Kjellbom *et al.,* 1999). We are currently investigating plasma membrane MIP protein phosphorylation in maize seedlings. In vitro phosphorylation experiment using purified plasma membrane, labels a major band with an apparent molecular mass of 29 kD. The protein is phosphorylated by a magnesium- and calcium-dependent kinase associated with the membrane fraction. The nature of the phosphorylated residue(s) was determined by acid hydrolysis of the radioactive band and analysis of the amino acids by 2-D thin layer electrophoresis. The data indicate the presence of radioactive phosphoserine. Protein purification, protease digestion, Edman degradation and amino acid sequencing allowed us to identify ZmPIP isoforms migrating at the same position as the 29 kD radioactive band. We are currently identifying which ZmPIP isoform(s) is phosphorylated.

5. CONCLUSION AND PERSPECTIVES

The abundance of MIPs in plant suggests a highly regulated expression, activity and cellular location of these proteins. Most of the MIPs tested in Xenopus oocytes are aquaporins. However, some of them including maize plasma membrane ZmPIP1a and 1b appeared to be inactive water channels. The observation of a positive cooperation between active and inactive aquaporins suggests that a fine-tuning of the P_f might be realized by interaction between different PIP isoforms. Another way to regulate MIP channel activity is the phosphorylation or dephosphorylation of the proteins: ZmPIP(s) appear to be phosphorylated both in vitro and in vivo. The function of this post-translational modification remains to be elucidated. Finally, the water permeability of the plasma membrane could also be regulated by the abundance of PIPs in the plasma membrane. In the principal cells of the renal collecting duct of mammals, vasopressin regulates the targeting of vesicles containing the aquaporin AQP2 (Borgnia *et al.*, 1999). The careful analysis of ZmPIP::GFP distribution in tobacco cells under different abiotic stress conditions could help to determine if this kind of regulation exists in plants.

ACKNOWLEDGEMENTS

This work was supported by a Pioneer Hi-Bred Research Award, the Belgian National Fund for Scientific Research and the Interuniversity "Poles of Attraction" Programme-Belgian State, Prime Minister's Office-Federal Office for Scientific, Technical and Cultural Affairs.

REFERENCES

Borgnia M Nielsen S Engel A and Agre P 1999 Cellular and molecular biology of the aquaporin water channels *Annu Rev Biochem* **68**: 425-458

Chaumont F Barrieu F Herman EM and Chrispeels MJ 1998 Characterization of a maize tonoplast aquaporin expressed in zones of cell division and elongation *Plant Physiol* **117**: 1143-1152

Chaumont F Barrieu F Jung R and Chrispeels MJ 2000 Plasma membrane MIP proteins from maize cluster in two sequence subgroups with differential aquaporin activity *Plant Physiol* **122**: 1025-1034

Kjellbom P Larsson C Johansson I Karlsson M and Johanson U 1999 Aquaporins and water homeostasis in plants *Trends Plant Sci* **4**: 308-314

Schäffner AR 1998 Aquaporin function structure and expression: are there more surprises to surface in water relations? *Planta* **204**: 131-139

Tyerman SD Bohnert HJ Maurel C Steudle E and Smith JAC 1999 Plant aquaporins: their molecular biology biophysics and significance for plant water relations *J Exp Bot* **50**: 1055-1071

4.4 Plant physiology

THE PERMEABILITY OF MIPS TO GASES

[1]Gordon J.Cooper, [2]Ram Dixit, [2]June Nasrallah and [3]Walter F. Boron
[1]School of Biological Sciences, G.38 Stopford Building, University of Manchester, Oxford Rd, Manchester, M13 9PT, UK,. [2]202 Plant Science bldg, Dept of Plant Biology, Cornell University, Ithaca, NY 14853. [3]Dept of Cellular and Molecular Physiology, Yale University, 333 Cedar St, New Haven, CT 06510, USA. E-mail: gordonjcooper@hotmail.com

1. BACKGROUND

The pioneering work of Overton at the turn of the last Century provided insight into how small neutral solutes enter cells, showing that the membrane permeability of a solute is proportional to solute's oil/water partition coefficient. This principle has been extended to gases, and it is commonly believed that gases diffuse freely through the lipid phase of cell membranes. However, in the last several years, a number of observations have challenged this dogma, and is becoming clear that not only can membranes have an extremely low gas permeability, but channels in some cases appear to augment gas permeability

2. MEMBRANES WITH LOW PERMEABILITY TO GASES

An observation made by Kikeri *et al.,*. in 1989 provided the first suggestion that the movement of gases across membranes might be more complex than originally believed. The thick ascending limb of the loop of Henle (TAL) reabsorbs NaCl in preference to H_2O and so plays a crucial role in controlling urine osmolality. In the TAL, the apical membrane provides the barrier to H_2O movement as well as being impermeable to H^+ (Guggino *et al.,* 1982, Hebert and Andreoli, 1984). On exposing the *basolateral*

Molecular Biology and Physiology of Water and Solute Transport
Edited by Hohmann and Nielsen, Kluwer Academic/Plenum Publishers, New York, 2000.

membrane of the TAL to NH_3/NH_4^+, Kikeri *et al.*, (1989) found that pH_i changes in the predicted manner (i.e. an alkalinisation followed by acidification (Boron and DeWeer, 1976), indicating a membrane freely permeable to NH_3 and NH_4^+. However, exposing the *apical* membrane to NH_3/NH_4^+ failed to produce even a transient pH_i increase; pH_i simply fell. Thus, the apical uptake of NH_4^+ dominates the pH_i changes. Whether the apical influx of NH_3 was in fact zero is not clear: Any tendency for NH_3 influx to increase pH_i would have been masked by the pH_i decrease due to NH_4^+ influx, which could not be fully blocked. Nevertheless, this work provided the first hint that a cell membrane may have a very low permeability to NH_3.

In 1994, work by Waisbren *et al.*, demonstrated conclusively that membranes can have an extremely low permeability to gases. The chief and parietal cells of the gastric gland are presented with a very hostile environment, a pH as low as 0.7 and the presence of pepsin. The apical membrane of these gland cells provides the barrier and prevents the cells from digesting themselves. Single micro-dissected gastric glands were perfused in the same manner as first used in renal tubules (Burg *et al.*,, 1966). Exposing the "bath" to 1% CO_2/5mM HCO_3^- caused the expected pH_i decrease, indicating that the basolateral membrane is CO_2 permeable. However, exposing the lumen to even 100% CO_2/22 mM HCO_3^- produced *no change* in pH_i. Moreover, exposing the bath to 20 mM NH_3/NH_4^+ at pH 7.4 caused a significant pH_i increase, whereas exposing the lumen to 135 mM NH_3/NH_4^+ at pH 8.0 did not elicit a change in pH_i, even though luminal [NH_3] was 25× greater than bath [NH_3]. Thus, the apical membranes of chief and parietal cells have no demonstrable permeability to CO_2 or HCO_3^-, nor to NH_3 or NH_4^+ (Waisbren *et al.*,, 1994)

3. CONTRIBUTION OF CHANNELS TO MEMBRANE GAS PERMEABILITY

It is clear that specialised cells can construct membranes impermeable to gases, but a question that arises is whether cells can use channels to increase gas permeability? Three recent studies indicate that the movement of gases through channels can contribute to membrane gas permeability.

3.1 Aquaporin 1 is permeable to Carbon Dioxide

To investigate the hypothesis that channels might be permeable to gases Cooper and Boron (1998) examined the effect of expressing AQP1 on the

CO_2 permeability of oocytes. Figure 1A shows experimental traces from three oocytes expressing different amounts of AQP1. We see that an increase in AQP1 expression is paralleled by an increase in CO_2 permeability. The level of AQP1 expression was judged from the time taken for the oocytes to lyse when exposed to de-ionised water: the faster the lysis time the greater the expression of AQP1. The rate of CO_2-induced acidification is an index of CO_2 permeability, with a higher acidification rate indicating a higher CO_2 permeability. From Figure 1B, which summarises the data from 34 such experiments, it is clear that the correlation between lysis time and CO_2 permeability is good. These results are consistent with AQP1 being permeable to CO_2, but two other possibilities cannot be excluded: (i) the oocyte contains a native gas channel that is induced by AQP1 and (ii) over-expression of AQP1 alters the membrane lipid composition, making the membrane leakier and thus more permeable to CO_2

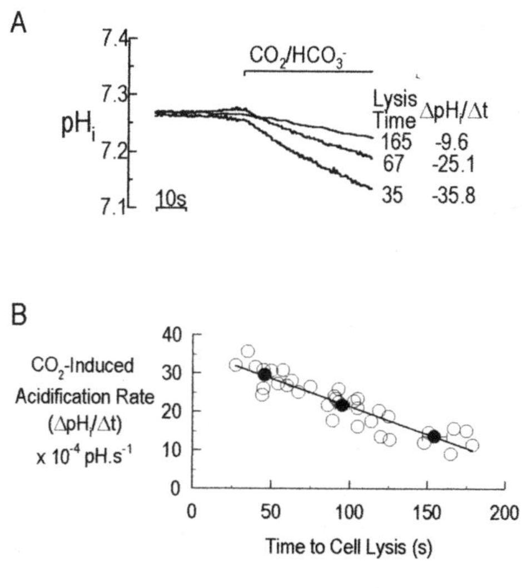

Figure 1. Effect of expressing AQP1 on oocyte CO_2 permeability. (A) Three single oocytes with increasing levels of AQP1 expression (as judged by cell lysis time in seconds when exposed to de-ionised H_2O) were exposed to CO_2 and the rate of acidification ($\times$ 10^{-4} $pH{\cdot}s^{-1}$) measured. (B) Shows the relationship between lysis time and CO_2-induced acidification rate for 34 oocytes expressing AQP1. The open circles indicate data from individual oocytes and the closed circles represent means for oocytes with lysis times of 0-60, 61-120 and 121-180s. The line plotted is the result of a linear regression [$y=(-0.144x+35.8)\times 10^{-4}$ $pH{\cdot}s^{-1}$]. Reproduced with permission of the American Physiological Society.

To distinguish among these three options, Cooper and Boron examined the effect of the mercurial pCMBS on the AQP1-dependent increase in CO_2 permeability. Mercurial agents such as $HgCl_2$ or pCMBS prevent the movement of H_2O through AQP1 (Preston *et al.*,, 1992). In a similar way, pCMBS inhibited the AQP1-dependent increase in CO_2 permeability (Figure 2A). As pCMBS does not interact with membrane lipids the third option – that AQP1 increases CO_2 permeability by altering lipid composition – can be discounted. However, the possibility that AQP1 induces a native gas channel can not as yet be ruled out. To address this issue, the authors exploited a mercurial-insensitive mutant of AQP1. Mercurials inhibit H_2O transport through AQP1 by interacting with the cysteine at residue 189, which lies close to the opening of the predicted channel pore. Indeed, the AQP1 mutant in which C189 was converted to a serine (C189S) has a water permeability that is normal but no longer sensitive to mercurials, (Preston *et al.*,, 1993). Expressing C189S increases oocyte CO_2 permeability to the same degree as does the wild type channel. However, the C189S-dependent increase in CO_2 permeability is no longer inhibited by pCMBS (summarised in Figure 2B). Thus, as well as being permeable to H_2O, the AQP1 channel also acts as a conduit for the entry of CO_2.

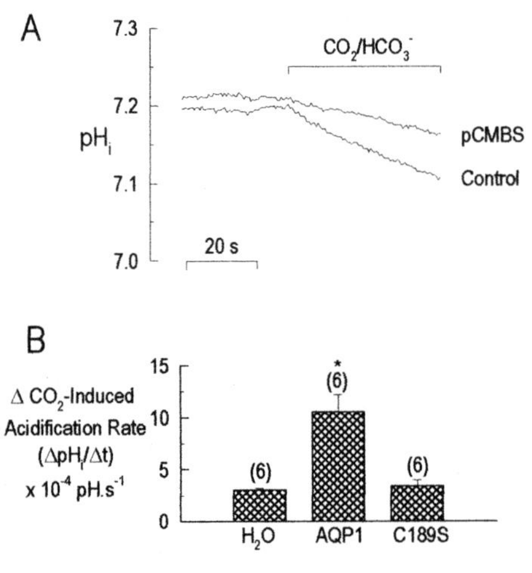

Figure 2. Effect of pCMBS on the CO_2 permeability of AQP1. (A) The two traces represent a paired experiment on a single oocyte exposed to 1.5%CO_2/10mM HCO_3^- before and after a 15 minute exposure to 1mM pCMBS. (B) summarises the changes in CO_2-induced acidification rate produced by pCMBS. Each bar represents the mean of paired differences before and after exposure to pCMBS. * indicates that the change in rate was significantly larger compared to the other 2 groups as judged by one-way ANOVA. Reproduced by permission of the American Physiological Society

3.2 The CO_2 permeability of red blood cells is inhibited by DIDS

The red blood cell membrane has an extremely high permeability to CO_2, a property that goes hand in hand with the major role of these cells (i.e., the transport of respiratory gases in the blood). The stilbene derivative DIDS, a compound commonly used as an inhibitor of HCO_3^- transporters, inhibits the CO_2 transport of red blood cells by about 90% (Forster et al.,, 1998). However the inhibitory action of DIDS on red blood cell CO_2 permeability does not involve an interaction with Band 3 (i.e., the red-cell Cl-HCO_3 exchanger) or intracellular carbonic anhydrase. Moreover DIDS does not interfere with membrane lipid composition. Thus, the authors concluded that the transport of CO_2 into the red blood cell involves a protein-mediated pathway. The water channel AQP1 is abundant in the red blood cell and it is intriguing to speculate that this protein may account for part of the protein-mediated transport of CO_2 in these cells.

3.3 The NH_3 permeability of the peribacterial membrane is protein-mediated

In legumes, biological nitrogen fixation (i.e., $N_2 + 3H_2 \rightarrow 2\ NH_3$) occurs in specialised root nodules that play host to the bacterium *rhizobia*. The *rhizobia* are enclosed in a compartment bounded by the peribacteroid membrane. The symbiotic arrangement found in the root nodules of these plants is of particular interest with respect to gas transport. High levels of O_2 inhibit the nitrogenase enzyme in rhizobia, which is responsible for nitrogen fixation. For efficient nitrogen fixation the peribacterial membrane needs to limit O_2 influx whilst maintaining a high permeability to N_2 and NH_3. Recently it was demonstrated that the about half of the movement of NH_3 across the peribacterial membrane is temperature dependent and also can be inhibited by the organic mercurial pCMBS (Niemietz, and Tyerman, 2000). Both of these properties are characteristics of a protein mediated process and the authors suggested that nodulin-26, found at high levels in the peribacteroid membrane, might account for the protein mediated fraction of NH_3 transport.

4. ARE MIPS PERMEABLE TO OTHER GASES?

The finding that AQP1 is permeable to the gas CO_2 raises the question of whether MIP proteins are permeable to other gases. In particular we have focused upon the gas NH_3 and the channels nodulin-26 and MIP-MOD.

4.1 Nodulin-26

As mentioned above, Nodulin-26 is expressed at high levels in the peribacteroid membrane of the root nodules of legumes. Although nodulin-26 is permeable to H_2O and certain other small solutes (Rivers *et al.*,, 1997), its precise role has yet to be determined. We believe that nodulin-26 may facilitate the movement of NH_3 and NH_4^+ across the peribacterial membrane.

4.2 MIP-MOD

In the self-incompatibility response of *Brassica campestris* (the field mustard plant), epidermal cells of the stigma (located at the surface of the female reproductive structure) recognize self-related pollen, resulting in the arrest of self-pollen tube development at the stigma surface. The *MIP-MOD* gene is thought to be involved in orchestration of the self-incompatibility response, because low expression of the gene correlates with the breakdown of the stigma self-incompatibility response (Ikeda *et al.*,, 1997). However, the precise role of the MIP-MOD protein is unknown, and it is possible that MIP-MOD transports pollen-inhibitory substance(s) to the stigma epidermal surface or removes from the stigma surface substance(s) needed for normal pollen-tube development. Characterisation of the transport properties of MIP-MOD is hence essential for understanding the nature of the participation of MIP-MOD in the self-incompatibility response.

4.3 Effect of nodulin-26 and MIP-MOD on the permeability of oocytes to NH_4^+ and NH_3

When control H_2O-injected oocytes are exposed to a Na^+-free solution at pH_o 8.0 containing 20 mM NH_3/NH_4^+ (19.5 mM NH_4^+ and 0.5 mM NH_3), pH_i briefly decreases (consistent with NH_4^+ influx) and then slowly increases (consistent with NH_3 influx). In oocytes expressing nodulin-26, the same manoeuvre causes a rapid and sustained acidification, suggesting that nodulin-26 is permeable to NH_4^+. Because the large NH_4^+ induced acidification could be masking any changes in pH_i associated with increased NH_3 influx, we raised the extracellular pH to 10, and lowered NH_3/NH_4^+ to 2 mM (0.5 mM NH_4^+ and 1.5 mM NH_3). In H_2O-injected oocytes, adding NH_3/NH_4^+ under these conditions causes a sustained acidification (indicative of NH_4^+ influx). However, in nodulin-26 expressing oocytes, the initial pH_i decrease is followed by a large increase, indicating an increase in NH_3 permeability.

In oocytes expressing MIP-MOD, exposure to NH_3/NH_4^+ under the high-pH conditions yielded results similar to those observed in nodulin-26-expressing oocytes (i.e., a decrease in pH_i, followed by a large increase). Thus, MIP-MOD also increases oocyte NH_3 permeability.

Because mercurial agents reduce the permeability of many MIP proteins, we examined the effect of the pCMBS on the increased NH_3 permeability of oocytes expressing MIP-MOD. We pre-incubated oocytes in pCMBS, and then exposed them to NH_3/NH_4^+ solutions at high pH. Whereas pre-incubation in pCMBS had no effect upon the pH changes observed in H_2O-injected oocytes, in oocytes expressing MIP-MOD, pCMBS abolished the secondary increase in pH_i and the cells acidified to the same extent as the H_2O-injected oocytes.

These results indicate that the MIP-MOD-dependent increase in membrane NH_3 permeability is due to permeation through a channel, and are consistent with our hypothesis that MIP-MOD, and maybe also nodulin-26, are permeable to the gas NH_3.

In the current study we have illustrated that the presence of MIPs can increase membrane permeability to both NH_4^+ and NH_3. The work supports the hypothesis that the gas NH_3 can cross cell membranes by a channel mediated-process. The ability of the mercurial pCMBS to prevent the MIP-dependent increase in NH_3 movement suggests that NH_3 is crossing the cell membrane by passage through these channels.

5. CONCLUSIONS

The most important implication from the recent studies on the transport of gases across membranes is the finding that gas transport can occur by protein-mediated processes. We have demonstrated that the channel AQP1 is permeable to the gas CO_2, although the physiological relevance of this property still has to be discerned. We have also found that nodulin-26 and MIP-MOD increase membrane NH_3 permeability. The movement of NH_3 may play an important role in biological nitrogen fixation, whereas the movement of NH_3 through MIP-MOD could play an important role in the prevention of self pollination.

REFERENCES

Boron WF and De Weer P 1976 Intracellular pH transients in squid giant axons caused by CO_2 NH_3 and metabolic inhibitors *J Gen Physiol* **67**: 91-112

Burg M Grantham J Abramow M and Orloff J 1966 Preparation and study of fragments of single rabbit nephrons *Am J Physiol* **210**: 1293-1298

Cooper GJ and Boron WF 1998 Effect of pCMBS on the CO_2 permeability of Xenopus oocytes expressing aquaporin 1 or its C189S mutant *Am J Physiol* **275**: C1481-C1486

Forster RE Gros G Lin L Ono Y and Wunder M 1998 The effect of 44'-diisothiocyanato-stilbene-22'-disulfonate on CO_2 permeability of the red blood cell membrane *Proc Natl Acad Sci USA* **95**: 15815-15820

Guggino WB Stanton BA and Giebisch G 1982 Regulation of apical potassium conductance in the isolated early distal tubule of the *Amphiuma* kidney *Biophys J* **37**: 338-338

Hebert SC and Andreoli TE 1984 Control of NaCl transport in thick ascending limb *Am J Physiol* **246** F745-F756

Ikeda S Nasrallah JB Dixit R Preiss S and Nasrallah ME 1997 An aquaporin-like gene required for the Brassica self-incompatibility response *Science* **276**: 1564-1566

Kikeri D Sun A Zeidel ML and Hebert SC 1989 Cell membranes impermeable to NH_3 *Nature* **339**: 478-480

Niemietz CM and Tyerman SD 2000 Channel-mediated permeation of ammonia gas through the peribacteroid membrane of soybean nodules *FEBS Lett* **465**: 110-114

Preston GM Carroll TP Guggino WB and Agre P 1992 Appearance of water channels in Xenopus oocytes expressing red cell CHIP28 protein *Science* **256**: 385-387

Preston GM Jung JS Guggino WB and Agre P 1993 The mercury-sensitive residue at cysteine 189 in the CHIP28 water channel *J Biol Chem* **268**: 17-20

Rivers RL Dean RM Chandy G Hall JE Roberts DM and Zeidel ML 1997 Functional analysis of nodulin 26 an aquaporin in soybean root nodule symbiosomes *J Biol Chem* **272** 16256-16261

Waisbren SJ Geibel J Modlin IM and Boron W 1994 Unusual permeability properties of gastric gland cells *Nature* **368**: 332-335

4.5 Plant physiology

PHYSIOLOGICAL AND GENETIC ANALYSIS OF PLASMA MEMBRANE AQUAPORIN FUNCTIONS IN *ARABIDOPSIS*

[1]Patricia Gerbeau, [1]Hélène Javot, [1]Virginie Lauvergeat, [2]Fabrice Martin-Laurent, [2]Josette Güçlü, [1]Véronique Santoni, [1]Colette Tournaire, [3]Pierre Ripoche and [1]Christophe Maurel
[1]Biochimie et Physiologie Moléculaire des Plantes, AGRO-M/INRA/CNRS UMR 5004, 2, place Viala, 34 060 Montpellier Cedex 01, France; [2]Institut des Sciences Végétales, CNRS, Avenue de la Terrasse, 91 198 Gif-sur-Yvette Cedex, France; [3]Service de Biologie Cellulaire, Département de Biologie Cellulaire et Moléculaire, CEA Saclay, 91 191 Gif-sur-Yvette Cedex, France. E-mail: gerbeau@ensam.inra.fr

1. INTRODUCTION

Terrestrial plants critically depend on a supply of water for their growth and development and display elaborate strategies to achieve water balance in the most adverse environments. The recent discovery in a large variety of organisms of water channel proteins named aquaporins has provided tremendous insights into the molecular mechanisms of transmembrane water transport and its regulation at the cellular level. Aquaporins belong to the large MIP family of membrane proteins, with members transporting water and/or small neutral solutes. Despite decades of physiological and biophysical research on water relations of plants, the discovery of aquaporins in both the intracellular and plasma membranes of plant cells was unexpected. This discovery revealed that the water and solute permeability of cell membranes may be more critical for plants than initially thought. The importance of aquaporins in plants is supported by their abundance, diversity and fine regulation at the gene and/or protein levels. Recent genetic and physiological evidence point to a crucial role for aquaporins in various processes including plant reproduction, cell elongation, root development and water uptake, and the adaptative responses of plants to dehydration and salinity.

Molecular Biology and Physiology of Water and Solute Transport
Edited by Hohmann and Nielsen, Kluwer Academic/Plenum Publishers, New York, 2000.

Studies performed in our laboratory aim at a comprehensive understanding of the role of cellular membranes in the water relations of higher plants, and should assess in particular the importance of aquaporins in plants. For this, we combine molecular, physiological, and genetic approaches in the model plant species *Arabidopsis thaliana* to assay water transport from cloned water channel molecules to native plant membranes and organs. These approaches include (i) the molecular cloning of plant aquaporins and their structure/function analysis after expression in *Xenopus* oocytes (Maurel *et al.,* 1993, Maurel *et al.,* 1995) (ii) membrane vesicle purification and stopped-flow spectrophotometry to achieve accurate water and solute permeability measurements on well-defined subcellular membrane fractions (Gerbeau *et al.,* 1999, Maurel *et al.,* 1997) (iii) cell pressure probe techniques to measure *in situ* the whole cell hydraulic conductivity of plant cells (iv) the isolation of knock-out aquaporin mutants and their phenotypic analysis.

The present study was focused on the aquaporins of the plant plasma membrane, the so-called Plasma membrane Intrinstric Proteins (PIPs), because these proteins are supposed to play a critical role both in cell osmoregulation and in water transport across plant tissues (Maurel, 1997). We report here two types of complementary approaches to address the plasma membrane aquaporin functions.

2. CALCIUM-DEPENDENT REGULATION OF WATER CHANNELS IN THE *ARABIDOPSIS* PLASMA MEMBRANE

The plasma membrane of plant cells exhibits a large range of water permeability values depending on either the cell type or the technique used for the measurements (Steudle, 1989). For instance, plasma membrane vesicles purified from tobacco suspension cells or wheat roots exhibited a reduced water permeability together with a lack of active water channels (Maurel *et al.,* 1997, Niemetz and Tyerman, 1997) whereas evidence for active water channels in *Arabidopsis* leaf protoplasts was provided using plants expressing a *pip1b* antisense transgene (Kaldenhoff *et al.,* 1998). In the present work, we used *Arabidopsis* suspension cells to investigate the mechanisms which may allow the regulation of plant plasma membrane permeability to water. This question was first addressed at the cell level by means of pressure probe measurements. For these, suspension cells were impaled with a micropipette coupled to a mechanical device and a pressure transducer. This set up allows to primarily measure or clamp the cell turgor pressure. Water movement generated upon imposition of a sudden hydrostatic gradient can be followed through the associated pressure relaxation. This type of measurements allows the determination of the cell hydraulic conductivity (L_p). Using this technique, we found that the L_p of *Arabidopsis* suspension cells can vary up to 8-fold depending on the measuring conditions. When cells were impaled with millimolar concentrations of KCl and $CaCl_2$ in the pipette and in the bathing solution, cell L_p was reduced 4-fold as compared to control values obtained in the same medium but in the absence of calcium. In contrast, a nearly 2-fold increase in L_p was observed when chloride ions were replaced by fluoride ions, the latter presumably acting as a non specific protein phosphatase inhibitor.

To investigate these changes in permeability at the membrane level, plasma membrane vesicles were purified by aqueous two phase partitioning either in standard conditions or in the presence of NaF and chelators of divalent cations such as EDTA. The latter conditions are referred to as protected conditions. The water permeability of purified vesicles was measured by stopped-flow spectrophotometry and a low activation energy indicated the presence of active water channels. Vesicles purified in standard conditions exhibited a reduced water permeability and no water channel activity. A 5-fold higher water permeability associated to the presence of active water channels was observed in plasma membrane vesicles purified in protected conditions. Furthermore, addition of calcium to these purified membrane vesicles reduced their water permeability down to the basal level of membranes isolated in standard conditions. A half reduction in membrane water permeability was observed for a free calcium concentration in the range of 20 nM. The cellular and molecular mechanisms of this plasma membrane-delimited regulation of *Arabidopsis* water channels by calcium are under investigation.

3. REVERSE GENETIC ANALYSIS OF PLASMA MEMBRANE AQUAPORIN FUNCTIONS IN *ARABIDOPSIS*

In plants, MIP homologues define a large gene family with for instance more than 30 members in *Arabidopsis* (Kjellbom *et al.,* 1999) 13 of them being in the PIP subfamily. Six of the *Arabidopsis* PIPs (PIP1a, PIP1b, PIP1c, PIP2a, PIP2b and RD28) have been expressed in *Xenopus* oocytes and shown to function as aquaporins (Daniels *et al.,* 1994, Kammerloher *et al.*, 1994). It is likely that the other PIPs also transport water and/or small neutral solutes. The expression profile of these and other PIP homologues in different plant species has also been described in some details. For instance, it has been suggested that PIP homologues may mediate the alterations in root hydraulic conductivity observed during day/night cycles or after changes in anionic nutriment availability (Carvajal *et al.*, 1996, Henzler *et al.*, 1999). However, direct informations about the function of PIPs at the whole plant level are still scarce. Kaldenhoff *et al.*, (1998) used a *pip1c*-derived antisense construct to target several genes of the *pip1* subfamily in transgenic *Arabidopsis*. While the aerial parts of the transgenic lines were similar to those of controls, the fresh weight of antisense plant roots was increased up to 4-fold, suggesting that antisense plants compensated a reduced root L_p by an increase in root surface area.

In the present work we have also addressed the function of PIP aquaporins by a reverse genetic approach, but by using a different strategy. For this, we have isolated and characterised single *pip* gene knock-out mutants in *Arabidopsis*. We searched for aquaporin insertion mutants in a library of more than 20,000 *Arabidopsis* lines independently transformed with an *Agrobacterium tumefaciens* transferred DNA (T-DNA). This library was screened by PCR using a combination of primers specific for *pip* genes and T-DNA border sequences, respectively. The approach reported here has only recently become applicable in plant research after the acquisition of large numbers of T-DNA or transposon insertion mutant lines. This approach has important advantages over the more classical antisense and co-supression strategies. It normally yields, when an insertion within the gene coding sequence can be identified, to a knock-out mutations. In addition, it should avoid cross-interference with the expression of several homologous genes when a specific member of a multi-gene family is targeted. After single T-DNA insertion mutants have been obtained, the possibility remains to combine different mutations, for a controlled knock-out of specific members in the gene family.

Using this strategy, six T-DNA insertion mutants were identified; the targeted genes belong to the *pip1* subfamily (2 mutants) and to the *pip2* subfamily (4 mutants). Whereas some of these mutants still need to be genetically characterised, plant lines homozygous for a single T-DNA insertion within a *pip* gene could be obtained and their phenotypic characterisation is in progress. In soil, hydroponic and *in vitro* cultures were used to screen a broad range of culture conditions and environmental stresses. The parallel study of *pip* gene expression level will help in revealing specific functions for these genes. One of the *pip* mutants showed a normal growth and development when grown in soil in standard conditions. However, this mutant showed an altered root growth, as compared to the wild type, when grown *in vitro* in the presence of NaCl. These preliminary observations need to be confirmed in a second, independently isolated mutant carrying a T-DNA insertion within the same *pip* gene. They suggest a role for aquaporins in the response of plants to salinity.

If phenotypes can be resolved in these mutant plants, this approach will be helpful in deciphering the specific function of single aquaporin genes. This will surely help us in understanding the meaning of the diversity of aquaporin genes in plants.

ACKNOWLEDGMENTS

The work from our laboratory was supported by the Centre National de la Recherche Scientifique (UPR0040, ATIPE) and the Commissariat à l'Energie Atomique (Direction des Science du Vivant, segment 28).

REFERENCES

Carvajal M Cooke DT and Clarkson DT 1996 Responses of wheat plants to nutrient deprivation may involve the regulation of water-channel function *Planta* **199**: 372-381

Daniels MJ Mirkov TE and Chrispeels MJ 1994 The plasma membrane of *Arabidopsis thaliana* contains a mercury-insensitive aquaporin that is a homolog of the tonoplast water channel protein TIP *Plant Physiol* **106**: 1325-1333

Gerbeau P Güçlü J Ripoche P and Maurel C 1999 Aquaporin Nt-TIPa can account for the high permeability of tobacco cell vacuolar membrane to small neutral solutes *Plant J* **18**: 577-587

Henzler T Waterhouse RN Smyth AJ Carvajal M Cooke DT Schäffner AR Steudle E and Clarkson DT 1999 Diurnal variations in hydraulic conductivity and root pressure can be correlated with the expression of putative aquaporins in the roots of *Lotus japonicus Planta* **210**: 50-60

Kammerloher W Fischer U Piechottka GP and Schäffner AR 1994 Water channels in the plant plasma membrane cloned by immunoselection from a mammalian expression system *Plant J* **6**: 187-199

Kaldenhoff R Grote K Zhu JJ and Zimmermann U 1998 Significance of plasmalemma aquaporins for water-transport in *Arabidopsis thaliana Plant J* **14**: 121-128

Kjellbom P Larsson C Johansson I Karlsson M and Johanson U 1999 Aquaporins and water homeostasis in plants *Trends in plant Science* **4**: 308-314

Maurel C Reizer J Schroeder JI and Chrispeels MJ 1993 The vacuolar membrane protein γ-TIP creates water specific channels in *Xenopus oocytes EMBO J* **12**: 2241-2247

Maurel C Kado RT Guern J and Chrispeels MJ 1995 Phosphorylation regulates the water channel activity of the seed-specific aquaporin α-TIP *EMBO J* **14**: 3028-3035

Maurel C 1997 Aquaporins and water permeability of plant membranes *Annu Rev Plant Physiol Plant Mol Biol* **48**: 399-429

Maurel C Tacnet F Güçlü J Guern J and Ripoche P 1997 Purified vesicles of tobacco cell vacuolar and plasma membranes exhibit dramatically different water permeability and water channel activity *Proc Natl Acad Sci USA* **94**: 7103-7108

Niemetz CM and Tyerman SD 1997 Characterization of water channels in wheat root membrane vesicles *Plant Physiol* **115**: 561-567

Steudle E 1989 Water flow in plants and its coupling to other processes: an overview *Methods Enzymol* **174**: 183-225

4.6 Plant physiology

A NUCLEAR MAGNETIC RELAXATION STUDY OF WATER UPTAKE IN BARLEY KERNELS

[1]Marco L.H. Gruwel, [1]Brock Chatson, [2]Xiang Yin and [1]Suzanne Abrams
[1] National Research Council, Plant Biotechnology Institute, 110 Gymnasium Place, Saskatoon, SK, S7N 0W9 Canada. [2] Prairie Malt Ltd./Cargill Malt Americas, Box 1150, Biggar, SK, S0K 0M0 Canada. E-mail: marco.gruwel@nrc.ca

1. INTRODUCTION

Water uptake in seeds is in general a triphasic process with the first phase represented by imbibition. The second phase shows the onset of major metabolic events while the third phase is concurrent with radicle elongation. Imbibition, however, occurs in all seeds, viable or not. Hydration is clearly a necessary, but not sufficient, condition for the initiation of biological activity in seeds. The rate at which water uptake occurs should provide information on seed viability.

The localization and measurement of water uptake in barley is often obtained using invasive and destructive techniques (Proudlove *et al.*, 1990). The most commonly used techniques to monitor water uptake are based on the use of ^3H_2O, iodine staining or use Mn^{2+} as a tracer added to the steep water (Davies *et al.*, 1989). Magnetic resonance (NMR) offers the possibility to study water uptake non-invasively. NMR has been used to study water transport and distribution in plants and seeds (Chudek *et al.*, 1997, Belton *et al.*, 1985). Most often the water distribution in different plants or seeds was studied by NMR micro-imaging (Morris *et al.*, 1990, Kano *et al.*, 1990). Studies of the dynamics of water uptake are less common (Ratkovic *et al.*, 1987, Bacic *et al.*, 1992). However, the non-invasive nature of the technique allows the study of water uptake with time on the same seed or plant in order to obtain information on the dynamics of water molecules in the seeds.

Molecular Biology and Physiology of Water and Solute Transport
Edited by Hohmann and Nielsen, Kluwer Academic/Plenum Publishers, New York, 2000.

Here we describe the use of a NMR technique based on the water proton (^1H) spin-lattice relaxation rate (R_1) to monitor the dynamics of water imbibition in whole barley kernels. Modification of the quantity of water and its physicochemical state within the kernels may in turn affect the actual water uptake rate and thus the time dependence of R_1. After imbibition the second phase in the water uptake process, involving water re-distribution within the kernel, could also be detected using the R_1 measurements.

2. EXPERIMENTAL SETUP

2.1 Sample preparation

Screened Harrington barley (Hordeum vulgare, cv. Harrington) kernels were used for the imbibition study, containing 10.7% protein and 11.8% moisture by weight. Small batches of 40-50 kernels were imbibed at room temperature in distilled water using a 50 mL beaker, while shielded from air and light. At selected intervals a single kernel was removed and blotted with filter paper to remove any excess water attached to the husk. The blotted kernels were centred in a capped sample container of 2 cm length and 0.7 cm o.d. and inserted into the NMR probe, kept at room temperature. For every NMR experiment a new kernel was used and NMR relaxation time measurements were performed up to three times per imbibition time point used in the NMR experiments.

2.2 Water uptake model for the first phase

We will use the theoretical approximation of a semi-infinite system representing the barley kernel which is considered large enough as for diffusion not to completely cover this volume during the course of the experiment. Considering the time scales involved in the barley imbibition experiments described here (*vide infra*), this is a good approximation. The kernel is assumed to be surrounded by a stirred layer of fluid which is present at the interface in a constant concentration. The diffusion of water with a constant D into the kernel is accordingly described by a diffusion equation. It can be shown that the number of water molecules N(t) present in the kernel at time t is given by $N(t) \approx \sqrt{\otimes Dt})$.

2.3 Two-Phase model for the interpretation of NMR data

In a complex heterogeneous sample such as a barley kernel, the observed R_1 of water can often be described by a single exponential (Harris 1976, Packer 1977). However, in such systems water is present in different states of mobility. These different water molecules are, in most cases, in fast exchange with each other. Each *type* of water has its own typical R_1 but due to the exchange, which is assumed to be faster than the individual relaxation rates, a single exponential decay of the water longitudinal magnetisation is observed.

In general the heterogeneous water distribution is assumed to show the characteristics of a two component system. One of the components represents mobile water while the other component represents molecules which experience hindered mobility. The latter molecules can be found near the surface of polyelectrolytes or membranes. This particular two-phase model has been successfully used in the interpretation of water R_1 in various systems (Packer 1977, Woessner *et al.*, 1970, Fung *et al.*, 1974) and results in,

$$R_1(obs) = p_m R_1^{m} + (1-p_m) R_1^{lm}. \qquad (1)$$

The fraction of mobile water is represented by p_m while indices *m* and *lm* refer to the mobile and less mobile fractions while $R_1(obs)$ represents the observed relaxation rate measured in an actual experiment.

Including the number of absorbed water molecules ($N(t) \approx \sqrt{\otimes Dt})$) in the description of the observed relaxation rate results in

$$R_1(obs) = R_1^{lm} + [\sqrt{t}/(\sqrt{t} + A)]\,(R_1^{m} - R_1^{lm}), \qquad (2)$$

with A representing a constant depending on the sample specific diffusion constant D.

3. RESULTS AND DISCUSSION

The imbibition time dependence of the water spin-lattice relaxation rate R_1 is shown in Figure 1. During the initial phase of imbibition a rapid increase in R_1 was observed which levelled off at ≈ 6-8 hours of imbibition after which a slow but continuous decrease in the relaxation rate was observed up to 50 hours of imbibition. The first ≈ 6-8 hours of water uptake represents the physical absorption of water into the kernel. During this process $R_1(obs)$ increased from 1.8 s^{-1} prior to imbibition to 4.0 s^{-1} after seven hours of imbibition.

From the fit of the imbibition data to equation (2) for $R_1(obs)$, shown in the inset in Figure 1, R_1^{m} and R_1^{lm} were determined to be $10.01 \pm 3.03\ s^{-1}$ and $1.75 \pm 0.09\ s^{-1}$, respectively. The two different molecular environments used to describe water mobility in barley thus approximately differ by a factor 5.

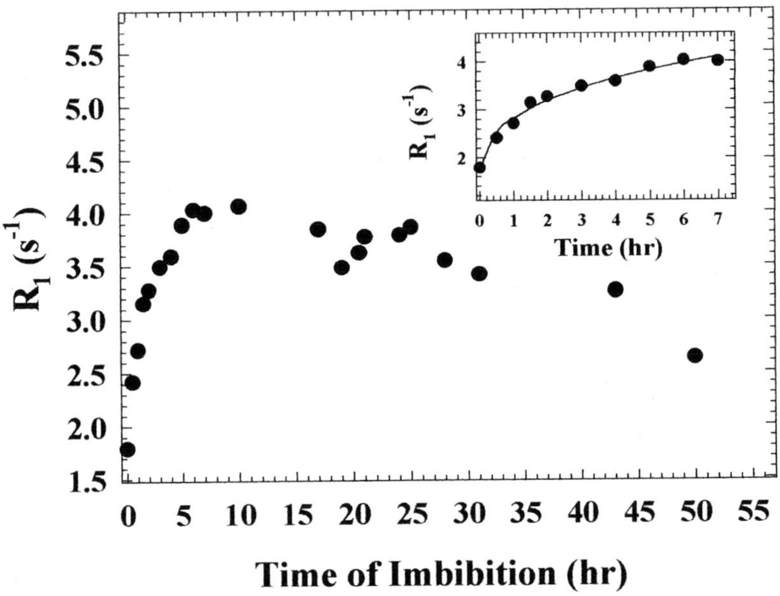

Figure 1. Imbibition time dependence of the ^1H water NMR spin-lattice relaxation rate. Each relaxation rate experiment was performed on a different kernel. The inset shows the fit of the data to the two-phase model

After ≈ 6-8 hours of imbibition, $R_1(obs)$ slowly starts to decrease while at the same time the spin-spin R_2 continues to decrease (not shown). These results indicate a continuous increase in water mobility. This increase, however, is not related to an increase in total absorbed water by the kernel. On the contrary, it is known that the total amount of water in the whole kernel after ≈ 6-8 hours of imbibition remains relatively constant (Yin *et al.*, 1996). After this initial period of rapid water uptake the kernels start to re-distribute water over the different kernel compartments. Apparently the onset of this process can clearly be followed by NMR R_1 measurements. It should be mentioned that, due to the anaerobic imbibition conditions, the last phase of the triphasic imbibition process during which radicle elongation occurs did not initiate.

Translational diffusion is a slow process in barley. Considering the values of the observed relaxation rates, translational diffusion cannot be considered the driving mechanism for relaxation. The main mechanism for relaxation appears to be provided by the rotational motion (Harris 1976). It can be shown that in pure water the rotational contribution to the observed ^1H spin-lattice relaxation rate is approximately twice as effective as the translational component. For Artemia cysts, Trantham *et al.,* (1984) observed a large reduction of the translational contribution to the total relaxation rate and an overall reduction of $R_1(Obs)$ by ≈50 times compared to pure water. A reduction in water translational motion is a key feature of many biological systems (Packer 1977).

Thus the translational freedom of water is highly restricted in barley, causing spin-lattice relaxation to be dominated by rotational motions. Additional support for this assumption is provided by water uptake in barley as measured by gravimetry and NMR imaging (Yin *et al.,* 1996). NMR micro imaging is an ideal technique to study the heterogeneity of the water distribution during imbibition. A typical NMR micro-image of a barley kernel after 6 hours of imbibition is shown on the left-hand side in Figure 2. After 6 hours of imbibition, water penetration into the kernel is still restricted to the embryo and scutellum. A small fraction of the total absorbed water occurs in the aleurone layer. Hydration of the endosperm occurs later, as shown by a typical micro-image obtained after 48 hours of imbibition, shown on the right-hand side in Figure 2. It should be noted that the images were acquired without saturation of the lipid resonance, however, the total lipid content of barley is low (≈2–4%, depending on the variety) and their contribution to the ^1H images can therefore be ignored.

Water viscosity for the two different phases can be calculated from the obtained relaxation data using a model such as the Stokes-Einstein-Debye model (SED). This model is based on the hypothesis of a continuous and homogeneous solvent and thus not directly applicable to small molecules such as water. However, the SED expression has been shown to provide a good first order approximation for many small molecules (Zwanzig *et al.,* 1985, van der Maarel *et al.,* 1987). SED relates the microscopic rotational correlation time of a probe to the macroscopic viscosity, η, of the solvent. We obtained η_m=37 mPas and η_{lm}=325 mPas.

A **B**

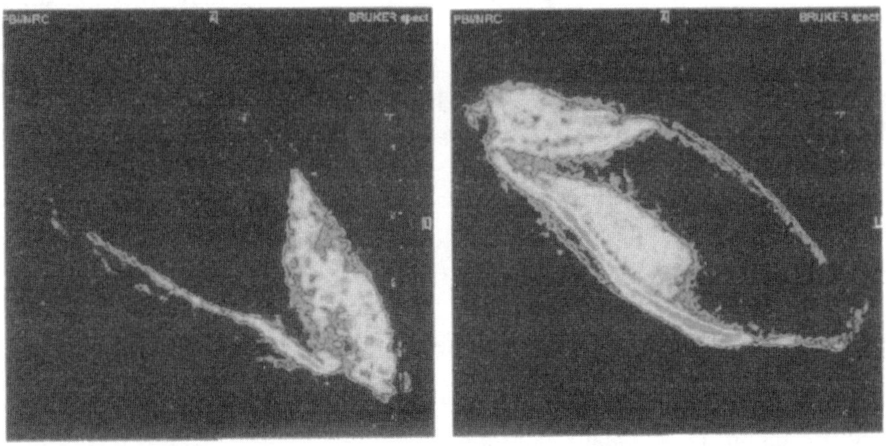

Figure 2. Typical grey-scale NMR micro-images of Harrington barley (two different kernels) obtained after 6 hours (A) and 48 hours (B) of imbibition. Both images were obtained in 17 min. with a spectral width (SW) of 100 kHz, using a Hahn echo sequence with a 2 ms SINC3 90° pulse, an echo time of 6.35 ms and a repetition time of 1.0 s. The field of view (FOV) is 1.0 x1.0 cm and consists of 256 x 256 points (39 μm pixel resolution) and a slice thickness of 0.6 mm. In total 4 scans were accumulated. Note that Figure A was slightly expanded compared to Figure B.

Comparing this with a water viscosity of 1 mPas at room temperature, a strong reduction of water mobility for the less mobile phase is clear. The actual overall water viscosity during the first phase of imbibition is given by a value in between these two limits and depends on the number of molecules in each phase, exactly as for R_1(obs). After seven hours of imbibition, η(obs) has dropped from 325 mPas to approximately 240 mPas (≈30% reduction), indicating that the less mobile water fraction, after 7 hr of imbibition, remains the dominant water fraction (≈70%). At this point the kernel has absorbed ≈24% water by weight (Yin *et al.*, 1996). Apparently imbibition of water causes not only an increase in the number of mobile water molecules but also the number of less mobile molecules.

Experiments on barley samples with different viabilities should indicate whether the constant A in eqn. (2) could be used to discriminate these viability differences.

ACKNOWLEDGEMENTS

Part of this research was funded by a grant from the Agricultural Development Fund (ADF Project 97000335) of the government of Saskatchewan, Canada. Prairie Malt Ltd. of Biggar, Saskatchewan kindly provided Harrington Barley.

REFERENCES

Bacic G Srejic R Lahajnar G Zupancic I and Ratkovic S 1992 Water and lipids in maize seed embryos: a proton NMR relaxation and diffusion study *Seed Sci and Technol* **20** 233-240

Belton PS and Ratcliffe RG 1985 NMR and compartmentation in biological tissues *Progr NMR Spectr* **17** 241-279

Chudek JA and Hunter G 1997 Magnetic resonance imaging of plants *Progr NMR Spectr* **31** 43-62

Davies NL 1989 Patterns of hydration in barley endosperm during steeping and their relation to malting performance In "EBC proceedings of the 22nd Congress"Zuerich CH pp 221-228

Fung BM and McGaughy TW 1974 The state of water in muscle as studied by pulsed NMR *Biochim Biophys Acta* **343** 663-673

Harris RK 1976 "Nuclear magnetic resonance spectroscopy" Pitman London UK p66-93

Kano H Ishida N Kobayashi T and Koizumi M 1990 1H-NMR imaging analysis of changes of free water distribution in barley and soybean seeds during maturation *Japan J Crop Sci* **59** 503-509

Morris PG Darceuil HE Jasinski A Jha AK McIntyre DJO and Northcote DH 1990 NMR microscopy of the germinating castor bean *Phil Trans R Soc London A* **333** 487-493

Packer K1997 The dynamics of water in heterogeneous systems *Phil Trans R Soc London B* **278** 59-89

Proudlove MO Davies NL and McGill W 1990 Quality requirements for malting barley and their measurement *Ferment* **3** 173-176

Ratkovic S 1987 Proton NMR of maize seed water: the relationship between spin- lattice relaxation time and water content *Seed Sci and Technol* **15** 147-154

Trantham EC Rorschach HE Clegg JS Hazelwood CF Nicklow RM and Wakabayashi N 1984 Diffusive properties of water in Artemia cysts as determined from quasi-elastic neutron scattering spectra *Biophys J* **45** 927-938

van der Maarel JRC de Boer HRWM de Bleijser J Bedaux D and Leyte JC 1987 On the structure and dynamics of water in AlCl3 solutions from HD and 17O and 27Al nuclear magnetic relaxation *J Chem Phys* **86** 3373-3379

Woessner DE and Snowden BS Jr 197 Pulsed NMR study of water in agar gels *J Colloid and Interface Sci* **34** 290-299

Yin XS Kendall EJ Fernets WJ Laycock Abrams GD and Abrams SR 1996 NMR imaging of water uptake and diffusion in barley *Master Brewers Assoc of the Americas Tech Quart* **33** 236-240

Zwanzig R and Harrison K 1985 Modification of the Stokes-Einstein formula *J Chem Phys* **83** 5861-5862

4.7 Plant physiology

LENTIL SEED AQUAPORINS

Phosphorylation , Oligomerization and Secondary Structure

[1]Pol Harvengt, [1]Ariane Vlérick, [1]Bruno Fuks, [2]Ruddy Wattiez, [3]Jean-Marie Ruysschaert and [1]Fabrice Homblé
[1]Laboratoire de Physiologie Végétale CP 206/2, Université Libre de Bruxelles,Brussels, Belgium: [2]Laboratoire de Biochimie, Université de Mons-Hainaut, Mons, Belgium. [3]Laboratoire de Chimie Physique des Macromolécules aux Interfaces, CP 206/2, Université Libre de Bruxelles, Brussels, Belgium. E-mail: pharveng@ulb.ac.be

1. INTRODUCTION

Aquaporins belong to the MIP superfamily of transmembrane channel proteins. AQP-1 from red blood cells and renal proximal tubules has been extensively characterized and structural studies have led to a common structural model for all members of the aquaporin family: aquaporins exhibit at least six closely associated α-helices spanning the lipid membrane. The loops respectively located between helices 2 and 3 and helices 5 and 6 each include a highly conserved Asn-Pro-Ala motif that could be essential for the protein activity. It is likely that this motif lines the water permeation path throughout the protein.

Plant aquaporins have been identified in both plasmalemma and tonoplast. In plants, aquaporins are thought to regulate the water flow through membranes during growth, development and stress responses. Actually, gene expression pattern is highly regulated by developmental or physiological programs. For instance, specific isoforms are expressed in seeds, in elongating zones of roots and shoots and in vascular tissues of shoots and roots (for review, Maurel, 1997).

During seed maturation, prior to dessication, storage proteins synthesized in the endoplasmic reticulum are sequestred in small spherical organelles (Protein Storage Vacuoles or PSV) of parenchyma cells of both developping

Molecular Biology and Physiology of Water and Solute Transport
Edited by Hohmann and Nielsen, Kluwer Academic/Plenum Publishers, New York, 2000.

cotyledons and endosperm (Herman and Larkins, 1999). These proteins provide the nitrogen required for the seedling growth during the early stage of germination, as the seed rehydrates. Exchanges of water and solutes across the tonoplast of PSV are required in order to make these nitrogen reserves available for the cellular metabolism. It has been shown that PSV membranes of *Phaseolus* seeds contained one major integral protein named α-TIP (Tonoplast Intrinsic Protein). α-TIP is a member of the aquaporin protein family (Johnson *et al.,* 1990). Early studies on native tonoplast have shown that α-TIP is phosphorylated by an unidentified calcium-dependent membrane-bound protein kinase (Johnson and Chrispeels, 1992). Heterologous expression of site-specific mutant of α-TIP in *Xenopus* oocyte has permitted to demonstrate that phosphorylation is required to activate the water transport (Maurel *et al.,* 1995).

Two homologous TIP cDNA have been cloned from bean (Johnson *et al.*, 1990) and *Arabidopsis* (Höfte and Chrispeels, 1992) seeds. They were respectively called α- and β-TIP suggesting that at least two TIPs are expressed in plant seeds. However, these two TIPs have not been purified so far. Moreover, β-TIP has not yet been identified in a tissue. Therefore, it is not known if these two homologous proteins are expressed simultaneously and targetted to the same organelles.

In this work, we isolated two TIP isoforms from the PSV membrane of *Lens* seeds. Chemical cross-linking experiments revealed that lentil TIP isoforms exist as hetero-oligomers in the membrane. We identified a 52 kDa magnesium-dependent protein kinase capable to phosphorylate each isoform

2. EXPERIMENTAL RESULTS

2.1 Purification and Identification of Lentil Seed Aquaporins

Most of the protein contained in *Lens* seeds were storage proteins of PSV. Although a large fraction of soluble protein was associated to the membrane fraction, two proteins were present in the membrane fraction only. The membrane fraction obtained after solubilization of the peripheral proteins with DOC at a 2/1 (w/w) detergent to protein ratio was significantly enriched in the two proteins. They accounted for 55 % of the membrane proteins and for about 5 % of the total PSV membrane proteins (before the DOC treatment) as estimated by scanner densitometry of SDS-PAGE gels stained with Coomassie blue. PSV membrane proteins were then solubilized in Genapol X-080 (at a detergent/protein ratio of 4). Solubilized proteins were mixed with hydroxyapatite. Both 25 and 26 kDa TIP proteins were co-eluted from hydroxyapatite after addition of K_2HPO_4 150 mM. The amino-terminal sequence of the 25 kDa protein (PRRAYGFGRADEATH) showed 87 % identity with the β-TIP of *Arabidopsis thaliana* and 60 % identity with α-TIP protein of *Phaseolus vulgaris*. The amino-terminal end of the 26 kDa polypeptide was blocked and an internal sequence (ATVAEFASTAIFLFAGEGS) was determinated after trypsinolysis. It shared a high homology with both *Arabidopsis* β-TIP (73 %) and *Phaseolus* α-TIP (78 %). 25 kDa and 26 kDa bands were blotted on a PVDF membrane and cleaved with CNBr. Comparison of the sequences of five internal peptides revealed a high homology of the two proteins. Such a strong homology has been already described in the case of *Phaseolus* α- and β-TIP cDNAs, which displayed 75 % sequence identity.

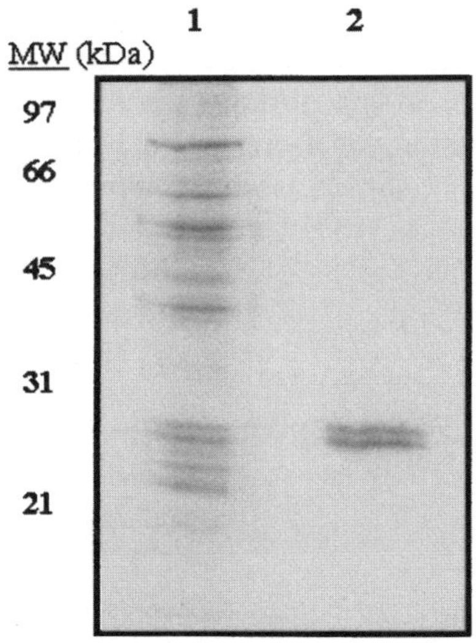

Figure 1. Purification of lentil seed aquaporins. Analysis by SDS-PAGE of polypeptides associated with PSV membranes (*lane 1*) and of th two TIP isoforms copurified after hydroxyapatite chromatography. (*lane 2*)

2.2 Phosphorylation of lentil seed aquaporins

For some of the plant and animal MIP homologs, it has been shown that the phosphorylation/dephosphorylation of serine residues regulated the protein activity. More precisely α-TIP of *Phaseolus* seeds is regulated by a membrane-bound protein kinase (Johnson and Chrispeels, 1992). This prompt us to determine whether or not both TIPs of *Lens* were phosphorylable and to investigate the regulation of the phosphorylation process. *In vitro* phosphorylation experiments were performed with the membrane fraction not treated with DOC. After incubation in the presence of [γ-^{32}P]ATP membrane proteins were separated on a 10.5 % acrylamide gel and the bands corresponding to the 25 kDa and 26 kDa TIP were cut out of the gel to quantify the radioactivity.

The time course of ^{32}P incorporation reached a plateau after 10 minutes. The radioactivity associated with the 25 kDa band was always higher than that of the 26 kDa band. For both proteins, the incorporation of ^{32}P was maximal at 30°C and at pH 6.5. The pH-dependence was slightly affected by the $MgCl_2$ concentration. Increasing the $MgCl_2$ concentration from 2 to 10 mM shifted the pH response curve to lower pH by about 0.5 unit.

In the absence of calcium ^{32}P incorporation into the 25 kDa TIP is magnesium-dependent. Saturation of the phosphorylation reaction occured at 10 mM magnesium whereas 1 mM $MgCl_2$ yielded to half saturation. However, in the absence of magnesium, calcium did not affect ^{32}P incorporation. This lack of effect also suggests that $MgCl_2$ stimulation of the phosphorylation reaction is not attributable to chloride. In our experiments, free Mg-ATP concentration is nearly constant over the range of Mg^{2+} concentration used. Thus, an excess of free Mg^{2+} is essential for kinase activation.

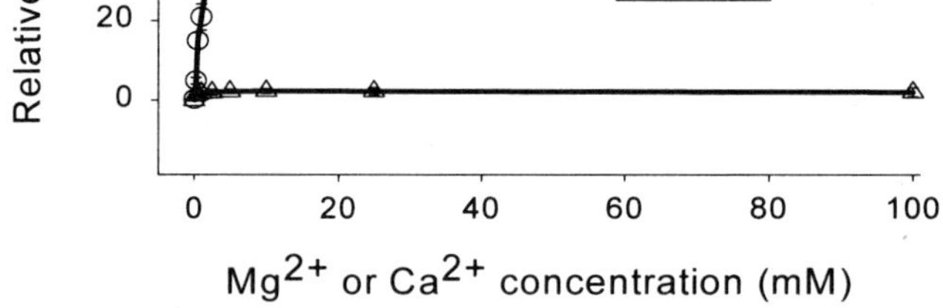

Figure 2. Effect of divalent cations on the kinase activity. PSV membranes were incubated at 30°C during 30 min., pH 6.5, in the reaction mixture containing either Mg^{2+} (o) or Ca^{2+} (Δ). Radioactivity was determinated by Cerenkov counting on excised SDS-PAGE gel slides.

At a constant magnesium concentration, the kinase activity is stimulated by micromolar amounts of free calcium. Maximal stimulation of ^{32}P incorporation was achieved at 1 and 2.5 μM free calcium with the 25 kDa and 26 kDa TIP, respectively. Kinase activity was further characterized by addition of kinase effectors and inhibitors. A concentration of 1 or 10 mM of Mn^{2+} resulted in a 30 % increase of the kinase activity. Phorbol ester did not affect the activity. Addition of W-7, a calmodulin antagonist also known as a CDPK inhibitor to the reaction mixture resulted in a 30 % loss in the kinase activity.

In order to detemine the apparent molecular weight of the kinase, we used an in-gel assay. Calf histone H1 or denaturated PSV membrane proteins were used as substrates for the kinase and mixed with the acrylamide solution of a 12 % SDS-PAGE gel containing 30 % less SDS. Membrane proteins were then separated in the gel by electrophoresis. The gel was soaked in a solution containing [γ-^{32}P] ATP. In the presence of [γ-^{32}P] ATP, kinase induced the phosphorylation of calf histone H1 (or PSV membrane proteins, data not shown) which permits its localization at 52 kDa by autoradiography. In a control experiment, it was checked that the lower SDS concentration did not affect the migration of proteins in the gel during electrophoresis.

2.3 Oligomerization of lentil seed aquaporins

Plant and animal aquaporins form tetrameric structures in their native membrane. In the case of lentil seed TIPs, the concommittant presence of two aquaporin isoforms in the PSV membrane raise question about their possible interaction during assembly of subunits into tetramers. Here we used chemical cross-linking experiments to show that close physical association occurs between lentil seed TIPs within PSV membrane. TIP-enriched membranes (after DOC treatment) were incubated with glutaraldehyde for increasing periods of time. SDS-PAGE revealed that as the incubation time increased, the amount of dimers, trimers and tetramers increased while the amount of monomers decreased. Similar results were obtained with DFDNB and DTSP.

To minimize any aspecific intermolecular coupling, membrane proteins were diluted in their lipidic environment and then incubated with glutaraldehyde. In this aim, liposomes were fused with the TIP-enriched membranes. Increasing the lipid to protein ratio delayed the formation of trimers and tetramers but not that of dimers. This suggest the specificity of the monomer-monomer interaction. Fusion efficiency was monitored by sucrose gradient ultracentrifugation. Detection of proteins in the different fractions of the gradient by SDS-PAGE revealed the presence of proteolipidic vesicles exhibiting a broad density range.

DTSP was used to distinguish homo-oligomers from hetero-oligomers. DTSP contains a disulfide bond which can be reduced, thereby breaking the covalent link between subunits. As expected, treatment with β-mercaptoethanol led to the disruption of cross-linked oligomers. This showed that each oligomer was made of both 25 kDa and 26 kDa TIPs.

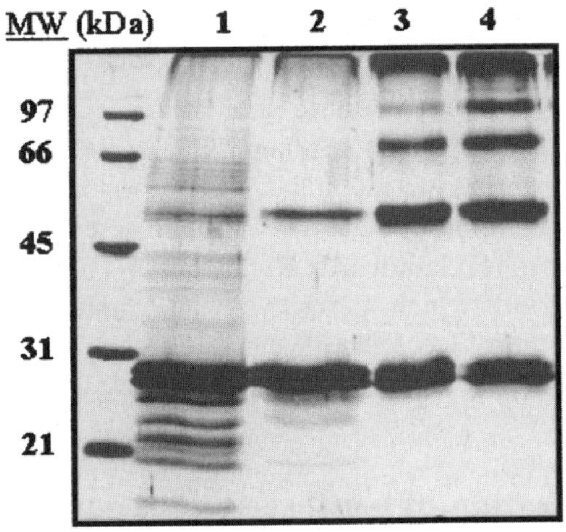

Figure 3. SDS-PAGE analysis of the DOC-treated PSV membranes after cross-linking with glutaraldehyde. TIP proteins were cross-linked in the presence of 0.125 % (v/v) of glutaraldehyde for 10 s. (*lane1*), 10min. (*lane 2*), 1 h (*lane 3*) and 3h (*lane 4*).

2.4 Reconstitution of Lentil Seed TIPs

Lentil seed aquaporins were purified by solubilization in Genapol X-080 and chromatography on hydroxyapatite. 200 μg of purified proteins were added to a mixture of 2 mg of purified asolectin LUVs in suspension in Tris 10 mM, N-Methyl-D-Glucamine 150 mM, NaN_3 0.02 % (w/v) pH 7.2. The mixture was incubated for one hour at 4°C under gentle agitation and the detergent was then removed by addition of 20 mg/ml (dry weight) of SM_2-Bio-Beads (extensively washed before use with methanol and distillated water). The mixture was kept under gentle agitation for 30 min. and Bio-Beads were then removed. Three additional cycles of incubation with Bio-Beads were then performed to ensure efficient detergent removal. Finally, proteoliposomes were harvested after ultracentrifugation (2 h at 180000 g and 4°C) and resuspended in Hepes 2 mM pH 7.2.

2.5 Secondary Structure Determination of Lentil Seed Aquaporin Hetero-oligomer

Attenuated total reflection infrared spectra (4 cm^{-1} resolution) were obtained on a Perkin-Elmer 1720X Fourier transform infrared (FTIR) spectophotometer as described elsewhere (Goormaghtigh *et al.*, 1994). Measurements were carried out at room temperature. Thin films were obtained by applying a sample to one side of an attenuated total reflection (ATR) plate and slowly evaporating the solvent under a nitrogen stream. Secondary structure determination was performed by analysis of the Amide I region of the infrared spectra. Fourier self-deconvolution (FSD) and curve fitting (Goormaghtigh *et al.*, 1990, 1994a,b) as well as partial least-square (PLS-ATR) with a basis set of circular dichroism (CD) and ATR-FTIR protein spectra (Cabiaux *et al.*, 1997).

Circular dichroism spectra were obtained on a Jasco J-600 spectropolarimeter. Detergent-solubilized proteins after hydroxyapatite purification were used for CD . Contribution of buffer and detergent were subtracted from the raw spectra before analysis.

ATR-FTIR spectra of lentil aquaporin hetero-oligomer reconstituted in asolectin vesicles show that the main component of the Amide I region is located at 1656 cm^{-1} and can be assigned to α-helical structure. A minor contribution is centered around 1627 cm^{-1} and has been interpreted as β-sheet. However, it must be noted that this component could also be due to the presence of non-helical structure, such as 3_{10}-helix or loops.

Analysis of Amide I absorption spectra carried out by the methods cited above gave helical content ranging from 43 to 50 % and a β-sheet content of 12-18 %. These results (*table 1*) are in perfect agreement with values obtained in our laboratory for the homologous AQP-1 (Cabiaux *et al.*, 1997).

Table 1. Determination of secondary structure of lentil seed aquaporins.

Proteins	Method	% α[1]	% β	% turns
Lentil seed TIPs (This paper)	FSD and curve fitting	(50 ± 5)	(18 ± 8)	(21 ± 1)
	PLS-ATR	(47 ± 7)	(13 ± 8)	(25 ± 1)
AQP-1 (Cabiaux *et al.*, 1997)	FSD and curve fitting	(48 ± 5)	(18 ± 8)	(21 ± 1)
	PLS-ATR	(48 ± 7)	(0 ± 8)	(11 ± 1)

[1]Values are presented with standard deviation between brackets

3. CONCLUSIONS

We show here that two proteins highly homologous are present in the membrane of PSV in lentil. The two TIPs identified in *Lens* are highly homologous but distinct: their apparent molecular weight are significantly different and after trypsinolysis, two peptides of each TIP are not fully identical. The occurence of two TIP proteins in seeds is consistent with the identification of two cDNA coding for TIP in plant seeds.

We investigated the in vitro phosphorylation properties of two TIPs from lentil seeds. Our data provide strong evidence that a protein kinase is associated with the PSV membrane. In *Phaseolus*, endogenous phosphorylation of α-TIP is activated by calcium and it has been therefore suggested that this kinase activity could be associated to a protein belonging to the CDPK family. Our results are consistent with this hypothesis : the kinase is activated by calcium, it requires Mg^{2+}, it has an apparent molecular weight between 50 and 60 kDa and it is inhibited by calmodulin antagonists.

Consistently with other reports on AQP-1, cross-linking experiments showed that TIPs from lentil seeds form hetero-oligomers. Physiological significance of the occurence of an hetero-oligomeric assembly in PSV membrane remains unclear. Coinjection of mRNA of each TIP isoform in oocytes would allow to investigate the role of the hetero-oligomer on water transport.

REFERENCES

Goormaghtigh E Cabiaux Vand Ruysschaert JM 1994aDetermination of soluble and membrane protein structure by Fourier transform infrared spectroscopy In Subcelular Biochemistry Vol 23 HJ Hilderson and GB Ralston editors Plenum Press New York pp329-450

Goormaghtigh E Vigneron L Scarborough G and Ruysschaert JM 1994b Tertiary conformational changes of the Neurospora crassa plasma membrane H(+)-ATPase monitored by hydrogen/deuterium exchange kinetics A Fourier-transform infrared spectroscopy approach J Biol Chem 269 27409-27413

Goormaghtigh E Cabiaux V and Ruysschaert JM 1990 Secondary structure and dosage of soluble and membrane proteins by attenuated total reflection Fourier-transform infrared spectroscopy on hydrated films Eur J Biochem 193 409-420

Herman E Mand Larkin BA 1999 Protein storage bodies and vacuoles Plant Cell 11 601-614

Hofte H and Chrispeels MJ 1992 Vegetative and seed-specific forms of tonoplast intrinsic protein in the vacuolar membrane of Arabidopsis thaliana Plant Cell 4 995-1004Johnson KD Herman EM and Chrispeels MJ 1989 An abundant highly conserved tonoplast protein in seeds Plant Physiology 91 1006-1013

Johnson KD Hofte H and Chrispeels MJ 1990 An intrinsic tonoplast protein of protein storage vacuoles in seeds is structurally related to a bacterial solute transporter (GIpF) Plant Cell 2 525-532
Johnson KD and Chrispeels MJ 1992 Tonoplast-bound protein kinase phosphorylates tonoplast intrinsic protein Plant Physiology 100 1787-1795
Mäder M and Chrispeels MJ 1984 Synthesis of an integral protein of the protein body membrane in Phaseolus vulgaris Planta 160 330-340
Marty F 1999 Plant vacuoles Plant Cell 11 587-599
Maurel C Kado RT Guern J and Chrispeels MJ 1995 Phosphorylation regulates the water channel activity of the seed-specific aquaporin a-TIP EMBO J 14 3028-3035

4.8 Plant physiology

EXPRESSION, LOCATION AND FUNCTION OF A PLASMA MEMBRANE AND CHLOROPLAST LOCATED AQUAPORIN FROM TOBACCO

Ralf Kaldenhoff, Beate Otto, Alexander Biela, Franka Siefritz, Martin Eckert
Universität Würzburg, Julius-von-Sachs-Institut, Molekulare Pflanzenphysiologie und Biophysik, Julius-von-Sachs-Platz 2, D-97082 Würzburg, Germany. E-mail: kaldenhoff@botanik.uni-wuerzburg.de

The characterization of a plant MIP from tobacco (*Nicotiana tabacum*) regarding its function as an aquaporin and role in the plant will be presented.

Heterologous expression in *Xenopus* oocytes and analysis of transport properties revealed that the protein facilitates the transport of water and small solutes like glycerol or urea. In contrast to the majority of plant aquaporins analyzed so far, the water permeability could not be inhibited by heavy metals. Addition of substances that increase c-AMP dependent protein kinase activity did not show any influence on water permeability of oocytes expressing this aquaporin. It is therefore concluded that regulation of water flux by aquaporins in the plant occurs at the transcription level. Expression of this aquaporin in plants was followed by northern-blot and in situ hybridization. The experiments indicate a high specific mRNA-level in roots and a moderate or low in shoots respectively leaves. Regulation of the corresponding gene was studied by transient promoter-GFP transformation into tobacco protoplast and revealed an induction of promoter activity by the phytohormones abscisic or gibberellic acid. Using immunolocalization with an antibody targeted to the N-terminus of the aquaporin, a localization at sites of expected high water transport rates from and to apo- or symplast could be demonstrated. An unexpected finding was a high concentration of the protein in rows of 10 to 20 petiole cells, which is interpreted as a prerequisite for a transcellular pathway of water transport. Possibly this water flux is connected with a diurnal leaf movement. To date, this specific way of water movement in plants was demanded just from theoretical conclusions, but experimental evidence was still missing.

Molecular Biology and Physiology of Water and Solute Transport
Edited by Hohmann and Nielsen, Kluwer Academic/Plenum Publishers, New York, 2000.

Separation of plant membranes and western-blotting indicated plasma membrane localization. In addition an antibody-signal could also be detected in proteins isolated from outer and inner chloroplast membranes. By electron microscopy studies the subcellular localization could be confirmed.

In order to analyze the function of this aquaporin in the plants' physiology and its contribution to water transport processes, tobacco antisense aquaporin plants were constructed. The subsequent reduction of aquaporin RNA-levels was confirmed by quantitative RT-PCR and northern analysis. The antisense aquaporin plants show specific morphological changes as a result of a reduced cellular water permeability. Differences to control plants could also be detected in osmotic values and salt concentrations as well as in hydraulic conductivity. The obtained data suggest a compensation mechanism for reduced water permeability, that is composed of specific morphological and physiological mechanisms.

ACKNOWLEDGMENTS

The work was supported by the Deutsche Forschungsgemeinschaft (SFB 251) and Fonds der Chemischen Industrie

REFERENCES

Biela A Grote K Otto B Hoth S Hedrich R and Kaldenhoff R 1999 The *Nicotiana tabacum* plasma membrane aquaporin NtAQP1 is mercury-insensitive and permeable für glycerol Plant J **18**: 565-570

Kaldenhoff R and Eckert M 1999 Features and function of plant aquaporins J Photochem Photobiol B: Biol **52**: 1-6

Otto B and Kaldenhoff R 2000 Cell specific expression of the mercury-insensitive plasma membrane aquaporin *NtAQP1* from *Nicotiana tabacum* Planta in press

4.9 Plant physiology

COORDINATION OF ROOT HYDRAULIC CONDUCTIVITY AND TRANSPIRATION IN HONEY LOCUST (*Gleditsia triacanthos* L.): A PROPOSED ROLE FOR AQUAPORINS

[1]Albino Maggio and [2]Robert J. Joly
[1]ENEA CR-Trisaia, S.S. Jonica Km. 419-500 – 75026 Rotondella (MT), Italy: [2]Department of Horticulture and Landscape Architecture, Purdue University, 1165 Horticulture Building, West Lafayette, IN-USA 47907-1165

1. INTRODUCTION

Plants exhibit rhythmic behavior in response to daily cycles of light and darkness. Such rhythms have been reported in leaf and petal movements, stomatal behavior, root resistance to water movement and solute transport in roots. It has been reported that when plants are transferred from recurring daily light and dark cycles to continuous darkness or continuous light, many such rhythms continue to be expressed for one or more periods. In constant light or darkness, rhythms generally have been observed to depart from 24-h periodicity (Sweeney, 1987). Several such rhythmic behaviors (e.g., stomatal movement) have been thoroughly characterized, and their physiological significance has been extensively documented. Many others have not been studied in sufficient detail, and our understanding of their physiological function is rather limited. For example, water homeostasis in plants is achieved by coordinated control of root water uptake from the soil and by transpiration through the stomata. Therefore, fine control of stomatal aperture and root hydraulic conductivity is essential for the maintenance of a favorable plant water status. Nevertheless, the rhythmic behavior of stomatal aperture and root resistance to water movement have never been functionally associated or explicitly studied as a potentially coordinated system that may regulate water homeostasis in plants.

Molecular Biology and Physiology of Water and Solute Transport
Edited by Hohmann and Nielsen, Kluwer Academic/Plenum Publishers, New York, 2000.

Henzler *et al.* (1999) recently sought to characterize physiological and molecular regulation of daily root water fluxes. Their report provided evidence that diurnal variations in root hydraulic conductivity in *Lotus japonicus* may be correlated with the expression of putative aquaporins. However, a functional linkage between aquaporin activity and diurnal rhythms of water flux is still missing. In the present study, we used honey locust (*Gleditsia triacanthos* L.), a leguminous tree species that exhibits a strong circadian rhythm in root water transport to investigate a possible relationship between water channel activity, circadian rhythm of root water flux and stomatal regulation. Here we show that: 1) the pattern of circadian root water flux (J_v) in honey locust is water stress-dependent; 2) root water flux in the ascendant region of the circadian rhythm is dramatically inhibited upon treatment with mercuric chloride, but completely reversed upon subsequent treatment with mercaptoethanol; 3) high and low values of root hydraulic conductivity (L_p) coincide with peaks and troughs, respectively, of the circadian water flux; 4) the circadian pattern of water vapor loss from leaves closely resembles that of root water fluxes although it is phase-shifted from that of the J_v rhythm by approximately 4 hours. Considering these results together, we conclude that diurnal variations in hydraulic conductivity of honey locust roots are associated with water channel activity, and we propose that coordination of aquaporin activity and stomatal conductance may be critical in alleviating cavitation and maintaining a favorable leaf water balance.

2. MATERIALS AND METHODS

2.1 Growth Conditions and Water Stress Experiment

Honey locust seedlings were grown in plastic pots (28 cm deep and 8 cm diameter) filled with a 2:2:1 (v/v/v) mixture of perlite, peat moss and top soil amended with $Ca(H_2PO_4)_2$, KNO_3 , $MgSO_4$, ground limestone and a standard trace element mixture. Seedlings were grown on benches in the Horticulture greenhouse at Purdue University, West Lafayette, Indiana under approximately 28° C daytime maximum and 23° C nighttime minimum temperature. Water stress was induced by withholding water when the seedlings were approximately 7 weeks old and 40 cm tall. Shoot water potentials were estimated using a pressure chamber (PMS Instruments Co.) immediately before measurement of root water flux rate.

2.2 Measurement of the Volume Flux of Water Through Roots

Shoots of the seedlings were severed 8 cm above the root collar, and root systems were mounted in the lid of a 48-L pressure chamber with the cut stumps protruding through a gasket. The root systems were immersed in half-strength Hoagland solution which was aerated and pressurized to 0.34 MPa by a compressed air cylinder. Exudate was delivered from each root system to an electronic balance via a computer-controlled automated system. Flux was measured during a 90 h period and was expressed on a root dry weight basis. For each treatment, 4 to 6 plants were analyzed.

2.3 Mercury Inhibition of Water Flux

Four one-liter plastic beakers were placed inside the pressure chamber described in 2.2 and filled with Hoagland solution. One root system was placed inside each beaker. After verifying that all 4 root systems presented a normal cycling behavior, $HgCl_2$ was added to two root systems at a final concentration of 0.5 mM, while the other two served as control. The injection of mercury was performed according to Maggio and Joly (1995) during the ascendant region of the circadian flux. Three hours after mercury injection, mercaptoethanol (ME) was added at a final concentration of 0.06 M. Water flux was recorded for more than 100 h. The experiment was replicated twice.

2.4 Transpiration Measurements

Honey locust seedlings were watered 12 hours before the experiment began. Single pots were sealed with polyethylene bags and aluminium foil to avoid water evaporation from the soil and then placed on an electronic balance connected to a computer. Water loss data were collected each 15 min through a 60 hour period for six plants.

3. RESULTS AND DISCUSSION

A circadian rhythm in volume flux (J_v) was observed in well-watered and water-stressed seedlings with maximal and minimal fluxes occurring approximately at the end of the dark period and continuing about 3-4 hours into early morning, and minimum flux coinciding with the beginning of the dark period (Figure 1). The severity of stress influenced the amplitude and trajectory of the rhythm as well as the absolute values of maximum and minimum fluxes. Seedlings subjected to a severe level of water stress (water potential of –2.8 MPa) exhibited an approximately 4-fold reduction in J_v at comparable points in the cycle.

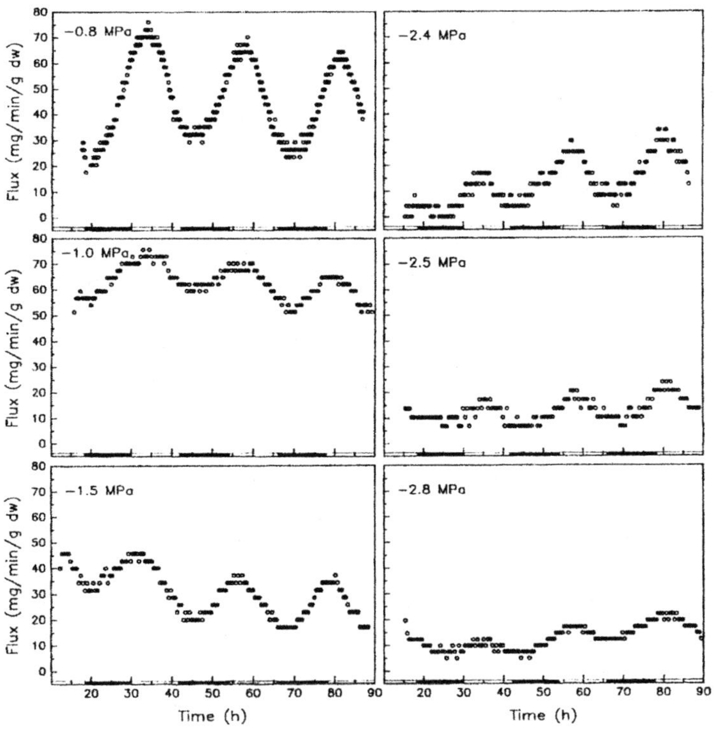

Figure 1. Circadian pattern in volume flux of water through root systems of a well-watered control (A, shoot water potential –0.8 MPa) and five water-stressed honey locust (B-F). Darkened bars on the X-axis represent the dark period.

The amplitude of the circadian water flux appeared to be very sensitive to water stress since it dropped approximately 50% at -1.0 MPa. In general, amplitude of the cycle and absolute minimum and maximum levels of water flux decreased at increasing level of stress.

Upon injection of $HgCl_2$ root flux decreased within 20 min (Figure 2). Flux continued to decline to zero within 3 hours. At this time ME was injected into the bathing solution. The addition of ME caused a relatively rapid response. Water flux increased in an exponential manner and it reached steady-state within 8 to 10 hours. At steady-state, the root water flux was almost identical to maximal values of flux measured prior addition of $HgCl_2$. However, the rhythm of water flux was abolished by $HgCl_2$ and ME treatments. Measurement of root hydraulic conductivity (L_p) at the time of maximum and minimum water flux (peaks and troughs of the circadian rhythm) revealed 50% reduction in root L_p at the end of the dark period (data not shown), in coincidence with the generally observed 50% reduction of root water flux.

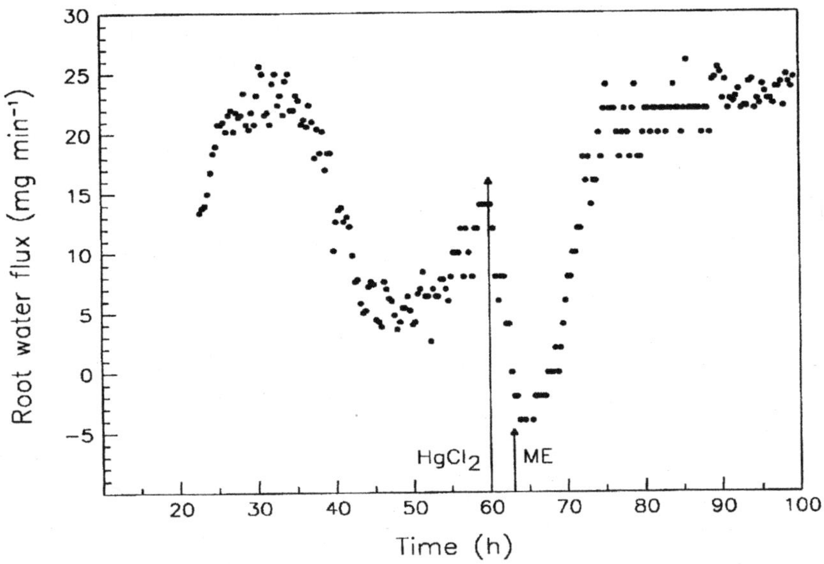

Figure 2. Pressure-induced water flux through a single $HgCl_2$- treated honey locust root system. Times of injection of $HgCl_2$ and ME are indicated by arrows. (Control plants, not shown, presented a typical water flux pattern as displayed in Figure 1A).

When the patterns of circadian rhythm of plant transpiration rates and root water flux rates were compared, two interesting features were apparent (Figure 3). First, the circadian pattern of water vapor loss from leaves closely resembled that of J_v. Second, the pattern of water vapor loss was phase-shifted forward by approximately 4 hours. During the period between sunset and sunrise, J_v increased linearly, reaching maximal values at approximately the time of sunrise and remaining high until late morning. The high J_v at sunrise (Figure 3) suggests that the period of lowest root resistance to water transport co-occurs with that of stomatal opening. High root L_p during the night and early hours after sunrise would allow for rapid entry of water across the root radius in response to a transpiration-induced water potential gradient in the hydraulic pathway.

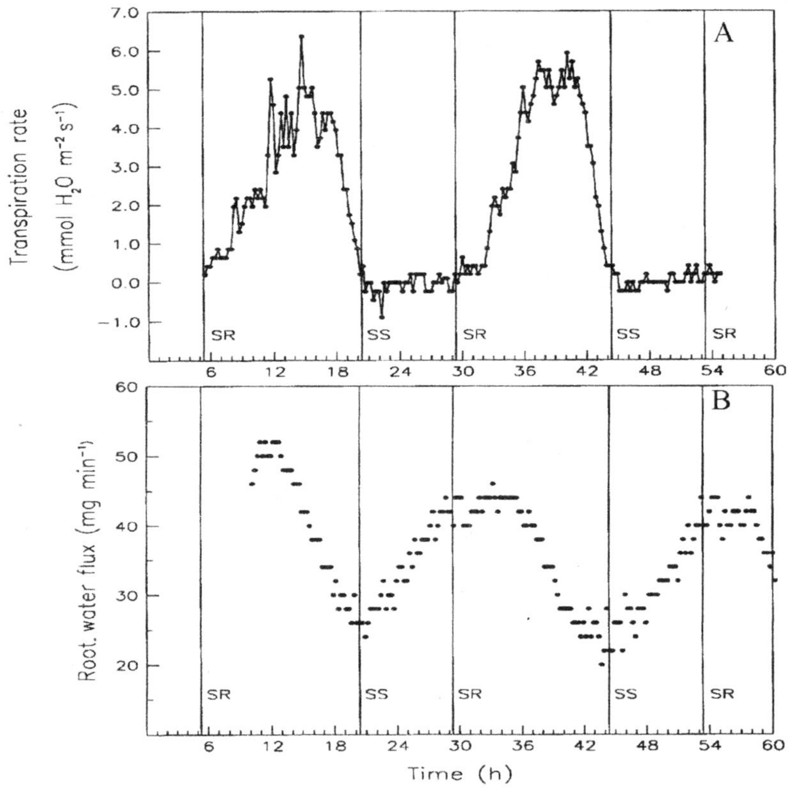

Figure 3. Comparison of the circadian rhythm of transpiration rate and root water flux in honey locust. A) whole-plant leaf transpiration rate. B) root water flux. SR and SS denote time of sunrise and sunset, respectively. Representative patterns of 6 independent measurements are shown.

This coordination between transpiration and root hydraulic conductivity may be of physiological significance in the prevention of xylem cavitation (Tyree and Sperry, 1989). For example, a decreasing root resistance to water entry into roots during the sunset-sunrise time interval may facilitate overnight replenishment of plant water reserves and, consequently, may allow for repair of embolisms that would otherwise break the continuity of the hydraulic path.

The diurnal patterns of root water flux shown here (Figures 1, 3) are remarkable in their clarity and repeatability. J_v is strongly reduced by mercury but can be recovered by addition of ME (although the rhythm is destroyed by this treatment). Further, L_p values estimated at the troughs of the circadian rhythm are reduced by approximately 50% from that estimated at the peaks. Considering these results together, we propose that root aquaporins modulate root permeability to water in a rhythmic manner. Specifically, we postulate that the circadian rhythm of root water flux is caused by opening and closure of water channels. The basal nighttime water flux (min J_v) observed (Figures 1, 3) may be explained by partial aperture of aquaporin pores or by the existence of a sub-population of constantly-open aquaporins that are not involved in the regulation of root resistances to water movement. Other possibilities to consider are the slippage of water molecules through lipid bilayers or the contribution of the apoplastic water flux.The model of an aquaporin-mediated circadian rhythm proposed here is consistent with observations made on water-stressed seedlings and supports the concept that root and leaf water fluxes are likely co-regulated. Lower leaf water potentials are associated with a concomitant reduction in both the average root water flux and the amplitude of the oscillatory response (Figure 1, compare A vs. C-F). We did not perform simultaneous measurements of stomatal conductance and root water fluxes at different levels of water stress. Thus, we do not know how water stress would have affected the relative phasing of stomatal aperture versus J_v. Experiments are needed to clarify whether a phase-shifted relation between root permeability and leaf conductance could play a role in the regulation of plant water status under irrigated and water-limited conditions. Lösch and Schulze (1995) have argued that control at the level of stomatal aperture is not sufficient, by itself, to stabilize plant water status. They have emphasized the idea that plants maintain a continuous yet regulated water flux, at a given water potential gradient between soil and atmosphere, by co-ordinating root water uptake and stomatal conductance. Lösch and Schulze (1995) summarize numerous reports where the observed relationships between leaf water potential, stomatal conductance and soil water deficit found a plausible explanation in an hypothetical coordination of root water uptake and transpiration rates (e.g., Berard and Thurtell, 1991; Meinzner and Grantz, 1990; Gollan *et al.*, 1986). Importantly, their review was written prior to the first reports of a role for aquaporins in plant roots.

4. FURTHER CONSIDERATIONS AND FUTURE WORK

A control of plant water homeostasis, through a combined regulation of root and stomatal water exchanges, may also have important physiological significance at moderate or advanced water stress. In contrast to the general pattern observed at moderately-high and severe levels of stress, analysis of root water flux in roots of mildly-stressed seedlings (-1.0 MPa) revealed that the average flux through the root system was not affected. Indeed, the minimum J_v was relatively higher in these plants compared to the non-stressed control (Figure 1A-B), suggesting that gating of root aquaporins may not be strongly affected under mild water deficit. The net effect may be to minimize resistances to water movement and maintain shoot hydration and leaf gas exchange.

This view may be interpreted in terms of feedforward-feedback regulation of plant water homeostasis (Schulze, 1993). If aquaporin expression or gating is indeed a feature of water stress perception and adaptation, then it also may have an important role in facilitating long distance transport of signal molecules, such as abscisic acid (Zhang *et al.*, 1987; Schurr, 1992). Accelerated or impeded delivery of such non-hydraulic signals would be expected to affect the timing of response of stomata, a critical factor in early adaptation to drought. However, to our knowledge, no studies have specifically evaluated the role of rhythmic fluctuations of root L_p or aquaporin expression on rates of delivery of signal molecules to leaves and subsequent effects on CO_2 and water vapor exchange rates.

REFERENCES

Berard RG and Thurtell GW 1991 The interactive effects of increased evaporative demand and soil water on photosynthesis in maize *Can J Plant Sci* **71**: 31-39

Gollan T Passioura JB and Munns R 1986 Soil water status affects the stomatal conductance of fully turgid wheat and sunflower leaves *Aust J Plant Physiol* **13**: 459-464

Henzler T Waterhouse RN Smyth AJ Carvajal M Cooke DT Schäffner AR Steudle E Clarkson DT 1999 Diurnal variations in hydraulic conductivity and root pressure can be correlated with the expression of putative aquaporins in the roots of *Lotus japonicus* *Planta* **210**: 50-60

Lösch R and Schulze E-D 1995 Internal coordination of plant responses to drought and evaporational demand In *Ecophysiology of Photosynthesis* E-D Schulze and M Caldwell (eds) Springer-Verlaag Berlin 571 pp

Maggio A and Joly RJ 1995 Effects of mercuric chloride on the hydraulic conductivity of tomato root systems: evidence for a channel-mediated water pathway *Plant Physiology* **109**: 331-335

Meinzer FC and Grantz DA 1990 Stomatal and hydraulic conductance in growing sugarcane: stomatal adjustment to water transport capacity *Plant Cell Environ* **13**: 383-388

Schulze E-D 1993 The regulation of transpiration: interactions of feedforward feedback and futile cycles In: *Flux Control in Biological Systems: From Enzymes to Populations and Ecosystems* E-D Schulze (ed) Academic Press San Diego 494 pp

Schurr U Gollan T Schulze E-D 1992 Stomatal response to drying soil in relation to changes in the xylem sap composition of *Helianthus annuus* II Stomatal sensitivity to abscisic acid imported from the xylem sap *Plant Cell Environ* **15**: 561-567

Sweeney BM 1987 *Rhythmic Phenomena in Plants* Academic Press San Diego 172 pp

Tyree M T and Sperry JS 1989 Vulnerability of xylem to cavitation and embolism *Annu Rev Plant Physiol Mol Biol* **40**: 19-38

Zhang J Schurr U and Davies WJ 1987 Control of stomata behaviour by abscisic acid which apparently originates in the roots *J Exp Bot* **38**: 1174-1181

4.10 Plant physiology

SOLUTE AND WATER PERMEATION ACROSS THE SYMBIOSOME MEMBRANE OF LEGUMES

Asymmetry of water and solute permeation through NOD26

[1]Stephen D. Tyerman and [1]Christa M. Niemietz
[1]School of Biological Sciences, Flinders University, Adelaide,GPO Box 2100 Adelaide 5001, Australia. E-mail: Steve.Tyerman@flinders.edu.au

1. INTRODUCTION

In legume nodules, nitrogen is fixed by *Rhizobium* bacteria encapsulated in a specialised organelle (symbiosome) in the plant cell cytoplasm. The symbiosome is delineated by the symbiosome membrane (SM) across which fixed nitrogen as NH_4^+ or NH_3 has to flow to be assimilated in the plant cytoplasm. Ammonium (NH_4^+) is supposed to enter the cytoplasm through specialised Ca^{2+}- and Mg^{2+}-regulated channels in the SM (Tyerman *et al.*, 1995; Whitehead *et al.*, 1998), while ammonia (NH_3) has been assumed to simply diffuse through the lipid phase of the membrane. Nodulin 26, a prevalent MIP in the SM that is only expressed when the bacteria infect a root cell, is a member of the aquaporin family. NOD26 confers increased permeability to water and some small molecules like formamide and glycerol (Rivers *et al.*, 1997). It has been suggested that the main function of NOD26 in the SM is that of osmoregulation (Dean *et al.*, 1999), the feature of its' glycerol-permeability being less understood since it is not known whether glycerol is used as an osmolyte in symbiosomes. If NH_3 uses the same pathway as glycerol or water then it could potentially permeate via NOD 26 and display similar permeation features.

Niemietz and Tyerman (2000) measured NH_3 permeability of SM vesicles from soybean by suddenly increasing the external NH_3 concentration in a stopped-flow spectrofluorimeter and measuring NH_3 influx by the increase in vesicle pH recorded using entrapped

Molecular Biology and Physiology of Water and Solute Transport
Edited by Hohmann and Nielsen, Kluwer Academic/Plenum Publishers, New York, 2000.

carboxyfluorescein. Ammonia (NH_3) permeability (8×10^{-3} cm s^{-1}) of SM vesicles was inhibited by up to 42% by $HgCl_2$ (EC_{50}= 2.9 μM, mercaptoethanol-reversible) and significantly reduced by ATP-pre-incubation. Such an effect of $HgCl_2$ was not observed for NH_3 permeation into lipid membrane vesicles formed from soybean lipid. The activation energy of NH_3 uptake (52 kJ mol^{-1}) increased (118 kJ mol^{-1}) with $HgCl_2$ treatment. Osmotic water permeability (P_{OS}) measured by inducing water efflux by a hyperosmotic gradient was also $HgCl_2$-sensitive (EC_{50}= 52.6 μM), but increased by ATP pre-incubation. However, we obtained significantly smaller P_{OS} (8.1×10^{-3} cm s^{-1}) to those reported for water efflux from soybean SM vesicles by Rivers *et al.*, (1997) (5×10^{-2} cm s^{-1}).

Nodulin 26 is phosphorylated by a SM-associated calcium dependent protein kinase (Weaver *et al.*, 1994), and voltage dependent gating of ion channel activity associated with NOD26 in artificial lipid bilayers is altered by phosphorylation (Lee *et al.*, 1995). Since ATP preincubation increased water permeability and decreased NH_3 permeability, we considered the possibility that water and NH_3 may take different pathways through NOD26. Alternatively NH_3 may permeate via a different protein in the SM.

The possibility that water and solutes may take different pathways through NOD26 was suggested by Rivers *et al.*, (1997) based on the apparent single file permeation of water through NOD26 and the fact that larger molecules such as glycerol also permeated. Our results also suggest that NOD26 may display phosphorylation-dependent gating for water and solute permeation. This may explain the difference between our P_{OS} values and those of Rivers *et al.*, (1997) if the phosphorylation status of NOD26 was different between the SM vesicle preparations. We have reexamined our P_{OS} values and performed additional experiments to examine why there can be such large disparity in P_{OS} for essentially the same SM preparation.

An aspect of gating of ion channels is that current flow in one direction is often enhanced by a gradient dependent change in open probability (Popen) of the channel. This leads to rectification of ion flow through some ion channels. Such rectification can also be caused by a molecule that blocks the channel from one particular side of the pore. It is not known if aquaporins form a continuously open pore (Popen=1), or if they open and close randomly in time (Popen<1) and if this Popen is altered. One test for gating would be to test for gradient dependent permeability and rectification of water flow. For the animal aquaporins AQP0-AQP5 water flow is not rectified (Meinild *et al.*, 1998). However, such tests have not been performed on plant aquaporins and also not for small solute permeation. Here we report on the directionality of water, glycerol and urea permeation across soybean SM and how mercury ions and ATP affects permeation in each direction unequally.

2. MATERIAL AND METHODS

2.1 Plant material and vesicle isolation

Soybean plants were grown and harvested at 6 weeks old for nodules as described in Niemietz and Tyerman (2000). SM vesicles were isolated by two-phase partitioning after first obtaining symbiosomes as described by Christiansen *et al.*, (1995). The SM vesicles were suspended in 0.33 M sucrose, 5 mM potassium posphate, pH 7.8 and stored at –80°C. After three partitioning steps, SM vesicles were obtained in the upper phase with little contamination from bacteroids as judged by glucansynthase II as a marker enzyme for the SM and cytochrome c oxidase as a marker for bacteroids. The sidedness of the SM was determined from ATPase latency assays (Christiansen *et al.*, 1995) and was found to be 70-80% inside out, that is, 70-80% of the vesicles had the bacteroid face of the SM facing to the outside.

2.2 Determination of water and solute permeability

Light scattering experiments were conducted as described by Niemietz and Tyerman (1997). For water permeability SM vesicles suspended in 0.6 M sucrose, 0.1 mM $CaCl_2$, 10 mM HEPES KOH pH 7.0 were simultaneously injected with a hypoosmotic solution that upon mixing yielded a 0.3 M sucrose gradient. The time course of vesicle swelling was followed as an increase in 90° light scattering at 500 nm. P_{OS} was calculated from the time-course of light scattering by comparing single exponential time constants fitted to simulated curves (Mathcad, rkfixed differential equation solver) in which P_{OS} was varied and with other conditions and vesicle dimensions as used in the experiments (see Eq. 1 in Lande *et al.*, 1995). An average SM vesicle diameter of 202 nm was measured by transmission electron microscopy.

Solute permeability was measured for efflux or influx by imposing 0.3 M concentration gradients of permeating solute (urea, glycerol, proline or alanine) while using sucrose (relatively impermeable) to maintain isosmotic conditions. Light scattering was measured as described above to obtained the kinetics of decrease in volume (solute efflux) or increase in volume (solute influx). For efflux, the vesicles were initially loaded with 0.6 M permeating solute and were then rapidly mixed with isosmotic sucrose so that an outwardly directed concentration gradient (0.3 M) of permeating solute occurred. For influx, the vesicles were initially loaded with sucrose isosmotic with an external mixing buffer of 0.6 M permeating solute giving a final inwardly directed gradient of 0.3 M permeating solute. P_{solute} was calculated from the time-course of light scattering by comparing single exponential time constants fitted to simulated curves (Mathcad, rkfixed differential equation solver) in which P_{Solute} was varied and with other conditions and vesicle dimensions as used in the experiments (see Eq. 2-6 in Lande *et al.*, 1995).

To check that imposed hypoosmotic gradients gave a linear response in light scattering, as would be expected if vesicles maintained osmotic integrity, we measured vesicle swelling in response to glycerol uptake from different concentrations.

3. RESULTS

3.1 Linear response to hypoosmotic gradients

Figure 1 shows that light scattering amplitude was linear with the imposed osmotic gradient for swelling vesicles. This indicated that for up to 0.3 M osmotic gradients the vesicles behaved as osmometers.

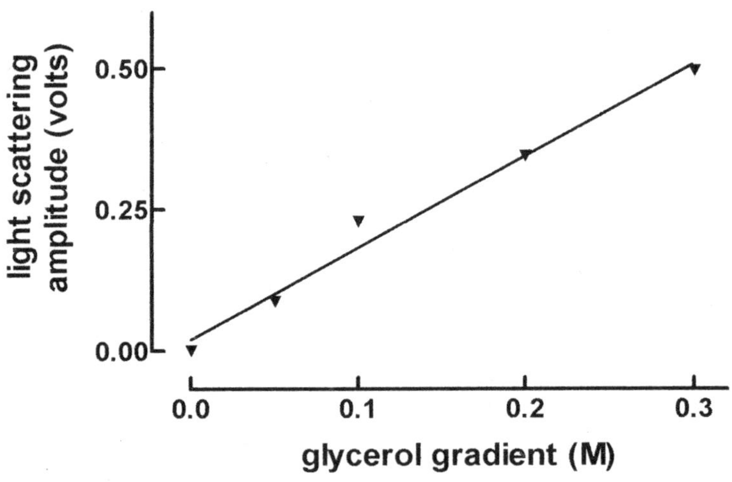

Figure 1. Light scattering amplitude in response to glycerol influx causing vesicle swelling.

3.2 Water permeability and $HgCl_2$ block

Water permeabilities of SM vesicles and the affect of $HgCl_2$ were previously measured by Niemietz and Tyerman (2000). These values were obtained from the kinetics of vesicle shrinkage and are shown in Fig. 2C. To examine why our values are in disagreement with Rivers *et al.*, (1997) we performed additional experiments where P_{OS} was determined from vesicle swelling kinetics (Fig. 2A). These values (Fig. 2C) are similar to those of Rivers *et al.*, (1997). We also tested swelling and shrinking kinetics on vesicles isolated from the same nodules (Fig. 2B) in case the difference between endosmotic P_{OS} and exosmotic P_{OS} was due to a long-term seasonal effect, since the two sets of experiments shown in Fig. 2C were obtained on different sets of nodules at different times of the year. The example in Fig. 2B shows that there appears to be an intrinsic rectification of water flow across the SM membrane. If the P_{OS} values were similar for both directions of flow we would expect a much larger rate constant for shrinking taking into account the effect of solute concentration on the kinetics.

Inhibition by $HgCl_2$ occurred for both directions of water flow (Fig. 2C). The concentration of $HgCl_2$ giving half maximal inhibition was not significantly different between swelling and shrinking.

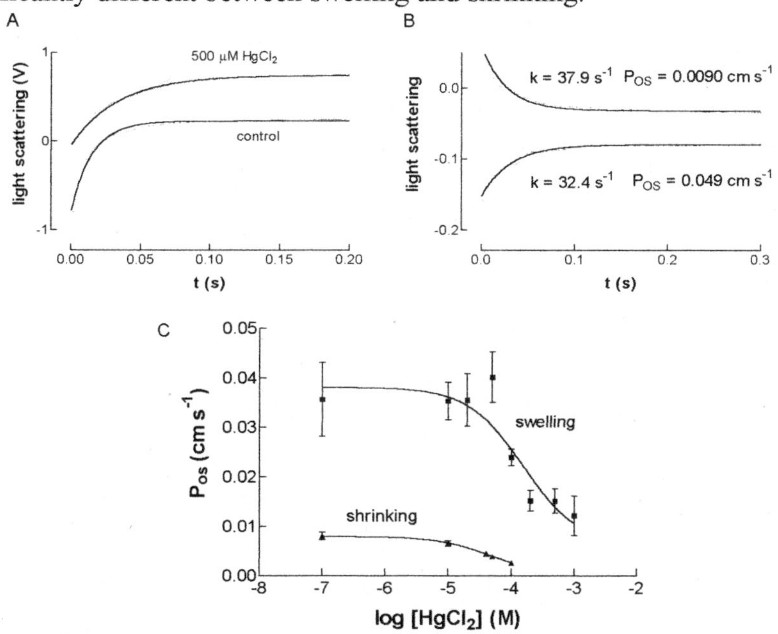

Figure 2. Water permeability measured from swelling or shrinking kinetics of SM vesicles subjected to $HgCl_2$ treatment. A. Example of light scattering kinetics for swelling and the effect of $HgCl_2$. B. Example of light scattering kinetics of SM vesicles from the same isolation batch subjected to hyper- and hypoosmotic gradients. C. Dose response curves for the effect of $HgCl_2$ on P_{OS} for swelling and shrinking. The values for shrinking were obtained from Niemietz and Tyerman (2000). The fitted dose-response curves give half-maximal inhibitions at 152 μM $HgCl_2$ for swelling and 53 μM $HgCl_2$ for shrinking.

3.3 Glycerol and urea permeability and rectification caused by HgCl2

Examples of glycerol influx and efflux kinetics are shown in Fig. 3A,B respectively. The effect of $HgCl_2$ was to inhibit glycerol efflux (Fig. 3B) but to stimulate glycerol influx (Fig. 3A). This resulted in strong rectification of glycerol permeation at high $HgCl_2$ concentrations (Fig. 3C).

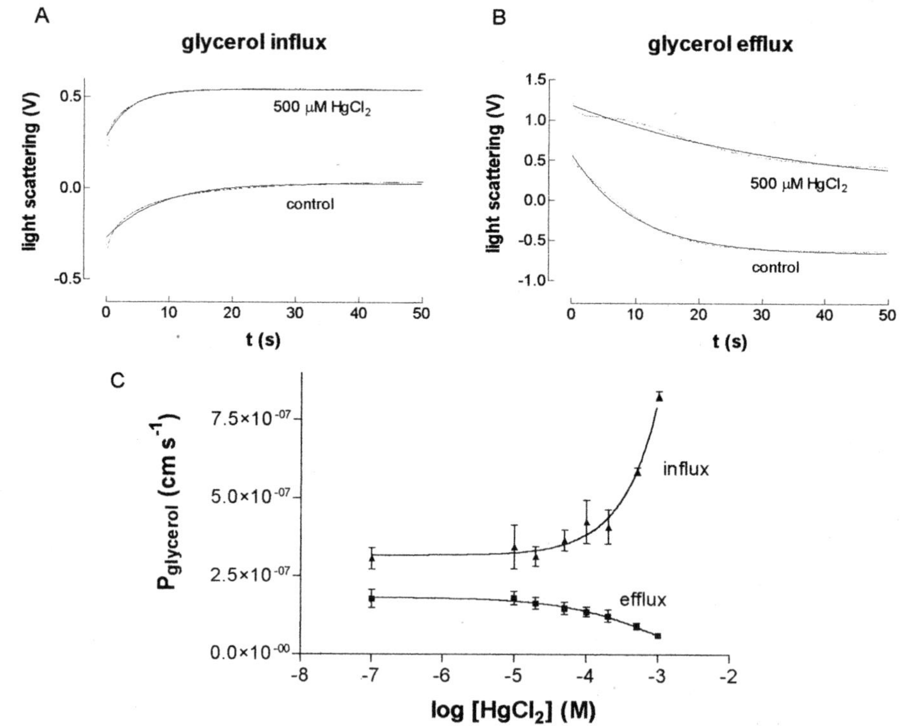

Figure 3. Glycerol permeability of SM vesicles and the effect of $HgCl_2$ treatment. A. Example of glycerol influx kinetics from light scattering and the stimulatory effect of $HgCl_2$. B. Example of glycerol efflux kinetics and the inhibitory effect of $HgCl_2$. C. Dose-response curves showing the significant degree of rectification at high $HgCl_2$ concentrations. For efflux, half maximal inhibition occurred at 965 μM $HgCl_2$.

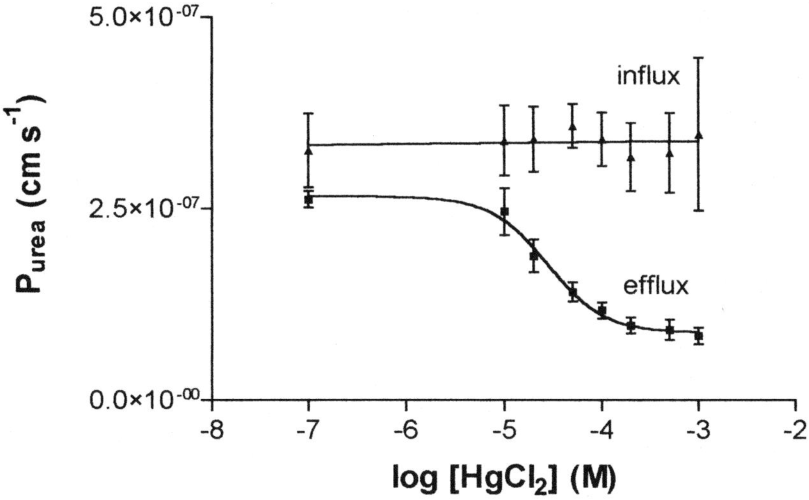

Figure 4. Urea permeability of SM vesicles and the effect of $HgCl_2$ treatment. The half maximal inhibition for urea efflux occurred at 31 μM $HgCl_2$.

Similar experiments were performed with urea as the permeating solute and the effect of $HgCl_2$ examined (Fig. 4). Efflux was strongly inhibited by $HgCl_2$, but in contrast to glycerol influx, there was no effect of $HgCl_2$ on urea influx.

3.4 Time dependent responses and effect of ATP

Rivers *et al.*, (1997) reported no apparent mercury inhibition of urea efflux from their SM vesicles. To reconcile our result (Fig. 4) with those of Rivers *et al.*, (1997) we tested the effect of overnight incubation at 4°C. This lead to a decrease in mercury sensitivity (Fig. 5) such that at 0.1 mM $HgCl_2$ (concentration used by Rivers *et al.*, 1997) there would be little apparent inhibition by mercury.

From our previous report of effects of ATP treatment on NH_3 permeability we tested the effect of ATP treatment on glycerol influx and efflux. Fig. 6 shows an example of the effect of ATP treatment for two periods on both influx and efflux. There was a clear inhibitory affect on glycerol efflux but very little effect on glycerol influx.

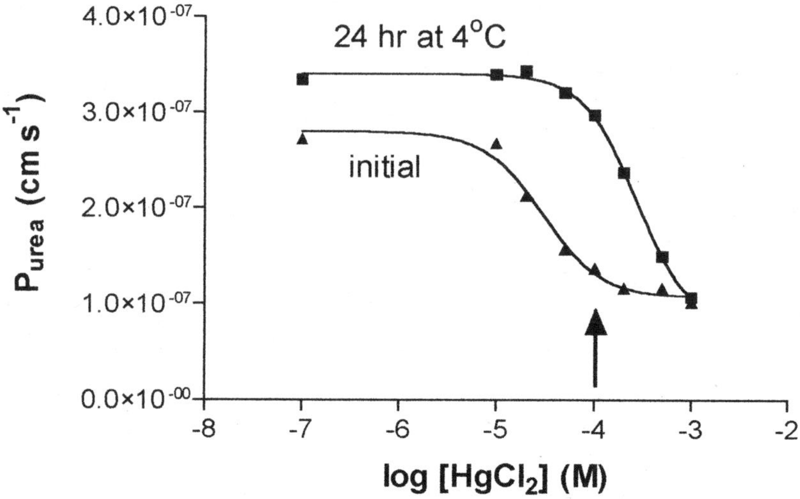

Figure 5. Effect of overnight incubation at 4°C on the mercury sensitivity of urea (efflux) permeability. The arrow indicates the concentration of $HgCl_2$ at which Rivers *et al.*, (1997) tested mercury sensitivity of their SM vesicles.

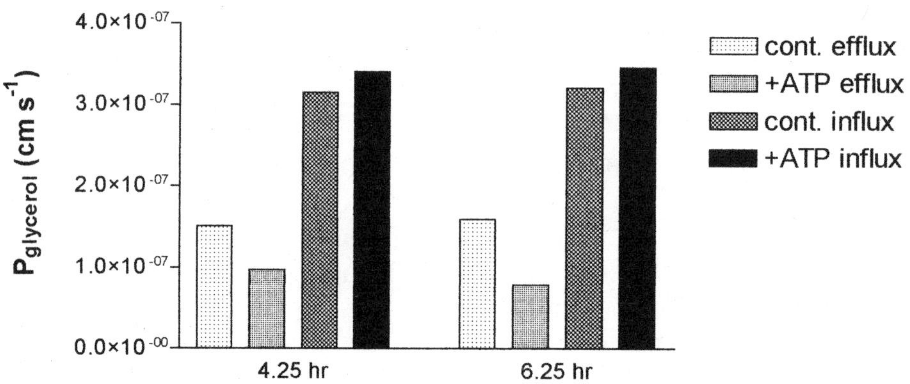

Figure 6. Effect of ATP incubation on the directionality of glycerol permeation in SM vesicles. The incubation medium contained 2 mM MgATP, 1 mM $CaCl_2$ in standard sucrose buffer.

4. DISCUSSION

4.1 Water permeation

Our results show an apparent rectification of water flow across the SM of soybean nodules. The P_{OS} values for water influx for SM vesicles, equivalent to water efflux from symbiosome to cytoplasm, are high and similar to those reported by Rivers *et al.*, (1997) for water efflux from SM vesicles. On the other hand our P_{OS} values for water efflux are significantly lower than those reported by Rivers *et al.*, (1997). One possible explanation for the disparity is that the sidedness of the vesicles used in the two studies was different. Perhaps the vesicles used by Rivers *et al.*, (1997) for efflux were predominantly in the opposite orientation to ours. Another possibility is that the water permeability of NOD26 is sensitive to osmotic pressure such that P_{OS} decreases as osmolarity increases. This has been observed for the hydraulic conductivity of *Chara* cells (Steudle and Tyerman, 1983) and could explain the rectification we have observed as well as the difference between our values and those of Rivers *et al.*, (1997), since they used lower bathing osmolarities for P_{OS} determinations.

The biological significance of rectification of water flow is unclear for the symbiosome. However, it does indicate that unlike AQP 0- AQP5 (Meinild *et al.*, 1998), NOD26 may not give symmetrical bidirectional water flow. This property of NOD26 makes it dissimilar to AQP 0 in an otherwise similar array of functional characteristics (Rivers *et al.*, 1997).

4.2 Glycerol and urea permeation

Rivers *et al.*, (1997) and Dean *et al.*, (1999) showed that NOD26 was permeable to glycerol and that glycerol permeation was inhibited by mercury. We also observed mercury inhibition but this clearly depended upon the direction of glycerol permeation across the SM membrane. Since mercury stimulated glycerol influx but inhibited glycerol efflux a significant degree of rectification occurred at high mercury concentrations. The vesicles remained osmotically competent at the higher mercury concentrations and showed the same magnitude of change in volume as the control vesicles. The rectification behaviour indicates that the NOD26 channel is somehow able to detect the gradient for glycerol permeation when treated with mercury. A similar effect of mercury was observed for urea permeation but stimulation of influx was not observed. Barone *et al.*, (1997) showed that mercury induced a conformational change in PMIP31, an aquaporin from *Beta vulgaris*. Perhaps such a conformational change in NOD26 can confer sufficient asymmetry as to allow rectification of solute flow. ATP treatment also resulted in greater rectification with efflux inhibited and influx stimulated. This indicates that phosphorylation of NOD26 may enhance rectification. The direction of permeation that is reduced by possible phosphorylation corresponds to solute influx to the symbiosome.

Our permeability values for glycerol are much smaller than those reported by Rivers *et al.*, (1997) for soybean SM vesicles. On the other hand our urea permeabilities are similar. The difference in glycerol permeability is difficult to reconcile. Perhaps there is an unknown gating factor of NOD26 that was different in our experiments. For example, both pH and Ca^{2+} have been shown to gate AQP 0 (Nemeth-Cahalan and Hall, 2000). Although pH was the same in our experiments as Rivers *et al.*, (1997), we had 0.1 mM Ca^{2+} present while Rivers *et al.*, (1997) had Mg^{2+} present.

On the basis of lack of inhibition by mercury, Rivers *et al.*, (1997) concluded that urea did not permeate NOD26. However, we observed that urea permeation of SM vesicles was very sensitive to mercury for urea efflux. Storage overnight at 4 °C significantly diminished the inhibition by mercury. Perhaps sidedness of the vesicles can also account for the difference in results as was suggested for water permeation, since urea influx was not affected by mercury.

5. CONCLUSION

Directionality of water and solute permeation occurs in the SM vesicles of soybean. Since NOD26 has been shown to account for both water and glycerol permeation in reconstituted proteoliposomes (Dean *et al.*, 1999) it is highly likely that NOD26 accounts for the rectification of permeation observed here in native SM vesicles. However, confirmation of these results is required using NOD26 expressed in oocytes to exclude permeation via other SM membrane proteins. Incorporation of purified NOD26 into proteoliposomes should be used to check its contribution to NH_3 permeation but may not reveal rectification if NOD26 has random orientation within the membrane. However, if water permeation through NOD26 is sensitive to solute concentration this would be revealed in proteoliposomes. Our results indicate that some aquaporins may have many other interesting facets of functional behaviour not unlike that observed for ion channels.

ACKNOWLEDGMENTS

This work was supported by a grant from the Australian Research Council to SDT. We wish to thank Wendy Sullivan for her expert technical assistance.

REFERENCES

Barone LM Shih C and Wasserman BP 1997 Mercury-induced conformational changes and identification of conserved surface loops in plasma membrane aquaporins from higher plants *J Biol Chem* **272**: 30672-30677

Christiansen JH Rosendahl L and Widell S 1995 Preparation and characterization of sealed inside-out peribacteroid membrane vesicles from *Pisum sativum* L and Glycine max L root nodules by aqueous polymer two-phase partitioning *J Plant Physiol* **147**: 175-181

Dean RM Rivers RL Zeidel ML and Roberts DM 1999 Purification and functional reconsititution of soybean Nodulin 26 An aquaporin with water and glycerol transport properties *Biochemistry* **38**: 347-353

Lande MB Donovan JM and Zeidel ML 1995 The relationship between membrane fluidity and permeabilities to water solutes ammonia and protons *J Gen Physiol* **106**: 67-84

Lee JW Zhang Y Weaver CD Shomer NH Louis CF and Roberts DM 1995 Phosphorylation of nodulin 26 on serine 262 affects its voltage-sensitive channel activity in planar lipid bilayers *J Biol Chem* **270**: 27051-27057

Meinild A-K Klaerke DA and Zeuthen T 1998 Bidirectional water fluxes and specificity for small hydrophilic molecules in aquaporins 0-5 *J Biol Chem* **273**: 32446-32451

Németh-Cahalan KL and Hall JE 2000 pH and calcium regulate the water permeability of aquaporin 0 *J Biol Chem* **275**: 6777-6782

Niemietz CM and Tyerman SD 1997 Characterization of water channels in wheat root membrane vesicles *Plant Physiol* **115**: 561-567

Niemietz CM and Tyerman SD 2000 Channel-mediated permeation of ammonia gas through the peribacteroid membrane of soybean nodules *FEBS Letters* **465**: 110-114

Rivers RL Dean RM Chandy G Hall JE Roberts DM and Zeidel ML 1997 Functional analysis of nodulin 26 an aquaporin in soybean root nodule symbiosomes *J Biol Chem* **272**:16256-16261

Steudle E and Tyerman SD 1983 Determination of permeability coefficients reflection coefficients and hydraulic conductivity of *Chara corallina* using the pressure probe: effects of solute concentrations *J Membrane Biol* **75**: 85-96

Tyerman SD Whitehead LF and Day DA 1995 A channel-like transporter for NH_4^+ on the symbiotic interface of N_2-fixing plants *Nature* **378**: 629-632

Weaver CD Shomer NH Louis CF Roberts DM 1994 Nodulin 26 a nodule specific symbiosome membrane protein from soybean is an ion channel *J Biol Chem* **269**:17858-17862

Whitehead LF Day DA and Tyerman SD 1998 Divalent cation gating of an ammonium permeable channel in the symbiotic membrane from soybean nodules *Plant J* **16**: 313-324

4.11 Plant physiology

ANALYSIS OF FOUR MIP-LIKE PROTEINS FROM THE DEVELOPING PEA SEED COAT

[1]Joost T. van Dongen, [1]Jolanda A.M.J. Schuurmans, [2]Peter M.T. Deen, and [1]Adrianus C. Borstlap
[1] Dept. of Plant Biology, Utrecht University, P.O. Box 80084, NL-3508 TB Utrecht, The Netherlands: [2] Dept. of Cell Physiology, University of Nijmegen, P.O. Box 9101, NL-6500 HB Nijmegen, The Netherlands. E-mail: J.T.vanDongen@bio.uu.nl

1. ASSIMILATE TRANSPORT IN PLANTS

Higher plants contain two specialised transport tissues: xylem and phloem. Tracheary elements in the xylem serve the transport of water and inorganic ions from the root system to the shoot, whereas the sieve elements in the phloem are responsible for the translocation of organic solutes (mostly sucrose and a mixture of amino acids) from 'source' (typically the full-grown, photosynthesising leaf) to 'sink' (young leaves, roots, fruits and seeds). Developing seeds store large amounts of starch, lipids and reserve proteins, which are synthesised from the organic solutes translocated in the sieve elements. Because this translocation is by a pressure-flow mechanism (Münch 1930), considerable amounts of water (Münch water) are co-translocated with the organic solutes, which may give rise to the circulation of water within the tissues of the 'sink' (Jenner *et al.*, 1988).

1.1 Transfer of nutrients in developing seeds

Seeds of legumes (pea, bean, soybean) consist of an embryo surrounded by a seed coat of maternal tissue. In the developing pea seed nutrients supplied by the mother plant arrive by way of the sieve elements in three vascular strands embedded in the parenchymal tissue of the seed coat (Fig. 1). Experiments with fluorescent tracers indicate that the contents of the sieve elements move symplasmically to the seed coat parenchyma cells from which they are released into the apoplasmic space facing the cotyledons of the embryo (Patrick *et al.,* 1995; Tegeder *et al.,* 1999).

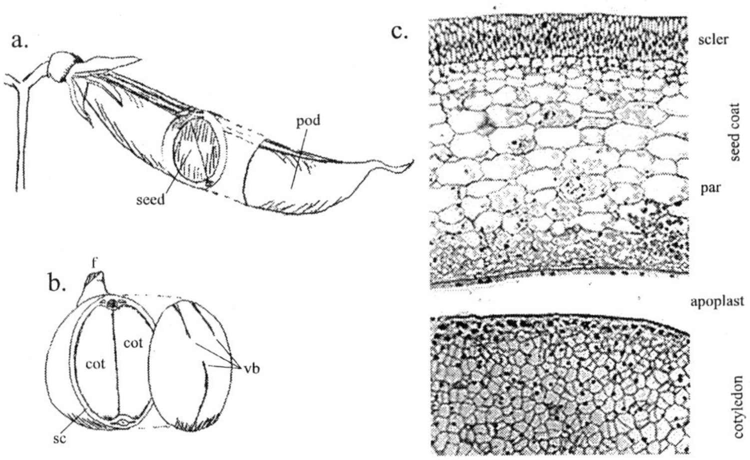

Figure 1. Anatomy of a pea fruit. a) Mature pod (30 days after flowering) carrying 2 to 6 seeds. b) Pea seed: the two cotyledons (cot) of the embryo are surrounded by a maternal seed coat (sc) with three vascular bundles (vb). The funiculus (f) connects the seed coat with the mother plant. c) cross section through a seed 14 days after flowering. The seed coat consists of two distinct layers: a wax-coated outer layer of thick walled sclerenchyma cells (scler) and an inner layer of parenchyma cells (par). The cotyledons and the seed coat are symplasmically isolated from each other.

Cotyledon transfer cells (specialised epidermal cells with cell wall invaginations that increase the membrane surface) take up the released nutrients (mainly sucrose and amino acids) from the apoplast. Recently, a range of H^+-symporters for sucrose and amino acids have been cloned from plants, including symporters that are expressed in the developing pea cotyledon (Tegeder *et al.,* 1999; 2000).

Because no symplasmic connections exist between seed coat and embryo, transport across the plasma membrane of seed coat parenchyma cells is a requisite step in nutrient unloading from the seed coat. Membrane transport in seed coat parenchyma cells has been investigated in our lab for sugars, amino acids and organic cations. Uptake experiments with ^{14}C-labelled substrates revealed a linear relationship between the influx and the external solute concentration, which suggests a diffusional transport mechanism. Estimated permeability coefficients for sucrose, mannitol, and several neutral amino acids were all in the range of 3 to 9 $nm \cdot s^{-1}$, which is four to six orders of magnitude larger than that of artificial lipid bilayers (De Jong *et al.*, 1996, 1997). Organic cations like L-lysine, L-histidine and choline have similar permeability coefficients at zero membrane potential, but their influx is accelerated at the resting membrane potential of about –50 mV (Van Dongen *et al.*, in prep.). The release of endogenous sugars and amino acids from seed coats as well as the influx of ^{14}C-labelled substrates was not affected by the protonophore CCCP and partially inhibitable by the organic mercury compound pCMBS. Taken together, these observations suggest that transport of sugars and amino acids across the membrane of seed coat parenchyma cells is facilitated by poorly selective, pore-forming transport proteins.

1.2 Possible roles of MIP-like proteins in pea seeds

Possible candidates for pore-forming proteins suggested to be involved in the release of sugars and amino acids from pea seed coats may be found in the family of the Major Intrinsic Proteins. Members of this family are particularly known for their ability to transport water and/or some small solutes like urea and glycerol (Van Os *et al.*, 1994; Maurel 1997a), but at least one MIP-like protein (AQP9 from rat liver and testis) seems to be a broad selectivity neutral solute channel facilitating the membrane transport of various neutral solutes such as carbamides, polyols, purines and pyrimidines (Tsukaguchi *et al.*, 1998). It has also been suggested that specific MIP-like proteins can facilitate the transport of ions. For example, Weaver *et al.*, (1994) showed increased permeability for ions after reconstitution of the MIP-like NOD26 in proteoliposomes. However, this observation has not been supported by heterologous expression experiments in *Xenopus* oocytes (Dean *et al.*, 1999).

Besides a possible role in assimilate unloading, two other functions of MIP-like proteins can be put forward in relation to seed physiology. First, it seems likely that aquaporins are involved in fluxing the Münch water through tissues of the developing seed. Secondly, it has been suggested that MIPs are of importance during germination. For example, ∀TIP cloned from bean might control the iso-osmotic status of the vacuoles during germination when starch and storage proteins are degraded (Maurel *et al.*, 1997b).

2. FOUR MIP-LIKES ARE EXPRESSED IN PEA SEED COAT TISSUE

2.1 Sequence alignment and similarity search

We recently cloned four genes coding for MIP-like proteins from a cDNA bank of the pea seed coat. One of these, PsTRG31 (EMBL Z18288) is a turgor-responsive gene previously cloned by Guerrero and Crossland (1993). The other three proteins we found are novel MIP-likes now registered as PsPIP2 (EMBL AJ243307), PsTIP (EMBL AJ243309) and PsNLM (EMBL AJ243308).

Table 1. Amino acid sequence homology between the four MIP-like genes isolated from pea seed coats and several other MIP-likes from plants. Highest homology percentages are highlighted in bold. Note that in the developing pea seed coat at least one member of each subgroup is expressed

		PsTRG31	PsPIP2	PsNLM	PsTIP
PIP1 group	PsTRG31	**100**	65	32	35
	AtPIP1a	**82**	66	27	34
	McMIPb	**83**	68	28	34
PIP2 group	PsPIP2	65	**100**	27	33
	AtPIP2b	69	**76**	30	35
	McMIPc	69	**78**	28	37
NLM group	PsNLM	32	30	**100**	35
	GmNOD26	28	28	**68**	35
	OsMIP	28	29	**70**	32
TIP group	PsTIP	35	33	35	**100**
	TrGTIPlp	32	33	35	**93**
	At(TIP	33	33	33	**80**

EMBL accession codes and plant species of the proteins: TRG31 (Z18288), PIP2 (AJ243307), TIP (AJ243309), NLM (AJ243308) from *Pisum sativum*, PIP1a (X75881), PIP2b(X75884), (TIP(X63552) from *Arabidopsis thaliana*, MIPb (L36097), MIPc (U73466) from *Mesembryanthemum crystallinum*, NOD26 (X04782) from *Glycine max*, MIP (D17443) from *Oryza sativa*, GTIPlp (Z29946) from *Trifolium repens*.

Amino acid sequence analysis revealed that all four subgroups as defined by Park and Saier (1996) and Schäffner (1998) are represented in the seed coat (Table 1). PsTRG31 and PsPIP2 belong to the PIP1- and PIP2-subgroups, respectively. PsTIP shows high similarity with the subgroup of tonoplast intrinsic proteins and PsNLM can be classified into the NOD26-like subgroup. NOD26 from soybean has been shown to be a glycerol facilitator after heterologous expression in *Xenopus* oocytes (Dean *et al.*, 1999) and induced an increased ion permeability after incorporation in lipid bilayers (Weaver *et al.*, 1994).

2.2 Functional analysis

Heterologous expression in *Xenopus* oocytes is currently used to screen the functionality of the selected proteins. Injection of 25 ng of polyadenylated and capped cRNA coding for PsPIP2 or PsTIP clearly resulted in an increased permeability of the oocyte membrane for water (Table 2). For PsPIP2, as little as 2 ng cRNA was sufficient to see a significant increase in water permeability. Injection of 25 ng of cRNA coding for PsTRG31 or PsNLM resulted in a small increase in water permeability only. It has not been tested yet whether the membrane permeability for different solutes will increase after heterologous expression, but these experiments will soon be carried out.

Table 2. Water permeability of the oocyte membrane after heterologous expression of the four MIP-likes. Ten oocytes were injected with 25 ng RNA coding for PsPIP2, PsTIP, PsTRG or PsNLM. After three days, the oocytes were transferred to 1/10 strength MBS and the time of bursting was recorded. Injection of the same volume (46 nl) of distilled water served as a control. All oocytes in this experiment came from one frog.

Time (min)	0	10	20	30	40	50	60
PsPIP2	0	3	9	**10**	**10**	**10**	**10**
PsTIP	0	4	5	8	**10**	**10**	**10**
PsTRG31	0	0	4	4	**10**	**10**	**10**
PsNLM	0	0	2	4	8	**10**	**10**
Water	0	0	2	2	5	8	**10**

2.3 Localisation and expression patterns

Northern blot analysis revealed that, within the pea fruit, PsNLM is only expressed in the seed coat; the localisation of the others was not organ specific (Figure 2). Schäffner (1998) has suggested that MIPs belonging to the same subgroup have the same intracellular localisation. This suggestion was based on the limited number of immunolocalisations of plant MIPs that was known at that time (Daniels *et al.*, 1994; Kammerloher *et al.*, 1994; Robinson *et al.*, 1995). More recent immunoassays confirmed this hypothesis for members of the TIP-group (Barkla *et al.*, 1999; Karlson *et al.*, 2000), but members of the PIP-group do not seem to be selectively located in the plasma membrane (Barkla 1999). To obtain information about the cellular and intracellular localisation antibodies are now being raised against specific epitopes of the four different MIPs cloned in our lab.

In seed coats, the expression level of PsTRG31, PsPIP2 and PsNLM increased during development whereas that of PsTIP decreased (data not shown). In the embryo (cotyledons) PsTRG31 was constitutively expressed, while RNA levels of PsPIP2 and PsTIP decreased during seed maturation. Johnson *et al.*, (1989) found an increase of the tonoplast protein TP 25 during maturation of bean cotyledons. The protein disappeared during germination. A highly related protein was detected in a pea cotyledon immunoblot using bean TP 25 antiserum. Later (Johnson *et al.*, 1990) this protein appeared to be a member of the MIP-like family and the protein was renamed as ∀TIP. Maurel (1997b) has suggested that ∀TIP may control the rehydratation of protein storage vacuoles during seed imbibition. Obviously, the TIP found in the present study, has another function, since it has a different expression pattern during development.

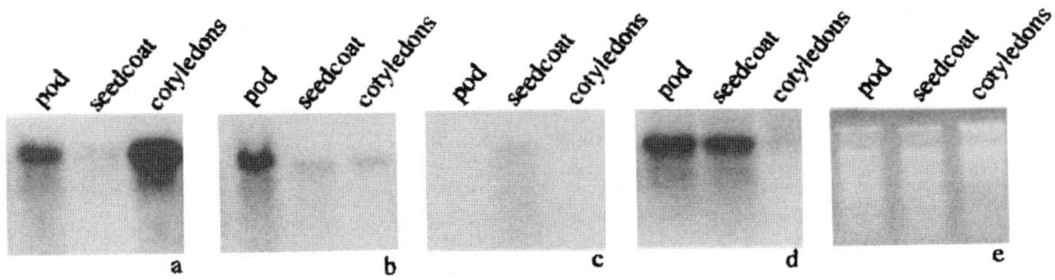

Figure 2. Northern blot analysis of a) PsTRG31, b) PsPIP2, c) PsNLM, d) PsTIP isolated from different pea fruit tissues. e) Ethidiumbromide stained RNA gels used as loading control.

Summarising, several MIP-like proteins are simultaneously present in the same organ. Their water permeabilities seem to differ and they do not have the same transcription patterns. Therefore, they will probably have different functions at various (sub)cellular locations. Whether one of them could function in assimilate unloading has to be investigated. Detailed analyses using antibodies raised against protein specific epitopes and further functional analysis using heterologous expression in *Xenopus* oocytes will, hopefully, help to solve this puzzle.

ACKNOWLEDGMENTS

We are indebted to Erik-Jan Kamsteeg (University of Nijmegen) for teaching JTvD how to perform the heterologous expression experiments, and Bas van Balkom (University of Nijmegen), Madeleine Wouterlood, Alex

Boonman and Gerco van Eikenhorst (Utrecht University) for their practical assistance.

REFERENCES

Barkla BJ Vera-Estrella R Pantoja O Kirch HH and Bohnert HJ 1999 Aquaporin localization – How valid are the TIP and PIP labels? *Trends Plant Sci* **4**: 86-88

Daniels MJ Mirkov TE and Chrispeels MJ 1994 The plasma membrane of *Arabidopsis thaliana* contains a mercury insensitive aquaporin that is a homolog of the tonoplast water channel protein Tip *Plant Physiol* **106**: 1325-1333

De Jong A Koerselman-Kooij JW Schuurmans JAMJ and Borstlap AC 1996 Characterization of the uptake of sucrose and glucose by isolated seed coat halves of developing pea seeds Evidence that a sugar facilitator with diffusional kinetics is involved in seed coat unloading *Planta* **199**: 486-492

De Jong A Koerselman-Kooij JW Schuurmans JAMJ and Borstlap AC 1997 The mechanism of amino acid efflux from seed coats of developing pea seeds as revealed by uptake experiments *Plant Physiol* **114**: 731-736

Dean RM Rivers RL Zeidel ML and Roberts DM 1999 Purification and functional reconstitution of soybean Nodulin 26 An aquaporin with water and glycerol transport properties *Biochemistry* **38**: 347-353

Guerrero FD and Crossland L 1993 Tissue-specific expression of a plant turgor-responsive gene with amino acid sequence homology to transport-facilitating proteins *Plant Mol Biol* **21**: 929-935

Jenner CF Xia Y Eccles CD and Callaghan PT 1988 Circulation of water within wheat grain revealed by nuclear magnetic resonance micro-imaging *Nature* **336**: 399-402

Johnson KD Herman EM and Chrispeels MJ1989 An abundant highly conserved tonoplast protein in seeds *Plant Physiol* **91**: 1006-1013

Johnson KD Höfte H and Chrispeels MJ 1990 An intrinsic tonoplast protein of protein storage vacuoles in seeds is structurally related to a bacterial solute transporter (GlpF) *Plant Cell* **2**: 525-532

Kammerloher W Fischer U Piechottka GP and Schäffner AR 1994 Water channels in the plant plasma membrane cloned by immunoselection from a mammalian expression system *Plant J* **6**: 187-199

Karlsson M Johansson I McCann MC Maurel C Larsson C and Kjellbom P 2000 An abundant TIP expressed in mature highly vacuolated cells *Plant J* **21**: 83-90

Maurel C 1997a Aquaporins and water permeability of plant membranes *Annu Rev Plant Physiol Plant Mol Biol* **48**: 399-429

Maurel C Chrispeels M Lurin C Tacnet F Geelen D Ripoche P and Guern J 1997b Function and regulation of seed aquaporins *J Exp Bot* **48**: 421-430

Münch E 1930 *Die Stoffbewegungen in der Pflanze* Fischer-Verlag Jena

Patrick JW Offler CE and Wang W-D 1995 Cellular pathway of photosynthate transport in coats of developing seed of *Vicia faba* L and *Phaseolus vulgaris* L I Extent of transport through the coat symplast J Exp Bot **46**: 35-47

Park JH and Saier Jr M H 1996 Phylogenetic characterization of the MIP family of transmembrane channel proteins *J Membrane Biol* **153** 171-180

Robinson DG Hoh B Hinz G and Jeong B-K 1995 One vacuole or two vacuoles: do protein storage vacuoles arise *de novo* during pea cotyledon development? *J Plant Physiol* **145**: 654-664

Schäffner AR 1998 Aquaporin function structure and expression: are there more surprises to surface in water relations? *Planta* **204**:131-139

Tegeder M Wang X-D Frommer WB Offler CE and Patrick JW 1999 Sucrose transport into developing seeds of *Pisum sativum* L *Plant J* **18**: 141-149

Tegeder M Offler CE Frommer WB and Patrick JW 2000 Amino acid transporters are localized to transfer cells of developing pea seeds *Plant Physiol* **122**: 319-326

Tsukaguchi H Shayakul C Berger UV Mackenzie B Devidas S Guggino WB Van Hoek AN and Hediger MA 1998 Molecular characterization of a broad selectivity neutral solute channel *J Biol Chem* **273**: 24737-24743

Van Os CH Deen PMT and Dempster JA 1994 Aquaporins: water selective channels in biological membranes Molecular structure and tissue distribution *Biochim Biophys Acta* **1197**: 291-309

Weaver CD Shomer NH Louis CF and Roberts DM 1994 Nodulin 26 a nodule-specific symbiosome membrane protein from soybean is an ion channel *J BiolChem* **269**: 17858-17862

4.12 Plant physiology

AQUAPORIN REGULATION UNDER SALT AND OSMOTIC STRESS IN THE HALOPHYTE *MESEMBRYANTHEMUM CRYSTALLINUM* L.

[1]Rosario Vera-Estrella, [1]Bronwyn J. Barkla, [1]Claudia Gallardo-Amarillas, [2]Hans J. Bohnert and [1]Omar Pantoja
[1]Departamento de Biología Molecular de Plantas, Instituto de Biotecnología, UNAM, A.P. 510-3, Colonia Miraval, Cuernavaca, Morelos, 62250, México, [2]Department of Biochemistry, University of Arizona, Tucson, AZ85721, USA. E-mail: rosario@ibt.unam.mx

1. INTRODUCTION

It is evident that plant aquaporins should play a dynamic role in maintaining cellular water homeostasis under conditions that necessitate modifications in water flux, including drought- and salt-stress. Changes in water uptake and allocation would be required to balance alterations in cellular osmotic potential and therefore, both plasma membrane and vacuolar membrane (tonoplast) aquaporin activity and/or expression must be tightly regulated.

1.1 MIP's and Water Transport in Plants

That MIP's play a role in plant water transport can be implied from their distribution in tissues with high water conductances (Kaldenhoff *et al.*, 1995 and 1998, Yamada *et al.*, 1995, Sarda *et al.*, 1997, Barrieu *et al.*, 1998, Grote *et al.*, 1998), and their expression during specific stages of plant development (Ludevid *et al.*, 1992, Yamada *et al.*, 1997, Fukuhara *et al.*, 1999). Environmental stimuli, resulting in an osmotic stress, have also been shown to regulate MIP expression. Transcripts and promotor-reporter gene expression, for both putative tonoplast and plasma membrane MIP, have been shown to be either induced or down-regulated by drought, dehydration, dessication and/or salinity stress (Daniels *et al.*, 1994, Fray *et al.*, 1994, Liu *et al.*, 1994, Yamada *et al.*, 1995, Mariaux *et al.*, 1998, Barrieu *et al.*, 1998. Pih *et al.*,., 1999). More direct evidence for MIP involvement in plant water transport comes from few studies in which the water channel (aquaporin) activity has been clearly demonstrated. Heterologous expression of several plant *Mip* mRNA's in *Xenopus laveis* oocytes resulted in an increased osmotic water permeability (P_f; Maurel *et al.*, 1993, Daniels *et al.*, 1994, Yamada *et al.*, 1995, Johansson *et al.*, 1998), and *Arabidopsis* constitutively expressing an antisense construct of the *AthH2* MIP, showed a reduced rate of protoplast osmotic swelling correlated to a reduction in AthH2 protein levels (Kaldenhoff *et al.*, 1995). Reconstitution of NOD26, a peribacteroid membrane specific MIP, in liposomes, resulted in a higher P_f than that measured in control liposomes (Dean *et al.*, 1999).

1.2 *Mesembryanthemum crystallinum* and MIP's

The halophyte, *Mesembryanthemum crystallinum*, presents itself as an ideal model plant system for the study of the role of MIP/aquaporins in salt-adaptation. To date, a total of 14 transcripts encoding different members of the MIP family have been identified, and nine full-length *M. crystallinum* MIP-like genes, with representative members from the PIP (plasma membrane intrinsic protein) and TIP (tonoplast intrinsic protein) subfamilies, have been characterized (Yamada *et al.*, 1995, Kirch *et al.*, 2000). Functional analysis of two *M. crystallinum* PIP (MIP-A and MIP-B), by heterologous expression in *Xenopus* oocytes, supported their role as aquaporins (Yamada *et al.*, 1995). In situ hybridization and immuno-cytological localization studies have demonstrated high expression of *M. crystallinum* MIP and aquaporins in tissues that contain a large percentage of expanding cells or show high water conductance, including phloem-associated cells, xylem parenchyma, and in the epidermis and endodermis of immature roots (Yamada *et al.*, 1995, Kirch *et al.*, 2000).

Using peptide specific antibodies, membrane localization of four of the *M. crystallinum* MIP was compared to that assigned from phylogenetic classification (Yamada *et al.*, 1995). Studies confirmed the TIP, MIP-F, as tonoplast located, whereas those grouped as PIPs, showed a complex membrane distribution which did not conform to that designated based on amino acid sequence analysis. MIP-A and -B were detected in tonoplast fractions and in fractions distinct from either tonoplast or plasma membrane and MIP-C, although abundant in the plasma membrane was also found in non-plasma membrane fractions (Barkla *et al.*, 1999a, Kirch *et al.*, 2000). These results raised the possibility that within the cell, MIP are mobile, and depending on the cellular requirement for water flux and the specific stimuli, trafficking of MIP between membranes may occur.

2. RESULTS

2.1 MIP Regulation in *M. crystallinum* by Salt- and Osmotic-Stress

M. crystallinum presents several adaptive mechanisms to combat salt- and drought-stress, including the accumulation and sequestration of sodium in the vacuole (Barkla *et al.*, 1999b), balanced by the synthesis of compatible solutes in the cytoplasm (Adams *et al.*, 1998). As a consequence, changes in osmotic potential drives the movement of water between cellular compartments and into the cell, suggesting a requirement for the regulation of aquaporin activity at both the plasma membrane and tonoplast.

Previously, transient reductions in amount of transcripts for *MipA*, *MipB* and *MipC*, were observed in roots upon exposure to osmotic stress (Yamada *et al.*, 1995). Western blot analysis of MIP protein from *M. crystallinum* plants under salt-stress conditions showed no salt-induced changes in expression for MIP-A or MIP-B, however MIP-C levels increased significantly (Kirch *et al.*, 2000). In contrast, MIP-F showed lower amounts in stressed tissue (Kirch *et al.*, 2000). Moreover, as NaCl concentrations were sequentially raised in cell suspension cultures of *M. crystallinum* (Vera-Estrella *et al.*, 1999), there was a steady decline in MIP-F protein amount as a function of the salt-concentration (Figure 1). Interestingly, this response was specific for salt-stress, as treatment of cell suspensions with mannitol resulted in a large increase in MIP-F protein (Figure 1), indicating a differential regulation of MIP-F by ionic- (salt) and osmotic-stress (mannitol).

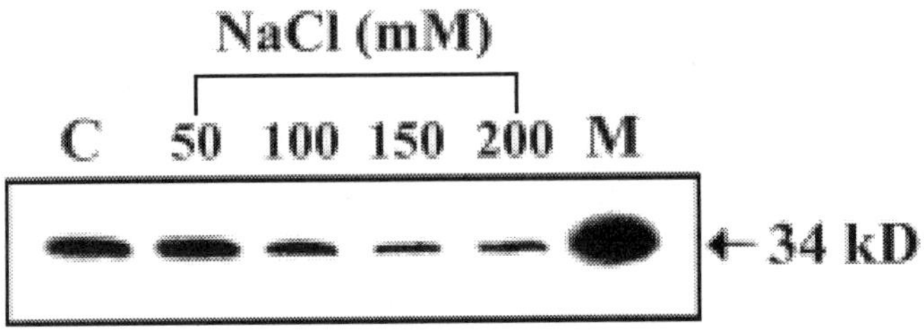

Figure 1. Western blot analysis of the effect of NaCl and mannitol (400 mM) on protein levels of MIP-F in purified tonoplast from cell suspensions of *M. crystallinum* treated for 5 days.

To investigate the possibility that environmental stimuli may alter the membrane distribution of specific MIP, as well as simply changing the amount of protein in a specific membrane fraction, localization of MIP-F protein in fractions from continuous sucrose gradients of microsomal membranes from control and mannitol-stressed cells of *M. crystallinum* were compared (Figure 2). As expected, there was an increase in MIP-F protein in the tonoplast (Figure 2, fraction #13), however, additionally MIP-F was now detected in membrane fractions of a higher density, overlapping with plasma membrane fractions, and the level of expression in these fractions was higher than that observed in the tonoplast (Figure 2, compare fractions #22 and #31 with #13).

2.2 Regulation of Tonoplast Water Permeability

In order to correlate the osmotic- and salt-induced changes in MIP expression at the protein level with changes in membrane water permeability, light scattering by tonoplast vesicles was measured using stopped flow spectrophotometry. As shown in Table 1, *M. crystallinum* tonoplast showed a high water permeability of 1,205 $\mu m\ s^{-1}$, a value two and 14 times larger than that described for *Nicotiana tabacum* and *Triticum aestivum* tonoplast, respectively (Maurel *et al.,* 1997, Niemitz and Tyerman 1997). From an Arrhenius plot we derived a low activation energy of 9 kJ $mol^{-1}\ K^{-1}$, indicating that the path for water movement across the tonoplast is facilitated by the presence of aquaporins in this membrane. Preliminary results have demonstrated that plants exposed to short term salt-stress showed an increase in leaf tonoplast osmotic water permeability.

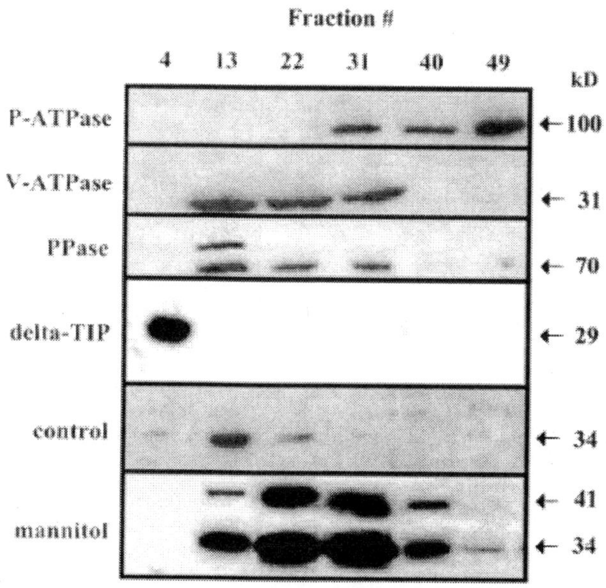

Figure 2. Immunological detection of MIP-F in selected membrane fractions of a continuous sucrose gradient from control and five-day mannitol-treated (400 mM) *M. crystallinum* cell suspensions. Membrane marker enzyme recognition in fractions from control cells are shown (P-ATPase, V-ATPase E subunit, PPase and δ-TIP).

Table 1. Tonoplast water permeability in *M. crystallinum*. Osmotic water permeability (P_f) and activation energy (E_a) were derived using stopped flow light-scattering in tonoplast vesicles isolated from leaves of control plants.

Species	P_f (μm s^{-1})	E_A (kJ mol^{-1} K^{-1})
M. crystallinum (leaves)	1,205	8.97
T. aestivum (roots)	86	23.32
N. tabacum (cell cultures)	690	10

3. DISCUSSION

In *M. crystallinum,* MIP-F regulation by salt- and/or osmotic stress was complex. Whereas, salt-stress caused a decrease in expression of MIP-F in tonoplast fractions, mannitol-induced osmotic stress resulted in a much higher level of expression of MIP-F protein compared to control untreated cell suspensions. The ability to distinguish between the ionic component of salt-stress and the osmotic changes induced by both salt- and mannitol-stress, indicates the presence of independent signalling pathways converging to differentially regulate MIP-F expression. Evidence for specific, distinct responses to ionic vs osmotic stress in the regulation of salt-adaptive mechanisms in *M. crystallinum* are accumulating, both the expression of IMT and V-ATPase activity have also been shown to respond differently to these two components (Barkla *et al.,* 1999b, Nelson *et al.,* 1998).

Concomitant with mannitol-induced changes in MIP-F abundance in the tonoplast, distribution of this protein in membranes was also altered. We suggest that within the cell, stress-induced alterations in MIP trafficking is occurring, in analogy to aquaporin cycling through animal endosomes (Ward *et al.,* 1999). MIP cycling and changes in abundance of protein may be fundamental for the control of cellular/tissue water homeostasis.

ACKNOWLEDGEMENTS

We are grateful to Drs. Ramon Serrano (Valencia, Spain), Phil Rea (Philadelphia, PA, USA), John Rogers (Pullwan, WA, USA) and Karl-Josef Dietz (Bielefel, Germany) for the antibodies against P-ATPase, PPase, δ-TIP and V-ATPase E subunit, respectively. Work in our lab was supported by grants from CONACyT (#25750 and #31794) and DGAPA (#IN232998).

REFERENCES

Adams P Nelson DE Yamada S Chmara W Jensen RG Bohnert HJ and Griffiths H 1998 Growth and development of *Mesembryanthemum crystallinum* (Aizoaceae) *New Phytol* **138:** 171-190

Barkla BJ Vera-Estrella R Pantoja O Kirch H-H and Bohnert HJ 1999a Aquaporin localization- how valid are the TIP and PIP labels? *Trends Plant Sci* **4:** 86-88

Barkla BJ Vera-Estrella R Maldonado-Gama M and Pantoja O 1999b Abscisic acid induction of vacuolar H^+-ATPase activity in *Mesembryanthemum crystallinum* is developmentally regulated *Plant Physiol* **120:** 811-819

Barrieu F Chaumont F and Chrispeels MJ 1998 High expression of the tonoplast aquaporin ZmTIP1 in epidermal and conducting tissues of maize *Plant Physiol* **117:** 1153-1163

Daniels MJ Mirkov ET and Chrispeels MJ 1994 The plasma membrane of *Arabidopsis thaliana* contains a mercury-insensitive aquaporin that is a homolog of the tonoplast water channel protein TIP *Plant Physiol* **106:** 1325-1333

Dean RM Rivers RL Zeidel ML and Roberts D M 1999 Purification and functional reconstitution of soybean nodulin 26 An aquaporin with water and glycerol transport properties *Biochem* **38:** 347-353

Fray RG Wallace A Grierson D and Lycett GW 1994 Nucleotide sequence and expression of a ripening and water stress-related cDNA from tomato with homology to the MIP class of membrane channel proteins *Plant Mol Biol* **24:** 539-543

Fukuhara T Kirch H-H and Bohnert HJ 1999 Expression of Vp1 and water channel proteins during seed germination *Plant Cell Environ* **22:** 417-424

Grote K von Trzebiatovski P and Kaldenhoff R 1998 RNA levels of plasma membrane aquaporins in *Arabidopsis thaliana Protoplasma* **204:** 139-144

Johansson I Karisson M Shukla VK Chrispeels MJ Larsson C and Kjellbom P 1998 Water transport activity of the plasma membrane aquaporin PM28A is regulated by phosphorylation *Plant Cell* **10:** 451-459

Kaldenhoff R Grote K Zhu J-J and Zimmermman U 1998 Significance of plasmalemma aquaporins for water-transport in *Arabidopsis thaliana Plant J* **14:** 121-128

Kaldenhoff R Kolling A Meyers J Karmann U Ruppel G and Richter G 1995 The blue light-responsive AthH2 gene of *Arabidopsis thaliana* is primary expressed in expanding as well as in differentiating cells and encodes a putative channel protein of the plasmalemma *Plant J* **7:** 87-95

Kirch H-H Vera-Estrella R Golldack D Quigley F Michalowski CB Barkla BJ and Bohnert HJ 2000 Expression of water channel proteins in *Mesembryanthemum crystallinum Plant Physiol* (in press)

Liu Q Umeda M and Uchimiya H 1994 Isolation and expression analysis of two rice genes encoding the major intrinsic protein *Plant Mol Biol* **26:** 2003-2007

Ludevid D Höfte H Himelblau E and Chrispeels MJ 1992 The expression pattern of the tonoplast intrinsic protein γ-TIP in *Arabidopsis thaliana* is correlated with cell enlargement *Plant Physiol* **100:** 1633-1639

Mariaux J-B Bockel C Salamini F and Bartels D 1998 Desiccation-and abscisic acid-responsive genes encoding major intrinsic proteins (MIPs) from the resurrection plant *Craterostigma plantagineum Plant Mol Biol* **38:** 1089-1099

Maurel C Reizer J Schroeder JI and Chrispeels MJ 1993 The vacuolar membrane protein ⊗-TIP creates water specific channels in *Xenopus* oocytes *EMBO J* **12:** 2241-2247

Maurel C Tacnet F Güçlü J Guern J and Ripoche P 1997 Purified vesicles of tobacco cell vacuolar and plasma membrane exhibit dramatically different water permeability and water channel activity *Proc Natl Acad Sci USA* **94:** 7103-7108

Nelson DE Rammesmayer G and Bohnert HJ 1998 Regulation of cell-specific inositol metabolism and transport in plant salinity tolerance *Plant Ce*ll **10:** 753-764

Niemietz CM and Tyerman SD 1997 Characterization of water channels in wheat root membrane vesicles *Plant Physiol* **115:** 561-567

Pih KT Kabilan V Lim JH Knang SG Piao HL Jin JB and Hwang I 1999 Characterization of two new channel protein genes in *Arabidopsis Mol Cells* **9:** 84-90

Sarda X Tousch D Ferrare K Legrand E Dupois JM Casse-Delbart F and Lamaze T 1997 Two TIP-like genes encoding aquaporins are expressed in sunflower guard cells *Plant J* **12:** 1103-1111

Vera-Estrella R Barkla BJ Bohnert HJ and Pantoja O 1999 Salt stress in *Mesembryanthemum crystallinum* L cell suspensions activates adaptive mechanisms similar to those observed in the whole plant *Planta* **207:** 426-435

Ward DT Hammond TG and Harris HW 1999 Modulation of vasopressin-elicited water transport by trafficking of Aquaporin 2-containing vesicles *Annu Rev Physiol* **61:** 683-697

Yamada S Katsuhara M Kelly WB Michalowski CB and Bohnert HJ 1995 A family of transcripts encoding water channel proteins: Tissue-specific expression in the common ice plant *Plant Cell* **7:** 1129-1142

Yamada S Nelson DE Ley E Marquez S and Bohnert HJ 1997 The expression of an aquaporin promotor from *Mesembryanthemum crystallinum* in tobacco *Plant Cell Physiol* **38:** 1326-1332

4.13 Plant physiology

GENOTYPICALLY DETERMINED ACTIN-REGULATED WATER PERMEABILITY OF TWO PLASMODEMATAL TRANSPORT CHANNELS

[1]Olga V. Volobuyeva, [1]Ludmila P. Khokhlova, [2]Gennady A. Velikanov, [3]Oleg A. Opanasyuk
[1]Department of Plant Physiology, Kazan State University,Kazan, Russia [2]Kazan Institute of Biochemistry and Biophysics in KSC RAS, Kazan, Russia [3]Department of Molecular Physics, Kazan State University, Kazan, Russia. E-mail: Olga.Volobuyeva@ksu.ru

1. INTRODUCTION

Plasmodesmata are channels ensuring connection of the next cells to similarly gap-junctions contacts of animal cells. On modern representation, plasmodesmata contain internal (desmotubule) and outside tubes, connecting accordingly vacuoles and cytoplasms of cells (Gamalei 1997), that has allowed us to enter concepts about symplasts in vacuoles and cytoplasm (Velikanov *et al.*, 1999).

Now study of plasmodesm regulation basically is focused on research of their ultrastructural features at use of electronic and immunofluorescence microscopy combined with cytochemical localization of various proteins (Olesonand Robards 1990, White *et al.*, 1994, Radford and White 1998, Blackman and Overall 1998). White *et al.*, (1994) have assumed that protein particles associated with desmotubule can be elements of cytoskeleton and, in particular, actin. This contractile protein is located along all length of plasmodesmata and is especially plentifully submitted in places of neck narrowings (Blackman and Overall 1998, Radford and White 1998). Probably, actin-myosin sphyncters submitted with both ends of plasmodesmata one (Radford, White 1998) are the basic regulator of intercellular interactions.

In the present work the attempt of water permeability study of two transport channels of plasmodesmata is undertaken at use of a NMR-method with pulse gradient of a magnetic field in winter wheat cultivar roots different on resistance after action of water stress of difference force. The method of NMR is unique for research of molecular mobility and allows to study alive tissues without essential disturbance of their state, that allows to avoid procedure of fixing of a plant material, which is the basic methodical complexity at research of functioning of plasmodesmata having fast wounded reaction even on the most sparing influences.

2. MATERIAL AND TECHNIQUE

The roots of 5-6-day old seedlings of three winter wheat cultivars differed in resistance Bezostaya 1 (low resistant), Mironovskaya 808 (resistant) and Albidum 114(very resistant) grown on distilled water were the object of investigation. Water stress of different force was created by placing roots of intact seedlings on 2,5 h in solution of different concentration of PEG-6000 (-0,3, -0,7, -1,5 MPa), creating accordingly weak, average and strong water stress. For revealing dependence of a water exchange of cells from a cytoskeleton structural conditions methodical reception based on modification of microfilaments *in vivo* by a 30-minute exposition of roots in a solution of cytochalasin B (5mkM, «Sigma») - inhibitor of actin protein polymerization after action of water stress was used.

For research of water transport a method of NMR-spin echo with pulse gradient of magnetic field was used (Tanner and Stejskal 1968). In the present work relative amplitude of an echo (R) depending on a square of a gradient of a magnetic field (g^2) at the fixed meanings of gradient impulse duration (δ) and time of diffusion supervision (t_d=100 ms) was registered.

New technical opportunities for study of dependence of R attenuation from g^2 in a range of three orders of its size at the expense of fast switching of sensitivity of the receiver, that enables to study the mechanism of this attenuation for plant tissues in the large range recently have been appeared. The measurements were carried out on NMR-diffusometer with working frequency of 60 MHz. The technique of registration of water self-diffusion coefficient (CSD) in researched roots is stated by us in article (Velikanov *et al.*, 1999). The application of membrane trope drug pipolphen, increasing permeability of membranes and method of paramagnetic doping have allowed to establish that is not cell distribution on the size but over-cellular and sub-cellular organization of a root are the reason of three CSD occurrence. Considering modern representation about uniform endoplasmatic continuum of plant organism (Gamalei 1997) and also taking into account presence of two transport channels of plasmodesmata connecting accordingly vacuoles and cytoplasms of the next cells (Overall *et al.*, 1982, Gamalei 1994) we have entered concepts of vacuolar and cytoplasmic symplasts and the nature of received CSD is explained. According to the offered hypothesis, D_3 characterizes the self-diffusion of water in system of vacuolar symplast and apoplast of root, D_2 - does it in system of cytoplasmic symplast, D_1 - characterizes the self-diffusion of water, at the initial moment of time registered in cellular membranes and/or water in vesicles, sub-molecular complexes and microbodies.

The roots were slightly dried by filter paper, were cut on segments of length 12 mm and placed on the bottom of thin-walled, calibrated on an internal diameter (8 mm) glass tubes. For measurements a zone of a stretching of a root was used cells of which have maximal density of plasmodesmata in radial section (Vakhmistrov 1985). Samples were placed in the gauge of diffusometer so, that the gradient of a magnetic field was directed in parallel to this section, thus, we registered the self-diffusion of water on radius of a root.

3. RESULTS

The form of diffusive attenuation (DA) of stimulated echo R (g^2) in seedling roots of a wheat has not exponential character (Fig.1). The extrapolation of a slow part of DA by the exponent (method of the least squares) and consecutive subtraction the exhibitor from total DA has allowed to carry out satisfactory diffusion of attenuation of an echo into three exhibitors with the coefficients of water self-diffusion (D_1, D_2, D_3), describing different on mobility a fraction of water.

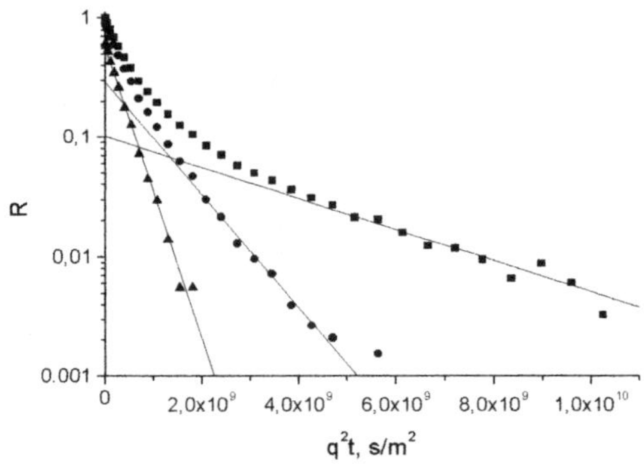

Figure 1. Display of decomposing of diffusion attenuation of echo on three exponents for control sample of wheat roots: 1--initial diffusion attenuation and straight characterizing D_1, 2-the sum of two exponents and straight D_2, 3-latter exponent, characterizing of the D_3.

The coefficients D_1, D_2 and D_3 (Fig.2) in roots of three research cultivars after influence of water stress and cytochalasin B were changed ambiguously. So, in control variant D_3, appropriative to a fast fraction, monotonously grew in a direction «no stress» → «strong stress» at low resistant cultivar Bezostaya 1 (Fig.2, A). At more resistant cultivars the course of curves had in the basic parabolic kind with a minimum at Mironovskaya 808 at weak stress and at Albidum 114 - on the average. The greatest D_3 after strong stress are common for researched cultivar.

The changes of average meaning on mobility of a fraction of water (D_2) of control variant in dependence of stress force were less expressed, than D_3. Small increase of D_2 at Bezostaya 1 was observed after average stress, at Mironovskaya 808 and Albidum 114 - after weak and average ones (Fig.2, B). Strong water loss of tissues at Bezostaya 1 caused fall of coefficient D_2, and at Albidum 114 - increase of it D_2 at Mironovskaya 808, having increased after weak stress, practically did not change after average and strong ones.

The amplitude of D_1 changes (least mobile fraction of water) in roots of control variant was even more insignificant in comparison with the fraction described above. Thus, in control variant the degree of CSD variability depending on the force of stress decreased in the line $D_3 \rightarrow D_2 \rightarrow D_1$, it means that the fast fraction of water with D_3 was characterized by the greater variability on the water loss especially in the resistant cultivars and less mobile one with D_1 least.

The action of cytochalasin B would not be reflected in D_3 at Bezostaya 1, as the changes were within the limits of the experiment mistake. The exposition of Mironovskaya 808 roots in solution of cytochalasin B would result in reduction of D_3 at after-action of weak and average stresses and in the reference of changes - after strong water loss. At Albidum 114 at all levels of water loss with different directions changes of D_3 are established.

The greatest effect of cytochalasin B would render on average on mobility fraction of water (D_2) especially at the resistant cultivars. The authentic increase (Mironovskaya 808 and Albidum 114) or tendency to increase (Bezostaya 1) of this coefficient in conditions without stresses and also after weak and average stresses was marked. The reaction of roots on cytochalasin B after strong stress would appear ambiguous. So, the increase of D_2 at Bezostaya 1, unauthenticity of changes - at Mironovskaya 808 and significant reduction - at Albidum 114 were registered. One must pay attention on the mirror appearance of curves D_2 and D_3 at the resistant cultivars which intensified by the action of cytochalasin B.

With the increase of cultivar resistance the action of cytochalasin B is directed to the party of mobility reduction of the slowest water fraction of a root, namely, the increase of factor D_1 at Bezostaya 1 would be accompanied by the absence of the cytochalasin B effect in these conditions at Mironovskaya 808 and decrease of the coefficient D_1 at Albidum 114.

4. DISCUSSION

In the light of the interpretation, stated in the methodical section, and the correlation of the coefficients D_1, D_2 and D_3 with the existing fractions of water in a root, it is possible to consider the increase of the factor D_2 after the action of weak and average water loss at all researched cultivars in the control variant (Fig.2, B) as the indication on the increase of water mobility in the system of the cytoplasmic symplast. First of all, it could be caused by the redistribution of the intracellular water. If we take into account that many organells are capable to swell with high amplitude, practically determined by an input and output of water. However, taking into account the decrease (Bezostaya 1 and Mironovskaya 808) or invariance (Albidum 114) of water proton relative part (data is not shown) of this water fraction at a level of moderate water loss in comparison with the norm, such assumption is flimsy. In this connection the most probable reason of the D_2 increase is the change of intercellular contact permeability. According to opinion Gamalei (1996), especially strong effect of the dehydration is shown on the plasmodesmata - the narrowest sites of the endoplastic network in which the general principle of its organization is repeated in a miniature. The small diameter of the plasmodesmata makes them the zone of the greatest vulnerability concerning the external influences - cold, water and the osmotic stresses, at the wounded reactions (Gamalei, 1996). In the structure of the plasmodesmata under the

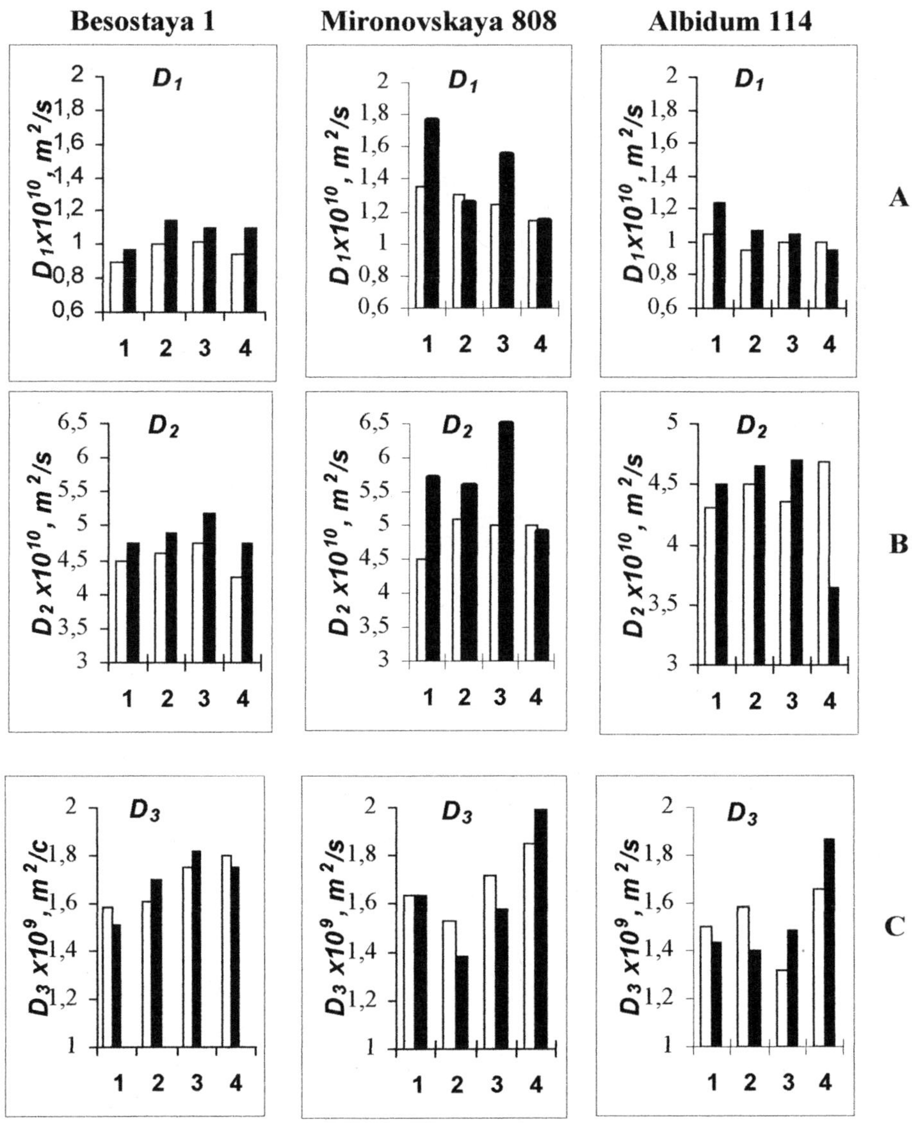

Figure 2. The influence of cytochalasin B on coefficients D_1 (A), D_2 (B) and D_3 (C) in winter wheat roots: 1 no stress, 2 weak stress, 3 average stress 4 strong stress. □ control (water), ■ cytochalasin B (5 mkM, 30 min).

action of the dehydration the structure of endoplasmic network is collapsed, up to the cores or the flat cisterns, thus increasing a gleam of the endoplasmic ring (Quander and Liebe 1995). Apparently, the reason of the D_2 increase consists in the strengthening of the cytoplasmic connection of the cells and the increase of the cytoplasmic channel contribution.

The significant genotipic differences were showed after the strong stress. On the basis of D_2 at Bezostaya 1 in the control it is possible to assume about the reduction of cytoplasmic water flow after the strong water loss. As Gamalei (1997) marks, the intercellular transport on plasmodesmata at a drought is slowed down or is stopped complitely. The decrease of the network endoplasmic turgor under the influence of the unsufficient water content results in the plasmodesmata closing and stop of all transport processes on the symplast. In our experiments the strong water stress reduced D_2 at the low resistant cultivar, practically did not change D_2 at the resistant cultivar and raised it at very resistant one. Thus, brightly expressed cultivar-specific strengthening of the intercellular contacts on the cytoplasmic continuum under the action of the strong water loss in the process of the cultivar resistance increase is observed. Apparently, the cultivar-specifity was showed in the work of the regulation mechanism of the intercellular connections on plasmodesmata.

Analyzing the dynamics of D_3 in the roots of the resistant cultivar it is necessary to note the decrease of this coefficient after the weak (Mironovskaya 808) and average (Albidum 114) stresses, that is opposite changes in relation to D_2. The modern model of plasmodesmata has actin-myosin sphincter inside the cytoplasmic ring supervising a diameter of desmotubule which unites the vacuoles of the next cells (Gamalei 1997). That's why one must expect the opposite changes of the coefficients D_2 and D_3 at the changes of the water permeability of one of these channels. Hence, the opposite orientation, found out in the experiment, of D_2 and D_3 changes allows to make the assumption of the diffusive water flows increase on the cytoplasmic symplast and the easing - on the vacuolar one and/or on the apoplast at the moderate water loss. The data on the parallel increase of D_2 and D_3 do not allow to make the similar assumption for the less resistant cultivar.

After the strong stress the increase of D_3, more significant at Mironovskaya 808 and Albidum 114 was marked. It would seem, it could be a consequence of strengthening of the vacuolar symplast permeability. Howewer, on a background of the strong stress, at which there are the processes of plasmolisis, cytorisis and devacuolization of the cells (Schimidt and Kaiser 1987) and also embolism of a plasmodesmata the strengthening of the water movement on the system of vacuolar symplast is improbable.

Therefore, in this case (at the strong stress) the coefficient D_3 in the greater degree characterizes the self-diffusion of the apoplast water. The marked greater increase of D_3 at Mironovskaya 808 and Albidum 114 can be caused, first, by the strengthening of the intercellular water outflow in the apoplast under the action of osmotic forces owing to the increase of the plasma membrane water permeability and/or, secondly, by the sharp receipt of water at the rehydration of the tissues and its localization in so-called «free space» of the cells, which includes the intercellular spaces and cell envelopes.

At the discussion of the data on the influence of the cytochalasin B on CSD it is necessary to note, that in the normal conditions and also on a background of weak and average stresses the inhibitor, as a rule, resulted in the opposite changes of D_2 and D_3 at the researched cultivars - in D_2 increase and D_3 decrease. The degree of their changes was more at Mironovskaya 808 and Albidum 114, than at Bezostaya 1. Hence, cytochalasin B in the norm and at the moderate level of water loss, apparently, on the one hand, increased the water conductivity of the cytoplasmic symplast and with another one - reduced the conductivity of the vacuolar symplast, and, probably, the transport of water in the apoplast.

The presence of both actin and myosin fibrillae in the cytoplasmic ring of plasmodesmata (White *et al.*, 1994, Blackman and Overall 1998; Radford and White 1998). Gamalei (1996) has presented the model of the plasmodesmata structure with actin-myosin sphincter inside the cytoplasmic ring supervising the diameter of endoplasmic tube. By the immunofluorescence and electronic microscopy at the use of contractile myosin protein inhibitor (2,3-butandion monoxim) the presence of actin-myosin sphincter in the neck narrowings of plasmodesmata is confirmed (Radford and White 1998).

From mentioned above it is possible to conclude that the destruction of the plasmodesmatal actin-myosin sphincters under the influence of cytochalasin B would result in the increase of water CSD cytoplasm. It is necessary to notice that the effect of the opposite D_2 and D_3 changes at the action of cytochalasin B would appear more precisely expressed in the conditions of water loss after-action in comparison with the normal conditions. And the resistant cultivar Mironovskaya 808 and Albidum 114 this effects of the inhibitor was showed more considerable than at the cultivar Bezostaya 1. Apparently, the water stress would strengthen the action cytochalasin B on plasmodesmata water conductivity and on the contrary it means is that synergism in operation of these two factors is marked. Hence, two mechanisms – the fall of the turgor of the plasmodesmatal central tube, on the one hand and the destruction of the actin-myosin sphincter – on the another hand, together thus render the stronger influence on a functional condition of plasmodesmata strengthening one another.

It is necessary to emphasize that D_2 and D_3 coefficients both in the control and cytochalasin B at the more resistant cultivars in comparison with not resistant one testifies about the higher sensitivity of intercellular transport ways and, first of all, through plasmodesmata at resistant cultivars.

After strong water loss the effect of inhibitor was changed, therefore D_2 was reduced and D_3 was increased in comparison with the control. The similar changes of CSD under the inhibitor influence in the greater degree were showed at the very resistant cultivar. At means that after the strong stress the structural-functional conditions of plasmodesmata under the influence of cytochalasin B would be modified in a direction of cytoplasmic channel closing and the strengthening of the water inflow in apoplast.

5. CONCLUSION

Summarizing the results of the above-stated researches, it is possible to note that the intercellular water transport regulation at a level of plasmodesmata in the roots of the resistant cultivars occurs on a basis of the compensator mechanisms. At the after-action of the moderate water loss the diffusive flows of water are increased on cytoplasmic symplast and decreased – on vacuolar one. After strong water loss alongside with the proceeding increase of the cytoplasmic symplast water conductivity and also inflow of water in the apoplast owing to the increase of plasma membrane permeability, probably, there is an almost complete termination of water transport on the vacuolar way. In roots of the low resistant cultivar these mechanisms are poorly expressed or completely absent. The additive action of the moderate water loss and cytochalasin B on the plasmodesmatal water conductivity, shown in the greater degree at resistant cultivars is revealed in comparison with low resistant one. After strong water stress the synergism was kept partially, as inhibitor blocking the cytoplasmic channel at the same time strengthened the effect of water loss raising the inflow of water and its localization in the apoplast.

REFERENCES

Badelet K White RG Overall RL Vesk M 1994 Ultrastructural specialization of the cell wall sleeve around plasmodesmata *Am J Botany* **81**: 1422-1427

Blackman LM Overall RL 1998 Immunolocalisation of the cytoskeleton to the plasmodesmata of *Chara corallina The Plant J* **14** : 733-741

Gamalei YV Vandel AJE Pakhomova MV Sjutkina AV 1994 Effects of temperature on the conformation of the endoplasmic reticulum and on starch accumulation in leaves with the symplastic minor-vein configuration *Planta* **194**: 443-453

Gamalei YV 1996 The outflow of photoassimilates in nature and experimental conditions *Rus J Plant Physiology* **43**: 328-347

Gamalei YV 1996 Over-cellular organization of plants *Rus J Plant Physiology* **44**: 819-846

Olesen P Robards AW 1990 The neck region of plasmodesmata: general architecture and some functional aspects In: *Parallels in Cell to Cell Junctions in Plants and Animals* Heidelberg: Springer Verlag 145-170

Overall RL Wolfe J Gunning BES 1982 Intercellular communication in Azolla roots: I Ultrastructure of plasmodesmata *Protoplasma* **111**: 134-150

Quader H Liebe S 1995 Actin Filament-Independent Formation of Tubular Endoplasmic Reticulum in Onion Epidermis Cells *J Plant Physiology* **188**: 332-339

Radford JE White RG 1998 Localization of a myosin-like protein to plasmodesmata *The Plant J* **14**: 743-750

Schimidt JE Kaiser WM 1987 Response of the Succulent Leaves of Peperomia magnoliaefolia to Dehydration *Plant Physiology* **83**: 190-194

Tanner TE Stejskal EO 1968 Restricted self-diffusion of protons in colloidal systems by the Pulse-Gradient Spin-echo Method *J Chem Phys* 49: 1768-1777

Vakhmistrov DB To the question about localisation of free space of barley roots *Rus J Plant Physiology* **14**: 397-401

Velikanov GA Volobuyeva OV Khokhlova LP 1999 Self-diffusion of water inside of a vacuolar and cytoplasic symplasts *The reports of Russian Academy of Scienses* **368**: 420-422

White RG Badelt K Overall RL Vesk M 1994 Actin associated with plasmodesmata *Protoplasma* **180**: 169-184

4.14 Plant physiology

ROLE OF AQUAPORINS DURING ELONGATION GROWTH OF CASTOR BEAN SEEDLINGS

Alfons R. Weig and Daniel A. Eisenbarth
Department of Plant Physiology, University of Bayreuth, D-95440 Bayreuth, Germany. E-mail: alfons.weig@uni-bayreuth.de

1. INTRODUCTION

When aquaporins were described as a molecular basis for transmembrane water flow in animal (Preston *et al.*, 1992) and plant cells (Maurel *et al.*, 1993), their physiological roles were often anticipated from their (subcellular) localisation with respect to the physiological process in the cell. However, these assumptions are difficult to prove and have been mainly restricted to gene knockout experiments in mice, yeast, and bacteria (reviewed in: Borgnia *et al.*, 1999). Gene knockouts in plants are laborious and only feasible in a few plants. Moreover, the large number of homologous genes in plant genomes makes the finding of a clear phenotype difficult. Instead of analysing a single gene or protein, we have decided to investigate the role of aquaporins in hypocotyl cell elongation because parameters involved in this process can be measured easily and the rate of the elongation can be modulated.

1.1 Advantages of the Castor Bean Seedling

During seed germination plants develop a more or less large hypocotyl (by definition the "stem" in between the cotyledons and the roots) whose main function is to expose the leaves to the light for photosynthesis. The hypocotyl tissue has a simple architecture: vascular bundles are embedded in parenchymatous tissue surrounded by an epidermal layer; the tissue is unbranched and the cells grow mainly in axial direction.

Molecular Biology and Physiology of Water and Solute Transport
Edited by Hohmann and Nielsen, Kluwer Academic/Plenum Publishers, New York, 2000.

We have selected this tissue for the analysis of aquaporin function because the cells show an endogenous on- and offset of net water uptake during development. Apical cells in the cell division zone (zone 1 in Fig. 1) are filled up with starch supplied via the phloem from source tissue (cotyledons and endosperm) but the cells remain quite small. At the hypocotyl hook the osmotically inactive starch granules are degraded and additional assimilates are unloaded from the phloem necessary for biosynthesis and turgor-driven cell elongation (zone 2 in Fig. 1). The osmotic potential rises from about -0.1 MPa in cells of the division zone to about -0.8 MPa in cells of the elongation zone (Balane, 1997). The cells of the elongation zone show the highest growth rates and therefore the highest net water uptake in the seedling. Cell elongation is later extremely reduced and these cells form the mature hypocotyl above the roots (zone 3 in Fig. 1). The involvement of transport proteins for sugar and water uptake and a regulation on gene and protein level has to be assumed, but the co-ordinated activity of the various membrane transport processes is widely unknown.

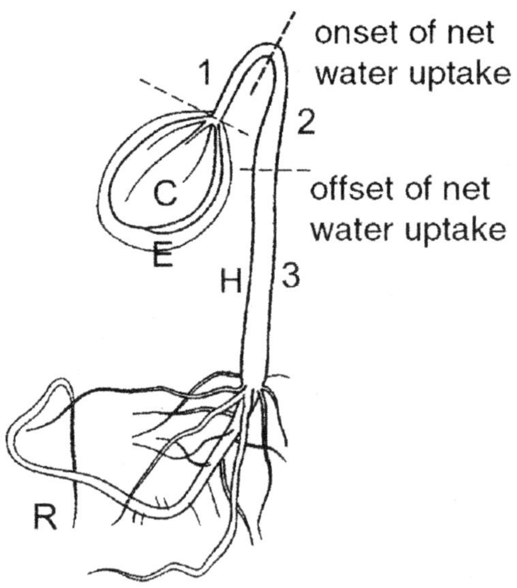

Figure 1. The castor bean seedling: Endosperm (E) surrounding the cotyledons (C); hypocotyl (H) consisting of the cell division zone (1), the elongation zone (2), and the mature hypocotyl (3); roots (R). Modified after Balane (1997).

The use of the castor bean (*Ricinus communis* L.) hypocotyl to study cell elongation has several advantages over other model plants. First, cells before, during, and after the elongation process are accessible and therefore regulations during the entire process can be studied on gene and protein level. Second, the seedlings grow within five to six days to a size which makes the hypocotyl tissue (5 to 10 cm in length) easily accessible without the need of a micromanipulator. Third, the growth rate (and thus the net water uptake into cells) can be manipulated in several ways: Illumination rapidly stops hypocotyl elongation, as it would occur naturally when a seedling reaches the light zone. Phloem loading at the source site is virtually accessible in castor bean seedlings which enables us to modify one of the prerequisites of cell growth, the assimilate supply to sink tissue.

The analysis of aquaporins and assimilate carriers on gene and protein levels along the castor bean hypocotyl axis could help us to answer the following questions: Are aquaporins necessary in the process of cell elongation which has relative low water transport rates? How is the activity of sugar carriers and aquaporins regulated at the gene and protein level in this economically important plant?

2. RESULTS

2.1 The Aquaporin Gene Family in *Ricinus communis* L.

Several cDNA clones for putative aquaporins were identified in castor bean seedlings by RT-PCR as described in Weig et al. (1997). Additionally, from an EST project on castor bean endosperm (van de Loo, 1995), two additional seed-specific sequences were obtained. Sequence comparison with known Arabidopsis aquaporins identifies them as members of the tonoplast intrinsic protein (TIP) or plasma membrane intrinsic protein (PIP) subgroup, respectively (Fig. 2).

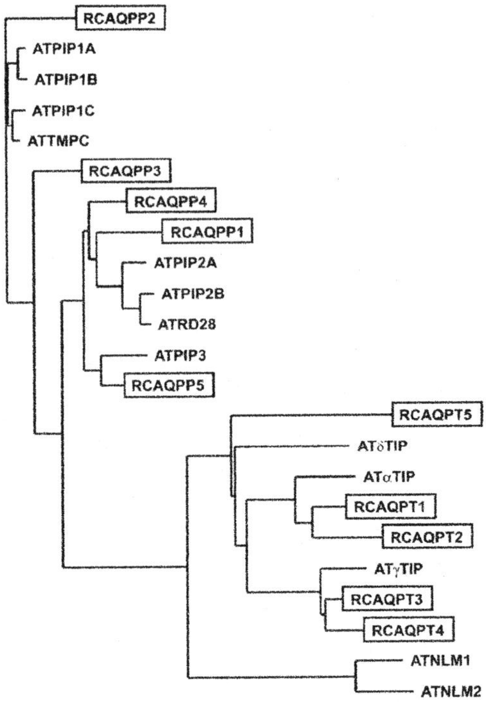

Figure 2. Sequence relationship of *Ricinus communis* (in boxes) and *Arabidopsis thaliana* MIP sequences.

2.2 Aquaporin and Sugar Carrier Gene Expression Analysis in Castor Bean Seedlings

RNA gel blot hybridisation showed that gene transcripts *RcAQPP2*, *RcAQPP5* (two putative plasmalemma aquaporins) and *RcAQPT4* (a putative tonoplast aquaporin) are the most abundant aquaporin gene transcripts among known castor bean MIP genes. Therefore, we focused on the analysis of these aquaporin homologs since changes in hydraulic conductivity along the hypocotyl axis should be mainly due to changes in the amount and/or activity of abundant aquaporins. In addition the expression profiles of two putative sucrose carriers and six putative monosaccharide carriers were compared with that of the MIP genes. The hypocotyl axis was divided into three parts as shown in Fig. 1: The cell division zone (1), the cell elongation zone (2), and the mature hypocotyl (3).

Expression analysis of the three putative aquaporin genes along the hypocotyl axis showed an increase in expression of each of these genes in the elongation zone (based on total RNA). While the *RcAQPP2* transcript remains high in fully elongated cells, *RcAQPP5* and *RcAQPT2* transcripts in the mature hypocotyl zone fall again to levels comparable to that of the cell division zone (Fig. 3). Similarly, *RcSCR1* sucrose carrier transcripts (Weig and Komor, 1996) are present in comparable amounts along the hypocotyl axis, while *RcSCR2* (Weig *et al.*, unpublished) and *RcHEX3* (Weig *et al.*, 1994) transcripts are transiently increased in the elongation zone.

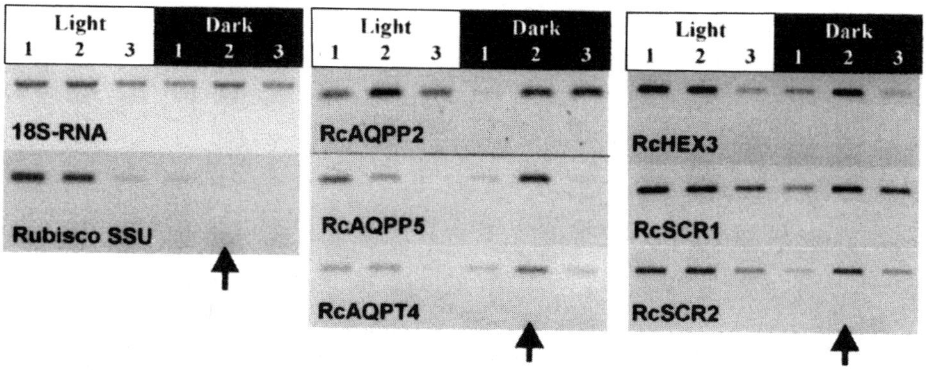

Figure 3. Aquaporin and sugar carrier gene expression along the castor bean hypocotyl (semi-quantitative RT-PCR with gene-specific primers): The numbers refer to the different zones along the hypocotyl axis as defined in Fig. 1. The arrow points to samples from the hypocotyl elongation zone of dark-grown seedlings. Light treatment was done at 350 $\mu E\ s^{-1}\ m^{-2}$ for 12 h. The comparable quality of the cDNA template was controlled with 18S-RNA and Rubisco primers.

Cell extension growth in plants is controlled mainly by three parameters: the cell wall extensibility, assimilate uptake and the water permeability of cell membranes. The cellular localisation of sugar carriers and aquaporins still has to be determined. However, the transient increase of membrane sugar and water transport capacity in elongating cells (as assumed from the increase in sugar carrier and aquaporin genes, respectively) indicates that at least two of the three parameters of cell extension growth are transiently increased in the hypocotyl elongation zone.

2.3 Changes in Membrane Transport Properties along the Hypocotyl Axis

We have measured the hydraulic conductivity of tonoplast membranes along the hypocotyl axis directly via osmotically driven swelling tests of vacuoles. Vacuoles were isolated from protoplasts out of the elongation and the mature hypocotyl zone and stained with neutral red (protoplasts of the cell division zone do hardly yield vacuoles, probably because the starch granules destroy most of the vacuoles during the isolation procedure). Vacuoles were then diluted with hypo-osmolar buffer and the swelling was observed under a microscope equipped with a video camera, which took a picture every two seconds. The increase in volume was analysed from the picture data and the osmotic water permeability was calculated. Vacuoles of the cell elongation zone showed a higher osmotic water permeability than vacuoles of the mature hypocotyl tissue. First experiments revealed that the water transport could be inhibited to a certain extent by mercury chloride. When dark grown seedlings were transferred to the light, the osmotic water permeability of vacuoles of the elongation zone was reduced significantly (Fig. 4).

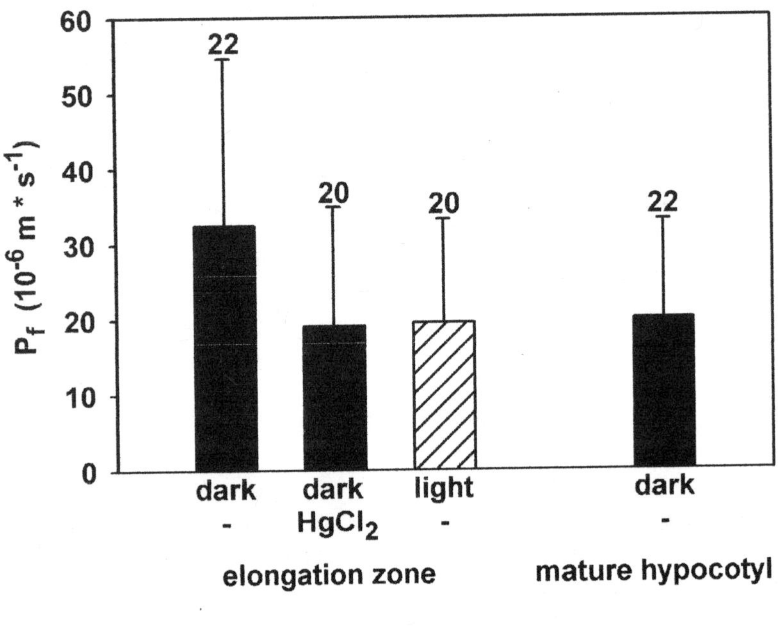

Figure 4. Osmotic water permeability of vacuoles of the elongation and mature hypocotyl zone: The growth conditions (darkness or transfer into the light) as well as the addition of mercury chloride is indicated. The number of experiments is given next to the error bar.

3. CONCLUSION

Ricinus communis like other plants contains a large number of MIP genes. Since most of them are closely related to plant aquaporins, they most likely encode water channels, too. Primary sequence analysis of the cDNA clones groups the castor bean clones together with the TIP and PIP subgroup of plant MIPs.

The expression profile of putative aquaporins and sugar carriers along the hypocotyl axis showed a transient increase in mRNA content. We assume that a co-ordinated increase in membrane transport capacity is necessary for cell elongation in this tissue. Since aquaporins are passive water channels, the membrane hydraulic conductivity is related to the amount of aquaporin molecules per membrane area (disregarding the possibility of activation via phosphorylation, membrane redistribution etc). The reduction of gene expression in the mature hypocotyl (Fig. 3) is probably underestimated because the signals relate to equal amounts of total RNA and not to comparable membrane areas (the cell size and membrane area increases but the amount of RNA per cell remains nearly unchanged).

The decrease in tonoplast osmotic water permeability from the elongation zone to the mature hypocotyl indicates that there is obviously a need for higher water transport capacities of vacuolar membranes during cell elongation. Whether this applies to the water permeability of each single cell or to that of the overall tissue has to be determined.

The results from seedlings, transferred into the light, clearly show the regulation of membrane transport proteins on the gene level. Sugar carriers and aquaporins that were highly expressed in the elongation zone of dark-grown seedlings are strongly induced in the newly formed growing zone, the former cell division zone next to the cotyledons.

The calculated Pf values for castor bean vacuoles seem to be an order of magnitude lower compared to vacuoles from other plants (Maurel *et al.*, 1997; Morillon and Lassalles, 1999). However, low hydraulic conductivity of castor bean hypocotyl cells has been measured previously with the cell pressure probe, too (Meshcheryakov *et al.*, 1992). Although the cell pressure probe integrates both the tonoplast and plasma membrane hydraulic conductivity (and therefore cannot be compared to our results directly), a lower hydraulic conductivity in hypocotyl cell membranes is likely.

The rate limiting parameter of cell elongation in the castor bean hypocotyl is not known in detail. The large drop in radial turgor pressure towards the outer layer of cells in the elongation zone (Meshcheryakov *et al.*, 1992) indicates a low tissue water conductivity. Thus it is now important to find out whether the radial tissue water conductivity in castor bean hypocotyl is low enough, that this might become rate limiting in hypocotyl elongation at least for peripheral cells.

ACKNOWLEDGMENTS

This work was supported by a DFG grant (We1680/4-1) to A.R. Weig and D.A. Eisenbarth. The authors thank E. Schwarz-Weig for carefully reading the manuscript.

REFERENCES

Balane SG 1997 Regulation des Stärke- und Zuckermetabolismus in der Wachstumszone des Hypokotyls des Rizinuskeimlings PhD thesis University of Bayreuth

Borgnia M Nielsen S Engel A and Agre P 1999 Cellular and molecular biology of the aquaporin water channels *AnnRev Biochem* **68**: 425-458

Maurel C Reizer J Schroeder JI and Chrispeels MJ 1993 The vacuolar membrane protein γ-TIP creates water specific channels in *Xenopus* oocytes *EMBO J* **12**: 2241-2247

Maurel C Tacnet F Güclü J Guern J and Ripoche P 1997 Purified vesicles of tobacco cell vacuolar and plasma membranes exhibit dramatically different water permeability and water channel activity *Proc Natl Acad Sci USA* **94**: 7103-7108

Meshcheryakov A Steudle E and Komor E 1992 Gradients of turgor osmotic pressure and water potential in the cortex of the hypocotyl of growing *Ricinus* seedlings - effects of the supply of water from the xylem and of solutes from the phloem *Plant Physiol* **98:** 840-852

Morillon R and Lassalles J-P 1999 Osmotic water permeability of isolated vacuoles *Planta* **210**: 80-84

Preston GM Carroll TP Guggino WB and Agre P 1992 Appearance of water channels in *Xenopus* Oocytes expressing red cell CHIP28 protein *Science* **256**: 385-387

van de Loo FJ Turner S and Somerville C 1995 Expressed sequence tags from developing castor seeds *Plant Physiol* **108**: 1141-1150

Weig A Deswarte C and Chrispeels MJ 1997 The major intrinsic protein family of *Arabidopsis* has 23 members that form three distinct groups with functional aquaporins in each group *Plant Physiol* **114**: 1347-1357

Weig A Franz J Sauer N and Komor E 1994 Isolation of a family of cDNA clones from *Ricinus communis* L with close homology to the hexose carriers *J Plant Physiol* **143**: 178-183

Weig A and Komor E 1996 An active sucrose carrier (Scr1) that is predominantly expressed in the seedling of *Ricinus communis* L *J Plant Physiol* **147**: 685-690

4.15 Plant physiology

FUNCTIONAL CHARACTERISATION OF *ARABIDOPSIS THALIANA* AQUAGLYCEROPORINS

Alfons R. Weig and Christiane Jakob
Department of Plant Physiology, University of Bayreuth, D-95440 Bayreuth, Germany. E-mail: alfons.weig@uni-bayreuth.de

1. INTRODUCTION

The major intrinsic protein (MIP) gene family encodes membrane transporters with aquaporin and glycerol permease activity as the predominant transport characteristics. However, functional expression of these proteins also showed that some of the MIPs are mixed-functional pores with transport activity for water, glycerol, urea and other small uncharged solutes (review: Borgnia *et al.,* 1999). In *Arabidopsis thaliana,* a model organism in plant biology, the MIP gene family consists of at least three major subgroups, the tonoplast intrinsic proteins (TIPs), the plasma membrane intrinsic proteins (PIPs), and the NOD26-like MIPs (NLMs; Weig *et al.,* 1997). While each of these subgroups contains functional aquaporins, the primary sequence analysis of the NLM cluster exhibits significant similarity to glycerol permeases of bacteria and yeast. Recently, aquaporin and glycerol permease activity of NOD26 from soybean has been demonstrated by heterologous expression in *Xenopus* oocytes (Rivers *et al.,* 1997) and functional reconstitution (Dean *et al.,* 1999).

Since a subcellular equivalent of the peribacteroid membrane of soybean root nodules (where NOD26 is integrated) is absent in *A. thaliana,* the physiological function of the AtNLM proteins cannot be deduced from results obtained for NOD26. In our contribution we will describe the *NLM* gene family of *A. thaliana* and present results on the functional expression analysis of some members of this gene family in *Saccharomyces cerevisiae.*

Molecular Biology and Physiology of Water and Solute Transport
Edited by Hohmann and Nielsen, Kluwer Academic/Plenum Publishers, New York, 2000.

2. RESULTS

2.1 Genomic Organisation of *NLM* Genes in *A. thaliana*

The first *NLM* gene transcript from *A. thaliana*, *AtNLM1*, has been identified in an EST database (Weig *et al.*, 1997). However, searching the DNA sequence databases from the Arabidopsis genome project results in at least seven additional genomic sequences with homology to *AtNLM1* (Fig. 1). The EMBO/GenBank accession numbers of the genomic clones referred to in Fig. 1 are AL021711 (*AtNLM1* and *AtNLM2*), AC005496 (*AtNLM3*), AC004077 (*AtNLM4*), AB016873 (*AtNLM5*), AC018907 (*AtNLM6*), AC011713 (*AtNLM7*), and AL049488 (*AtNLM8*).

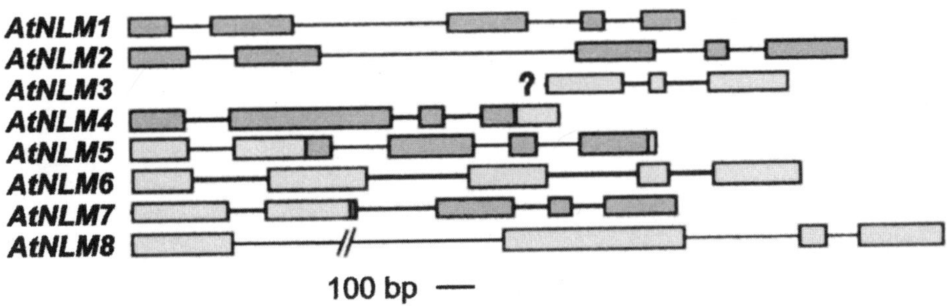

Figure 1. Structure of *NLM* genes in *A. thaliana*. Boxes represent exons, which have been verified by corresponding cDNA sequence analysis (dark) or which were deduced from computer analysis and manual inspection (grey). Lines connecting the exons are introns. In *AtNLM8*, the first intron (about 1.72 kb) is shortened for better viewing. Each figure represents the region in between the (putative) start and stop codons (except for *AtNLM3*).

Cluster analysis of the putative amino acid sequences clearly identified them as members of the NLM subgroup of plant MIPs. *AtNLM1*, *-2*, *-5*, *-6* and *-7* contain four introns in between the (putative) start and stop codons. In *AtNLM4* and *AtNLM8*, the second introns are missing while the other introns are still present as could be confirmed (at least for *AtNLM4*) from sequence analysis of a cDNA clone. *AtNLM3* seems to represent only a fragment of the genes since the first two exons could not be identified in the upstream region. With the exception of *AtNLM3*, putative start and stop codons could be identified for all the other *NLM* genes. Interestingly, the genes for *AtNLM1* and *AtNLM2* lie very close together (about 53 kb distance) on chromosome IV.

Five of the *NLM* genes are virtually transcribed: *AtNLM1* (Weig *et al.*, 1997) as well as *AtNLM2*, *-4* and *-5* were isolated as cDNA sequences in our lab (see also Fig. 2). Recently an EST sequence corresponding to *AtNLM7* has been added to the EMBO/GenBank database.

2.2 Amino Acid Signature of NLMs

A survey of more than 150 MIPs from different organisms revealed, that aquaporins and glycerol permeases could well be distinguished upon five specific amino acid residues (Froger *et al.*, 1998). AtNLM1 represents an intermediate form of MIPs since three of the five amino acid residues are similar to the aquaporin subclass while the other two are similar with residues of the glycerol permease group. Extended sequence analysis of all the *Arabidopsis* NLM proteins available so far clearly confirmed this finding (Fig. 2).

Interestingly, two aquaporins from *Nicotiana tabacum* (NtAQP1 and NtTIPa), for which water and glycerol transport activity has been shown in the *Xenopus* oocyte expression system (Biela *et al.*, 1999; Gerbeau *et al.*, 1999), do have clear 'aquaporin' residues at all five positions described by Froger et al. (1998). In addition, the two "highly conserved" NPA boxes, found in most MIPs, are less well conserved in NLMs. The first box reads 'NPS' in AtNLM6 and AtNLM8 (not shown) and the second box reads 'NPG', 'NPR', and 'NPV' in AtNLM1/2, AtNLM7, and AtNLM8, respectively (Fig. 2).

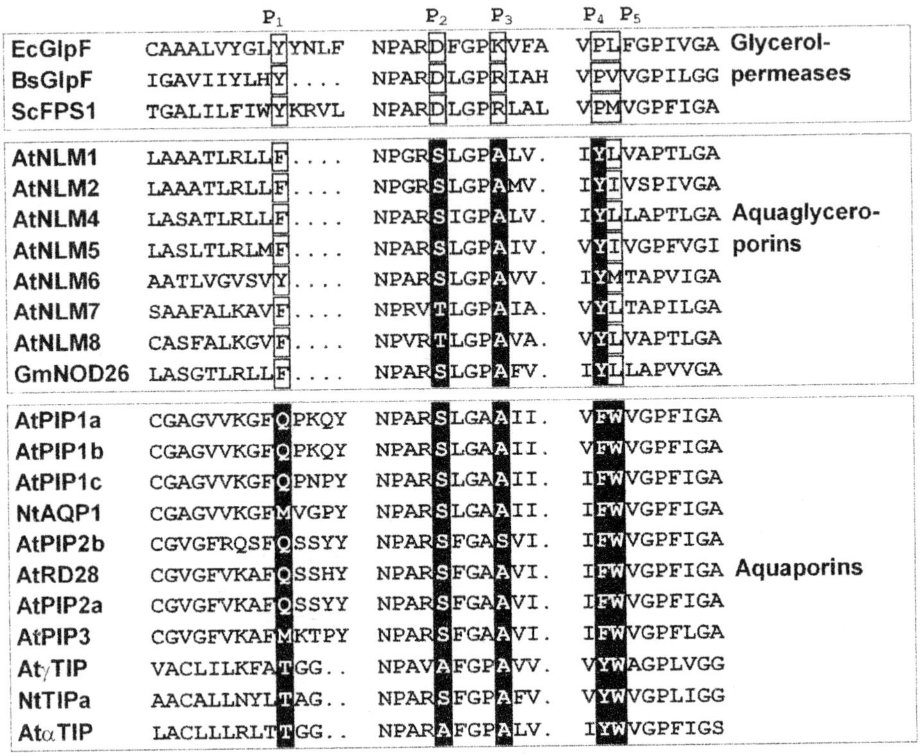

Figure 2. Amino acid signatures of glycerol permeases, aquaglyceroporins, and aquaporins: The significant residues of MIPs as defined by Froger et al. (1998) are indicated as P_1 to P_5. White boxes indicate amino acid residues found in glycerol permeases, black boxes mark amino acid residues found in aquaporins.

2.3 Expression Analysis of *Arabidopsis NLM* Genes

To analyse the expression profile of various *NLM* genes, we designed sequence specific primers for *AtNLM1* to *-5* corresponding to (putative) exons and performed RT-PCR experiments on cDNA prepared from different organs of adult *A. thaliana* plants (Fig. 3). With the exception of *AtNLM3*, each of the other four genes was expressed in root tissue. Moreover, *AtNLM1* and *AtNLM5* seem to be expressed in root cells exclusively. *AtNLM2* fragments were detectable in all organs except siliques, while *AtNLM4* fragments could be detected in cauline leaves and stem tissue. The identity of the PCR fragments as transcripts of the respective genes was verified by DNA sequence analysis. For *AtNLM3* no PCR product could be amplified in any of the organs tested.

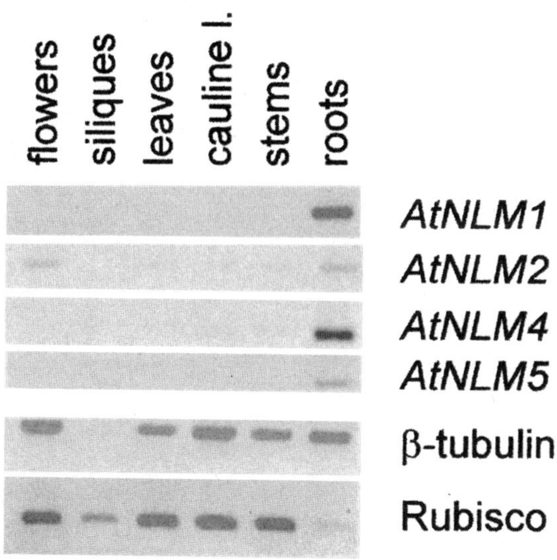

Figure 3. Expression profile of *NLM* genes in different organs of *A. thaliana* (semi-quantitative RT-PCR with gene-specific primers): The comparable quality of the cDNA template was controlled with β-tubulin and Rubisco primers.

2.4 Functional Expression of *NLM* Genes in *Saccharomyces cerevisiae*

Baker's yeast cells have been successfully used as heterologous expression systems for a variety of plant membrane transport proteins including sugar and amino acid carriers. We used a deletion mutant of FPS - a glycerol permease in *S. cerevisiae* (Tamas *et al.*, 1999) - to analyse glycerol transport activities of *AtNLM1* and *AtNLM2* after heterologous expression in these cells. Fig. 4A shows the glycerol permease activity of the two *Arabidopsis* NLM proteins expressed in the *fps1* mutant. The initial uptake rate is linear up to 100 mM glycerol in the extracellular medium (Fig. 4B).

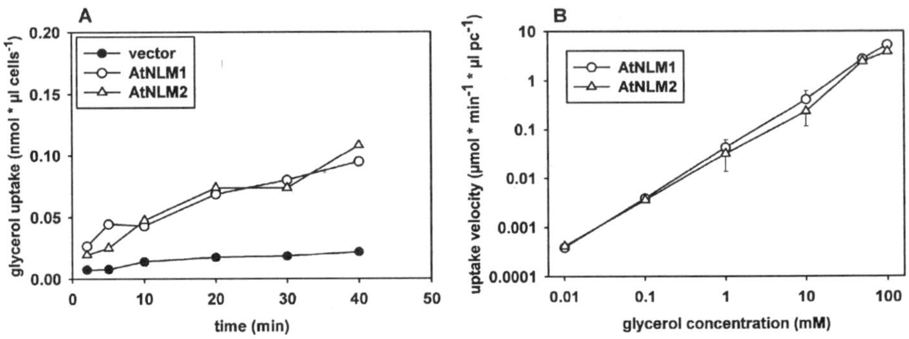

Figure 4. Glycerol transport characteristics of *AtNLM1* and *AtNLM2* expressed in *fps1* yeast mutants. A: Time course of glycerol uptake at 0.1 mM extracellular glycerol concentration; B: Initial uptake rates at different extracellular glycerol concentrations.

In plants, both passive and active transport systems for sugar transport have been identified in plasma membrane and tonoplast. But each of the cloned sugar carriers tested so far exhibits active, proton-coupled sugar transport activities. To analyse whether NLM proteins allow passive permeation of sugars, we expressed *AtNLM1* and *AtNLM2* in a multiple deletion mutant of baker's yeast, EBY.VW4000. This yeast strain is defective in more than twenty sugar carriers and two glucose sensors (Wieczorke *et al.,* 1999). None of the two NLM proteins allowed growth of the transformed yeast cells on glucose, fructose or galactose containing medium. However, this yeast strain transformed with an active glucose carrier from *Ricinus communis* (Weig *et al.,* 1994) grew on each of these carbon sources (Fig. 5).

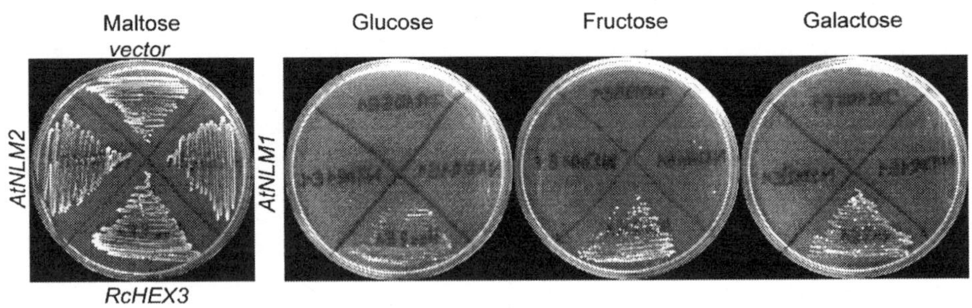

Figure 5. Growth of *AtNLM1* and *AtNLM2* and *RcHEX3* transformed EBY.VW4000 yeast cells on different carbon sources.

3. CONCLUSION

As of March 2000, the 'Arabidopsis Genome Initiative' (AGI) has annotated about 87% of the Arabidopsis genomic sequence data to public databases. This allowed us to identify not only the gene of *AtNLM1* but also gave us a good overview on other *NLM* genes in Arabidopsis.

Five out of eight *NLM* genes contain four introns in between the (putative) start and stop codons and five genes are virtually expressed in different tissues, mainly in roots. Inspection of the deduced amino acid sequences indicated signatures, which could be found in aquaporins and glycerol permeases. Heterologous expression of *AtNLM1* and *AtNLM2* in *S. cerevisiae* revealed glycerol transport activity of these proteins, which was not saturable up to 100 mM external glycerol concentration. However, sugars did not permeate at detectable rates through the pores. Therefore, we assume that NLMs do allow the passage of glycerol (and eventually other small solutes) but are not identical with the passive sugar carriers found in plant membranes.

Currently we are working on the cell specific expression of *NLM* genes and subcellular localisation of the aquaglyceroporins in order to uncover the physiological functions of these transport proteins in plants. Since most of the genes are expressed in roots where glycerol is often produced upon flooding or hypoxia/anoxia (Gerber *et al.*, 1988), it will be interesting to find out whether these proteins do play a role in the plant response to these types of stress.

ACKNOWLEDGEMENTS

The authors thank Stefan Hohmann (University of Göteborg, Sweden) and Eckard Boles (University of Düsseldorf, Germany) for providing the yeast mutants used in this work. This work is supported by a DFG grant (We1680/5-1) to A.R. Weig and C. Jakob. The authors thank E. Schwarz-Weig for carefully reading the manuscript.

REFERENCES

Biela A Grote K Otto B Hoth S Hedrich R and Kaldenhoff R 1999 The *Nicotiana tabacum* plasma membrane aquaporin NtAQP1 is mercury-insensitive and permeable for glycerol *Plant J* **18**: 565-570

Borgnia M Nielsen S Engel A and Agre P 1999 Cellular and molecular biology of the aquaporin water channels *Ann Rev Biochem* **68**: 425-458

Dean RM Rivers RL Zeidel ML and Roberts DM 1999 Purification and functional reconstitution of soybean nodulin 26 An aquaporin with water and glycerol transport properties *Biochemistry* **38**: 347-353

Froger A Tallur B Thomas D and Delamarche C 1998 Prediction of functional residues in water channels and related proteins *Protein Science* **7**: 1458-1468

Gerbeau P Guclu J Ripoche P and Maurel C 1999 Aquaporin Nt-TIPa can account for the high permeability of tobacco cell vacuolar membrane to small neutral solutes *Plant J* **18**: 577-587

Gerber DW Byerrum RU Gee RW and Tolbert NE 1988 Glycerol concentrations in crop plants *Plant Sci* **56**: 31-38

Rivers RL Dean RM Chandy G Hall JE Roberts DM and Zeidel ML 1997 Functional analysis of nodulin 26 an aquaporin in soybean root nodule symbiosomes *J Biol Chem* **272**: 16256-16261

Tamas MJ Luyten K Sutherland FW Hernandez A Albertyn J Valadi H Li H Prior BA Kilian SG Ramos J Gustafsson L Thevelein JM and Hohmann S 1999 Fps1p controls the accumulation and release of the compatible solute glycerol in yeast osmoregulation *Mol Microbiol* **31**: 1087-1104

Weig A Deswarte C and Chrispeels MJ 1997 The major intrinsic protein family of *Arabidopsis* has 23 members that form three distinct groups with functional aquaporins in each group *Plant Physiol* **114**: 1347-1357

Weig A Franz J Sauer N and Komor E 1994 Isolation of a family of cDNA clones from *Ricinus communis* L with close homology to the hexose carriers *J Plant Physiol* **143**: 178-183

Wieczorke R Krampe S Weierstall T Freidel K Hollenberg CP and Boles E 1999 Concurrent knock-out of at least 20 transporter genes is required to block uptake of hexoses in *Saccharomyces cerevisiae FEBS L* **464**: 123-128

5

WATER CHANNELS AND GYCEROL FACILITATORS IN MICROORGANISMS

5.1 Microbial MIP channels

THE MICROBIAL *ESCHERICHIA COLI* AQUAPORIN-Z WATER CHANNEL

Structural, functional and physiological properties

Giuseppe Calamita

Department of General and Environmental Physiology, University of Bari, via Amendola 165/A, 70126 Bari, Italy; E-mail: calamita@biologia.uniba.it

1. INTRODUCTION

Movement of water across plasma membranes is a necessary feature of many physiological functions of all living organisms. However, the molecular pathway of the osmotic flux of water across membranes became evident only after the discovery of the aquaporin water channels (Preston *et al.*, 1992), proteins belonging to the MIP family of transmembrane channels (Park and Saier, 1996). Aquaporins were first recognised in animals and plants where they are being found to play important physiological roles (Agre, 1998; Maurel, 1997). Recently, aquaporins have also been reported in microorganisms (Calamita *et al.*, 1995), a finding consistent with the striking ability of microbial cells to rapidly adapt to the changes in environmental water activity to which they are often exposed.

1.1 AqpZ as a model system for aquaporins

Since the discovery of the bacterial *Escherichia coli* Aquaporin-Z water channel (AqpZ) in 1995 (Calamita *et al.*, 1995), the number of known microbial aquaporin genes has rapidly increased due to extensive genome sequencing and homology cloning by polymerase chain amplification. At present, 30 aquaporin genes have been identified and partially characterized in Archaea, Eubacteria, Fungi and Protozoa. Most of the available biological information on microbial aquaporins derives from studies on the *E. coli*

AqpZ gene and protein. These studies have revealed the striking biochemical (Borgnia *et al.*, 1999) and structural (Ringler *et al.*, 1999; Scheuring *et al.*, 1999) homologies existing between AqpZ and the its counterparts in higher organisms, an observation in line with the molecular conservation anticipated by phylogenetic analyses (Calamita *et al.*, 1995; Park and Saier, 1996). AqpZ has been implicated in processes of particular importance for cell growth and survival while additional roles are under investigation (Calamita *et al.*, 1998). Although useful information will no doubt derive from studies on currently uncharacterized microbial aquaporins, definition of the biochemical, biophysical, structural and physiological features of AqpZ are already providing valuable insights into the importance of aquaporins.

2. MOLECULAR PROPERTIES OF AQPZ

Many molecular properties of the *aqpZ* gene and AqpZ channel protein have been elucidated due to the relative ease with which bacterial systems may be manipulated both biochemically and genetically. The structural organisation of the *E. coli aqpZ* gene has been defined (Calamita *et al.*, 1997). The *Xenopus laevis* oocyte expression system as well as functional reconstitution in proteoliposomes have shown the biophysical features of the AqpZ channel (Calamita *et al.*, 1995; Borgnia *et al.*, 1999). Biochemical and structural studies have been carried out by using the E. coli AqpZ protein produced by a convenient homologous expression system (Borgnia *et al.*, 1999).

2.1 The *E. coli aqpZ* gene

The *E. coli aqpZ* gene has a monocistronic organisation (Calamita *et al.*, 1997). This structure is shared with other known prokaryotic aquaporin genes but contrasts with that of the *aqpZ* bacterial paralogue, the glycerol facilitator gene, *glpF*, which is part of an operon coding for proteins concerned with glycerol uptake and metabolism (Sweet *et al.*, 1990). All known *aqpZ* genes have a chromosomal location except that of the pathogenic species *Shigella sonnei*, *ORF10S*, which is carried by a plasmid responsible for its bacterial virulence (Houng and Venkatesan, 1998).

The -35/-10 promoter region of the *E. coli aqpZ* does not entirely match the canonical consensus by suggesting that the active open complex formation at the *aqpZ* promoter may require an interaction with a protein activator (Calamita *et al.*, 1997). As observed by Northern blot analysis and confirmed by regulatory studies, *aqpZ* is maximally transcribed during the exponential phase of growth (Calamita *et al.*, 1998).

2.2 Biophysical features of the AqpZ channel

Heterologous expression in *Xenopus* oocytes and functional reconstitution in proteoliposomes showed that AqpZ is an orthodox aquaporin by being highly permeable to water (P_f= 10 x 10^{-14} cm^3 s^{-1} $subunit^{-1}$) but excluding small uncharged solutes such as glycerol, sorbitol and urea and lacking apparent ionic conductances (Calamita *et al*., 1995; Borgnia *et al*., 1999). Like other major water channels, the calculated Arrhenius activation energy for the movement of water through the AqpZ channel is low (E_A= 3.8 kcal/mol) (Calamita *et al*., 1995). AqpZ water permeability is unchanged at pH ranging between 5 and 7 (Borgnia *et al*., 1999). Functional reconstitution studies also revealed that AqpZ does not require contribution from additional proteins for its water transport activity, a feature shared with aquaporins in higher organisms.

Cryoelectron microscopy studies revealed *in vivo* that AqpZ is functionally expressed in *E. coli* where it mediates osmotic movements of water in both directions (Delamarche *et al*. 1999). The bi-directional water channel activity is also featured by multiple mammalian aquaporins likely representing a general property among aquaporins (Meinild *et al*., 1998).

2.3 Structure of the AqpZ channel protein

Borgnia *et al*. (1999) devised an efficient homologous expression system to produce milligram quantities of the *E. coli* AqpZ protein which was then used for biochemical and structural studies. Like the mammalian AQP1, the best-characterized eukaryotic aquaporin, AqpZ has a homotetrameric organisation in its native state. However, unlike other known aquaporins, AqpZ is a markedly hydrophobic and tightly packed protein. This may explain why the AqpZ tetramers are very stable by not being readily dissociated even by 1% SDS at neutral pH.

Electron crystallography and atomic force microscopy studies performed by Ringler *et al*. (1999) and Scheuring *et al*. (1999), respectively, confirmed the homotetrameric structure and demonstrated experimentally the six transmembrane domains topology of the AqpZ monomer, a structural feature shared with the higher eukaryotic aquaporins. As usual, the two typical Asn-Pro-Ala (NPA) aquaporin signature motifs of AqpZ were located in loops B and E that have been shown to fold back into the membrane to form a highly specific channel (*hourglass model*) (Jung *et al*., 1994).

2.4 Regulation of the *aqpZ* gene

Regulation of AqpZ at a genetic level was studied by using an *aqpZ::lacZ* gene fusion system whose expression was driven by a 532 bp region containing the promoter and *cis*-regulatory elements of the *E. coli aqpZ* (Calamita *et al*., 1998). The results of the regulatory studies provided hints into the understanding of the physiological roles of AqpZ.

The *aqpZ* gene is growth-phase regulated because its transcription rate peaks at the mid-logarithmic phase of growth while declining to a significantly lower level during the late stationary phase. *aqpZ* was also found to be osmotically regulated since its expression was significantly increased under long-term hypoosmotic conditions but strongly reduced in hyperosmolar media (Calamita *et al*., 1998). Because aquaporin genes responding to osmotic stimuli have also been observed in plants, mammals and fungi, it is likely that osmotic gene regulation may be a general feature among aquaporins.

3. PHYSIOLOGICAL ROLES OF AQPZ IN *E. COLI*

The physiological role of AqpZ has been investigated by characterizing the phenotype expressed by *E. coli aqpZ* mutant cells (Calamita *et al*., 1998).

A direct role for AqpZ in maintaining turgor pressure, a property needed for cell growth and survival, has been observed by cryoelectron microscopy (Delamarche *et al*., 1999). It was shown that the AqpZ channel mediates both the rapid outwardly and inwardly osmotic fluxes of water across the cytoplasmic membrane triggered by abrupt shifts in extracellular osmolarity. AqpZ also participates in the adaptive response of *E. coli* to prolonged hypoosmotic stress as indicated by the considerable low cell viability characterizing the AqpZ$^-$ *E. coli* cells grown in hypoosmotic media (Calamita *et al*., 1998). This finding is consistent with the presence of typical *aqpZ* genes in bacteria permanently living in fresh water such as cyanobacteria (Kaneko *et al*., 1996).

AqpZ may have a role in mediating the bulk water uptake needed for cell expansion during rapid growth (Calamita *et al*., 1998). This is suggested by the reduced viability characterizing the *aqpZ* mutant grown at 39 °C, a temperature where the growth rate is maximal, as well as by regulatory studies showing an increased expression of *aqpZ* during the exponential phase of growth. However, the precise physiological significance of AqpZ in rapidly dividing *E. coli* cells is unclear because is has been calculated that sufficient water for cell division may be absorbed by simple diffusion.

Interestingly, expression of *aqpZ*-like open reading frames seems to be one of the necessary requirements for the expression of cell surface antigens needed by enteric bacteria to invade the host epithelial cells. In fact, deletion of an *aqpZ*-like coding region, *ORF10S*, carried by the pathogenetic species *Shigella sonnei* eliminated the form I antigen and consequently its bacterial virulence (Houng and Venkatesan, 1998). A similar observation has been made with *Plesiomonas shigelloides* (serotype 017) where an *aqpZ*coding region, *ORF10P*, has been found to contribute to the expression of a cell surface O-antigen (Chida *et al.*, unpublished observation). At present, the specific role of the proteins coded by *ORF10S* and *ORF10P* in the mechanism of expression of the above-mentioned surface antigens is still unknown.

The yeast *Saccharomyces cerevisiae* AQY1-1 aquaporin may be implicated in the control of the osmotic balance which is critical for processes such as mating, budding, pseudohyphal formation and sporulation (Bonhivers *et al.*, 1998). The protozoan *Dictyostelium discoideum* WacA aquaporin has been suggested to mediate the extrusion of water that leads to prespore cell formation (Flick *et al.*, 1997).

4. OTHER MICROBIAL AQUAPORINS

Since identification of the *E. coli aqpZ*, 29 additional aquaporin genes have been reported variously distributed in Archaea, Eubacteria, Fungi and Protozoa (Fig. 1). Two of these genes, the slime mold *Dictyostelium discoideum wacA* (Flick *et al.*, 1997) and the yeast *Saccharomyces cerevisiae AQY-1* (Bonhivers *et al.*, 1998), have been proven to code for proteins featuring a water transport activity. The identity of the remaining genes has been assumed based on sequence analyses, gene organization and phylogenetic position relative to proven aquaporins (Kayingo *et al.*, 2000) and functional studies are therefore needed to confirm their aquaporin nature experimentally. As suggested by the phylogenetic analysis (Fig. 1), it is possible that the proteins coded by some aquaporin genes possess other biophysical properties alternatively/in addition to the water transport function. For example, the *Synechococcus* SmpX protein apparently plays a role in copper ion transport (Kashiwagi *et al.*, 1995) although it is phylogenetically related to the aquaporin cluster (Kayingo *et al.*, 2000)

It is interesting to note that except for *Klebsiella pneumoniae* and *Caulobacter crescentus*, whose genomes comprise at least two possible aquaporin genes, it can be seen that bacteria possess at most one aquaporin gene. Aquaporins seem to be more related to Gram-positive than to Gram-negative bacteria, as suggested by the fact that most Gram-positive species whose genome has been entirely sequenced do not possess canonical aquaporin genes. Although still unclear, this observation may be consistent with the fact that Gram-positive bacteria use different strategies to respond to changes in the environmental osmolarity (Poolman and Glaasker, 1998).

So far, no explanations have been provided to justify the apparent lack of canonical aquaporins and other MIP family channels in some microorganisms such as the archaea *Methanococcus jannaschii* and *Pyrococcus horikoshii* and the bacteria *Aquifex aeolicus*, *Helicobacter pylori*, *Mycobacterium tuberculosis*, *Treponema pallidum*, *Chlamydia trachomatis*, *Chlamydia pneumoniae*, *Rickettsia prowazekii*, *Neisseria meningitidis* and *Campylobacter jejuni*. It is possible that these microbes do not experience considerable changes in water or solute availability in their natural habitats and therefore they may not require water or solute channels. Alternatively, these microorganisms might possess other currently unknown transport mechanisms.

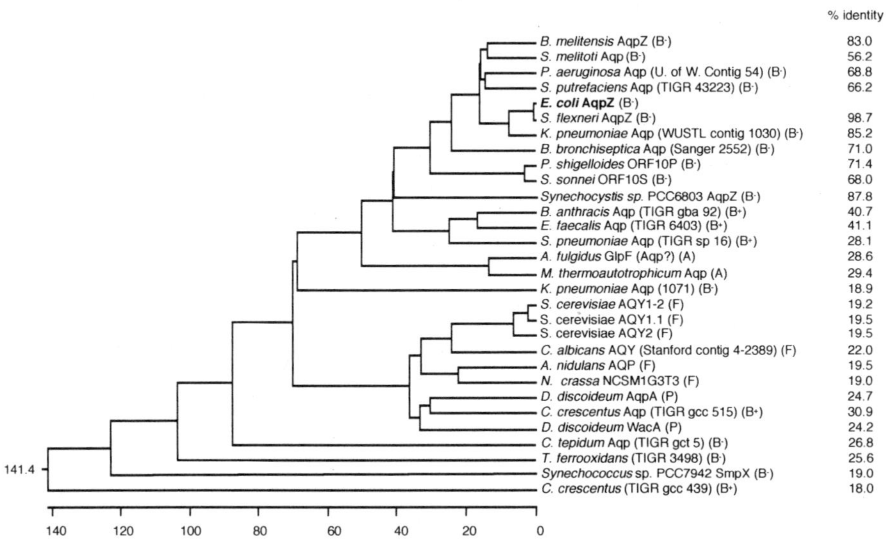

Figure 1. Phylogenetic analysis of *E. coli* AqpZ and other possible microbial aquaporins. Deduced amino acid sequences from the indicated aquaporins were aligned by using the Clustal method. Phylogenetic tree was constructed with Lasergene computer program. Branch lengths are proportional to evolutionary distances. Lower scale reports the genetic distances. The percent identity compared with the *E. coli* AqpZ (*bold*) is indicated. Preliminary sequence data was obtained from The Institute for Genomic Research website at http://www.tigr.org. Abbreviations are: A, Archaea; B+, Gram-positive Eubacteria; B-, Gram-negative Eubacteria; F, Fungi; P, Protozoa. U. of Washington (U. of W.): genome.washington.edu; WUSTL: www.genome.wustl.edu; Stanford: www.stanford.edu;

5. CONCLUSIONS AND PERSPECTIVES

Efforts are still needed to identify new homologues and to fully understand the biological and molecular properties of microbial aquaporins. Extensive functional studies are needed to assess the biophysical features of the proteins coded by the numerous as yet uncharacterized microbial aquaporin genes. Possible additional or alternative functions to the water channel activity are not excluded. Little is known about the molecular mechanisms by which microbial aquaporins are regulated both at genetic and protein level, a primary importance aspect which requires deeper investigation. Valuable information regarding the structure of aquaporin proteins might be obtained by means of X-ray diffraction studies using 3D AqpZ crystals. Although important roles have been already ascribed to AqpZ, additional ones may as yet be unknown. Phenotypic examination of microorganisms carrying disrupted aquaporin genes will help in fully defining the physiological functions of aquaporins in microorganisms. It is likely that AqpZ will continue to be one of the most viable options for understanding the molecular properties and biological importance of aquaporins.

ACKNOWLEDGMENTS

I thank Professors Peter Agre and Erhard Bremer and Doctors Mario J. Borgnia, Christian Delamarche, Stefan Hohmann and Daniel Thomas for collegial discussions.

REFERENCES

Agre P Bonhivers M and Borgnia M 1998 The aquaporins blueprints for cellular plumbing systems *J Biol Chem* **273:** 14659-14662

Bonhivers M Carbrey JM Gould SJ and Agre P 1998 Aquaporins in *Saccharomyces J Biol Chem* **273:** 27565-27572

Borgnia MJ Kozono D Calamita G Maloney PC and Agre P 1999 Functional reconstitution and characterization of AqpZ the *E coli* water channel protein *J Mol Biol* **291:** 1169-1179

Calamita G Bishai WR Preston GM Guggino WB and Agre P 1995 Molecular cloning and characterization of AqpZ a water channel from *Escherichia coli J Biol Chem* **270:** 29063-29066

Calamita G Kempf B Bonhivers M Bishai WR Bremer E and Agre P 1998 Regulation of the *Escherichia coli* water channel gene *aqpZ Proc Natl Acad Sci USA* **95:** 3627-3631

Calamita G Kempf B Rudd KE Bonhivers M Kneip S Bishai WR Bremer E and Agre P 1997 The Aquaporin-Z water channel gene of *Escherichia coli*: structure organization and phylogeny *Biol Cell* **89:** 321-329

Delamarche C Thomas D Rolland J-P Froger A Gouranton J Svelto M and Agre P 1999 Visualization of AqpZ-mediated water permeability in *Escherichia coli* by cryoelectron microscopy *J Bacteriol* **181:** 4193-4197

Flick KM Shaulsky G and Loomis WF 1997 The *wacA* gene of *Dictyostelium discoideum* is a developmentally regulated member of the MIP family *Gene* **195:** 127-130

Houng HS and Venkatesan MM 1998 Genetic analysis of *Shigella sonnei* form I antigen: identification of a novel IS630 as an essential element for the form I antigen expression *Microb Pathog* **25:** 165-173

Kaneko T et al (1996) Sequence analysis of the genome of the unicellular cyanobacterium *Synechocystis sp* strain PCC6803 IISequence determination of the entire genome and assignment of potential protein-coding regions *DNA Res* **3:** 109-136

Kashiwagi S Kanamaru K and Mizuno T 1995 A *Synechococcus* gene encoding a putative pore-forming intrinsic membrane protein *Biochim Biophys Acta* **1237:** 189-192

Kayingo G Bill R Calamita G Hohmann S and Prior BA 2000 Microbial water and glycerol channels In *Aquaporins* (Agre P Nielsen S and Hohmann S eds) Academic Press London in press

Jung JS Preston GM Smith BL Guggino WB and Agre P 1994 Molecular structure of the water channel through aquaporin-CHIP The hourglass model *J Biol Chem* **269:** 14648-14654

Maurel C 1997 Aquaporins and water permeability of plant membranes *Ann Rev Plant Physiol Mol Biol* **48:** 399-429

Meinild AK Klaerke D A Zeuthen T 1998 Bidirectional water fluxes and specificity for small hydrophobic molecules in Aquaporin 0-5 *J Biol Chem* **273:** 32446-32451

Park JH and Saier HM Jr 1996 Phylogenetic characterization of the MIP family of transmembrane channel proteins *J Membrane Biol* **153:** 171-180

Poolman B and Glaasker E 1998 Regulation of compatible solute accumulation in bacteria *Mol Microbiol* **29:** 397-407

Preston GM Piazza-Carroll T Guggino WB and Agre P 1992 Appearance of water channels in *Xenopus* oocytes expressing red cell CHIP28 water channel *Science* **256:** 385-387

Ringler P Borgnia MJ Stahlberg H Maloney PC Agre P and Engel A 1999 Structure of the water channel AqpZ from Escherichia coli revealed by electron crystallography *J Mol Biol* **291:** 1181-1190

Scheuring S Ringler P Borgnia MJ Stahlberg H Müller DJ Agre P and Engel A 1999 High resolution AFM topographs of the *Escherichia coli* water channel Aquaporin-Z *EMBO J* **18**: 4981-4987

5.2 Microbial MIP channels

STUDY OF FAST WATER MOVEMENTS IN BACTERIA BY CRYOELECTRON MICROSCOPY

[1]Alexandrine Froger, [1]Jean-Paul Rolland, [2]Maria C. Rodriguez, [2]Juan M. Garcia-Lobo, [1]Laurence Duchesne, [1]Isabelle Pellerin, [1]Annie Cavalier, [1]Jean-François Hubert, [1]Stéphane Deschamps, [1]Patrick Bron, [1]Christian Delamarche and [1]Daniel Thomas

[1]Equipe Canaux et Récepteurs Membranaires, UMR-CNRS 6026, Université de Rennes 1, Campus de Beaulieu, 35042 Rennes cedex, France. [2]Departamento de Biologia Molecular de la Universidad de Cantabria, Spain.

1. INTRODUCTION

Movement of water into or out of cells is a fundamental process of life found throughout nature. However the molecular pathway of this transport remained elusive until discovery of the aquaporins, a large family of water channel proteins. In bacteria, the osmotic movement of water across the cytoplasmic membrane is one of the mechanisms triggered to maintain the cell turgor, a function essential for growth and survival. The first known prokaryotic aquaporin water channel gene, *aqpZ*, has been reported in *Escherichia coli* (Calamita *et al.*, 1995) and functionally characterized (Borgnia *et al.*, 1999; Delamarche *et al.*, 1999). This indicates that, in spite of the high surface-to-volume ratio characterizing bacteria, the simple diffusion of water through the membrane lipids could not be always sufficient to preserve turgor. Presently, several microbial members of the MIP family have been identified by sequence homology, but only few microbial MIPs have been functionally studied (Maurel *et al.*, 1994).

Glycerol facilitators, predicted from sequence analysis, occur in bacteria for all subgroups whereas aquaporins have been found only in some Gram-negative bacteria and alleged in two archeal species (Hohmann *et al.*, 2000). Gram-positive bacteria seem to possess only one MIP gene, which forms a

Molecular Biology and Physiology of Water and Solute Transport
Edited by Hohmann and Nielsen, Kluwer Academic/Plenum Publishers, New York, 2000.

separate branch from glycerol facilitators and hence could bear unorthodox functional properties.

Xenopus oocytes have been largely used as an experimental system for expression and functional analysis of heterologous proteins, but this system is not always satisfactory. For example, for MIP proteins, reduced P_f values in injected oocytes could result from low expression or misrouting of the protein (Mulders *et al.,* 1997; Lagrée *et al.,* 1998).

To study the water transport properties of new bacterial MIPs, we propose a rapid method using *E. coli* as a heterologous expression system and cryoelectron microscopy to monitor water flux in *E. coli* cells. To date, most electron microscopic studies of cellular structures are based on conventionally prepared specimen where specimen are subjected to chemical fixation, dehydrated and embedded in resin. Such procedure may rise to severe structural artefacts. The method of preparing and observing specimens in a thin layer of vitreous ice, known as cryoelectron microscopy, enables biological specimens to be maintained in a close-to-life structural organization and thus to be imaged under physiological conditions (Dubochet *et al.,* 1988). Bacteria like *E. coli* are less than 1 µm thick and thus can be embedded *in toto* in vitreous ice. Moreover, cryoelectron microscopy can record different stages of a biological process (Lepault *et al.,* 1983), thus, making it possible to perform time-resolved studies. We have monitored the response to an osmotic shock of an MIP-deficient *E. coli* strain containing or not an exogenous *mip* gene. Using this approach, new bacterial aquaporins have been functionally characterized using *E.coli* as expression system.

2. METHODS

2.1 Bacterial strains and plasmids

In this study we have used *mip* genes from *Escherichia coli* (*aqpZ, glpF*), *Brucella abortus* (*aqpX*) (Rodriguez *et al.,* submitted) and *Lactococcus lactis* (*glaLac*) (Froger *et al.,* submitted). Different MIP ORF *aqpZ, aqpX, glaLac* and *glpF* were cloned into pUC19. The parental wild-type *E. coli* strain MM294 and the *aqpZ* and *glpF* null *E. coli* strain SK46 (Calamita *et al.,* 1998) transformed by the control plasmid pUC19 or its derivatives were used for osmotic challenges.

2.2 Osmotic shock and cryoelectron microscopy

E. coli cells were grown at 37°C until exponential phase. Bacteria were then pelleted and suspended in M9 medium (osmolality, 240 mosM) at room temperature. A 2.5 µl drop of cell suspension was placed directly on a copper grid coated with a thin carbon film rendered hydrophilic by glow-discharge. Osmotic up-shocks were induced by rapidly mixing 2.5 µl of a 1.2 M sucrose-M9 solution with the cell suspension rising the osmolality to 1,000 mosM. Immediately or after a delay of 10 s, the suspension was briefly blotted with filter paper to a film of water and immediately plunged into liquid ethane held at liquid nitrogen temperature. Specimens were examined at –170°C in a Philips CM12 microscope with a Gatan model 626 cryoholder. Micrographs were recorded under low dose conditions and at a nominal magnification of x6,300.

3. RESULTS AND DISCUSSION

We have already described the morphological appearance of osmotically unchallenged *E. coli* cells, when observed by cryoelectron microscopy: cells are sharply delineated by the bacterial envelope, the perisplasmic space is minimal, and no shrinkage of the cytoplasm is apparent (Delamarche *et al.*,, 1999). Figure 1 shows *E. coli* strain SK46 ($AqpZ^-$, $GlpF^-$) quickly frozen following a 10 s hyperosmotic shock. The *aqpZ* and *glpF* null bacteria, illustrated in figure 1a, doesn't display any obvious morphological alteration. In contrast SK46 *E coli* strain expressing AqpZ (figure 1b) and SK46 *E. coli* strain SK46 expressing the *Brucella* aquaporin AqpX (figure 1c) clearly show dramatic cytoplasm retractions forming plasmolysis spaces. Similarly, *E. coli* strain SK46 expressing Gla_{Llac} the *Lactococcus* MIP (figure 1d) shows significant shrinkage of the cytoplasm but less severe than in bacteria expressing an aquaporin. The same experiment was conducted on *E. coli* strain SK46 expressing GlpF but the osmotic challenge of 10s was without effect on the cell morphological appearance as expected (data not shown). The data presented in the present work provide a straight forward method to functionally characterize microbial MIPs expressed in bacteria. On one hand, like AqpZ, AqpX from *Brucella abortis* mediates fast and large osmotic flux and thus appears to be a genuine aquaporin. On the other hand Gla_{Llac} from *Lactococcus* significantly mediates water fluxes albeit with lower permeability than bona fide aquaporins.

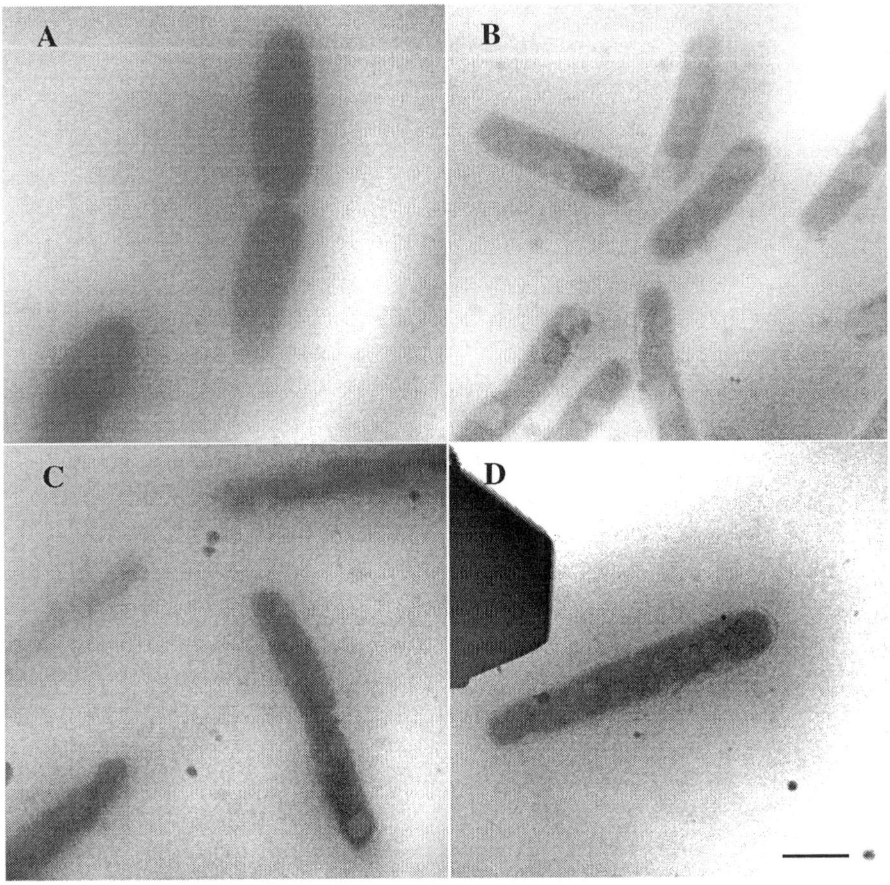

Figure 1. Cryoelectron micrographs of *E. coli* strain SK46 (AqpZ$^-$, GlpF$^-$) following a 10s hyperosmotic chock. (a) *aqpZ and glpF* null bacteria; (b) SK46 *E. coli* strain expressing AqpZ; (c) SK46 *E. coli* strain expressing AqpX; (c) SK46 *E. coli* strain expressing Gla_{Llac}. Bar, 1μm

ACKNOWLEDGEMENTS

This work was supported by La Fondation Langlois (Rennes, France).

REFERENCES

Borgnia MJ Kozono D Calamita G Maloney PC and Agre P 1999 Functional reconstitution of AqpZ the *E coli* water channel protein *J Mol Biol* **291**: 1169-1179

Calamita G Bishai WR Preston GM Guggino WB and Agre P 1995 Molecular cloning and characterization of AqpZ a water channel from *Escherichia coli J Biol Chem* **270**: 29063-29066

Calamita G Kempf B Bonhivers M Bishai WR Bremer E and Agre P 1998 Regulation of the *Escherichia coli* water channel gene aqpZ *Proc Natl Acad Sci* USA **95**: 3627-3631

Delamarche C Thomas D Rolland J-P Froger A Gouranton J Svelto M Agre P and Calamita G 1999 Visualization of AqpZ-mediated water permeability in *Escherichia coli* by cryoelectron microscopy *J Bacteriol* **181**: 4193-4197

Dubochet J Adrian M Chang JJ Homo J-C McDowall AW and Schultz P 1988 Cryo-electron microscopy of vitrified specimens *Q Rev Biophys* **21**: 129-228

Hohmann S Bill RM Kayingo G and Prior B 2000 Microbial MIP channels *Trends Microbiol* **8**: 33-37

Lagrée V Pellerin I Hubert J-F Tacnet F Le Cahérec F Roudier N Thomas D Gouranton J and Deschamps S 1998 A yeast recombinant aquaporin mutant that is not expressed or mistargeted in *Xenopus* oocyte can be functionally analyzed in reconstituted proteoliposomes *J Biol Chem* **273**: 12422-12426

Lepault J Booy F and Dubochet J 1983 Electron microscopy of frozen biological suspensions *J Microsc* **129**: 89-102

Maurel C Reizer J Schroeder JI Chrispeels MJ and Saier MHJr 1994 Functional characterization of the *Escherichia coli* glycerol facilitator GlpF in *Xenopus* oocytes *J Biol Chem* **269**: 11869-11872

Mulders SM Knoers NV Van Lieburg AF Monnens LA Leumann E Wuhl E Schober E Rijss JP Van Os CH and Deen PM 1997 New mutations in the AQP2 gene in nephrogenic diabetes insipidus resulting in functional but misrouted water channels *J Am Soc Nephrol* **8**: 242-248

5.3 Microbial MIP channels

AN AQUAPORIN GENE IN PLASMODIUM FALCIPARUM

Molecular cloning and functional expression

Martin Hansen, Eric Beitz and Joachim E. Schultz
Institute of Pharmacy, University of Tübingen, Morgenstelle 8, D-72076 Tübingen, Germany. E-mail: martin.hansen@uni-tuebingen.de

1. INTRODUCTION

Since the discovery of the first water channel in mammalian red blood cells by Peter Agre's group less than a decade ago, the aquaporin family dramatically extended to far more than 100 members today. These membrane spanning transport proteins proved to be important players in the maintenance of fluid homeostasis and were identified in a large number of organisms including bacteria, plants, insects, amphibians and mammals. No aquaporins from protozoa have been described so far. The Malaria Genome Project of Plasmodium falciparum provided the opportunity to screen the genetic information of this protozoan parasite for aquaporin related sequences. P. falciparum causes the most severe form of malaria, malaria tropica, of which still 1-2 million people die every year. Strong efforts are being undertaken to reveal weak sites of this parasite mainly focusing on metabolic properties and specific features of its variable coat in order to identify targets for a therapeutic approach.

Molecular Biology and Physiology of Water and Solute Transport
Edited by Hohmann and Nielsen, Kluwer Academic/Plenum Publishers, New York, 2000.

2. IDENTIFICATION OF AN AQUAPORIN IN PLASMODIUM FALCIPARUM

Comparative BLAST searches in the TIGR database of the P. falciparum genome using mammalian aquaporin sequences as query yielded overlapping fragments from chromosome 11 coding for a open reading frame of 774 bp. The sequence shows highest identity (32.1%) to an aquaporin from E. coli, the glycerol-uptake facilitator (glpF; GenBank acc. #P11244). We preliminary designated this first aquaporin sequence from P. falciparum PfAQP1.

2.1 Cloning and Expression of PfAQP1

Based on the Malaria Genome Project derived sequence data we designed specific primers spanning the complete open reading frame of PfAQP1.

Using genomic DNA from P. falciparum cultured in red blood cells as a PCR template we obtained a product of 792 bp with a coding region of 774 bp (for 258 amino acids; see Figure 1) that was identical to the database sequence. There was no evidence for introns on the DNA level or after sequence comparisons with other aquaporins on the protein level, which is not unusual for protozoan genes.

For a functional assay the PCR product was subcloned into pOG1, a vector designed for in vitro cRNA-transcription and subsequent protein expression in Xenopus laevis oocytes. Stage V and VI oocytes were injected with either PfAQP1-cRNA (5 ng per oocyte) or DEPC-water as control.

After 72 h of incubation in isotonic medium the oocytes were exposed to hypotonic stress and were video-monitored using a 2.5 x objective. The volume increase due to water influx was calculated from the area covered by the oocyte. PfAQP1-cRNA injected oocytes swelled rapidly and ruptured within 1-2 minutes whereas control oocytes showed only a marginal volume increase. Thus, we established the presence of a functional water channel protein in P. falciparum.

2.2 Properties of the PfAQP1 Sequence

Aquaporins are characterized by a topology consisting of six transmembrane spanning domains and two highly conserved asparagine-proline-alanine (NPA) motifs in both protein halves which are pivotal in the formation of the actual water pore. Quite obviously, the molecule contains an internal symmetry which is assumed to originate from an ancient gene duplication.

Computer-based topology calculation using the PHD software predicts a set of six membrane spans for PfAQP1 (Figure 1). This was somewhat expected due to reasonable sequence similarities with other aquaporins in these domains. However, both canonical NPA motifs are changed to NLA and NPS, respectively. This is similar to a previously characterized glycerol pore from Saccharomyces cerevisiae. In this yeast aquaporin the same amino acid exchanges are present but in an inverted arrangement, i.e. NPS and NLA (Van Aelst *et al.*, 1991, Tamás *et al.*, 1999). Whether these motifs are responsible for a directed transport of glycerol, as supposed by the authors, has still to be proven. The higher sequence identity of PfAQP1 with glycerol facilitators than with pure water pores, especially in the characteristic aspartic acid two residues downstream of the NPS motif, might indicate that we are dealing with a glycerol facilitator, too.

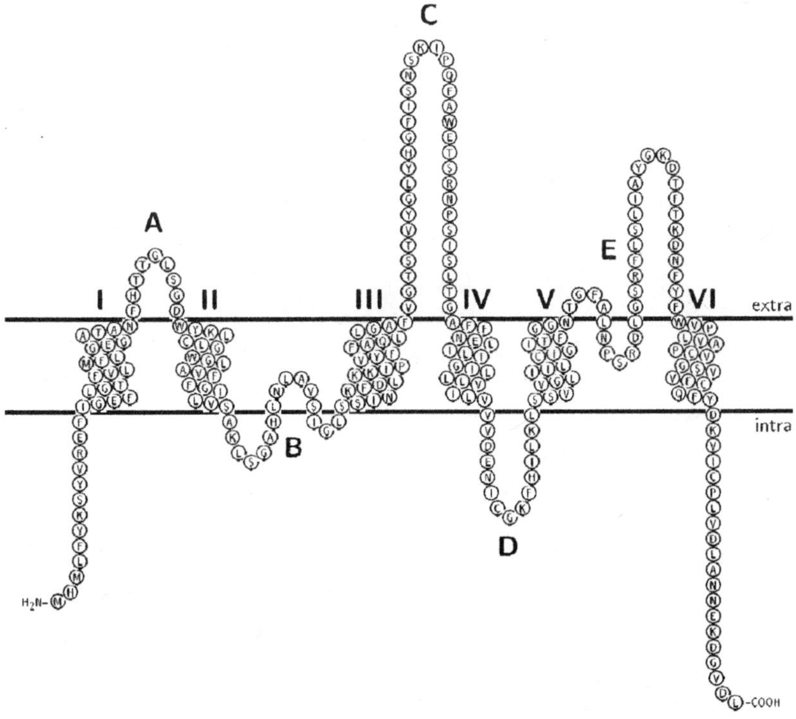

Figure 1. Topology plot of PfAQP1 based on the PHD prediction software

3. PERSPECTIVE

In addition to these initial studies we are going to further define the transport characteristics of the new potential glycerol facilitator PfAQP1. Does the unconventional amino acid arrangement in the pore region define a new subfamily of aquaporins with the capability of transporting other kinds of cargo? Such an assumption is speculative but this possibility should not be ruled out if one considers the recent finding that AQP6 constitutes a chloride channel (Yasui *et al.*, 1999). Trans-membranous transporters at the interface between the parasite and a host red cell likely have important functions in the exchange of nutrients and metabolites and thus call for further biochemical studies.

REFERENCES

Tamás MJ Luyten K Sutherland FCW Hernandez A Valadi H Li H Prior BA Kilian SG Ramos J Gustafsson L Thevelein JM and Hohmann S 1999 Fps1p controls the accumulation and release of the compatible solute glycerol in yeast osmoregulation *Mol Microbiol* **31**: 1087-1104

Van Aelst L Hohmann S Zimmermann FK Jans AWH and Thevelein JM 1991 A yeast homologue of the bovine lens fibre MIP gene family complements the growth defect of a Saccharomyces cerevisiae mutant on fermentable sugars but not its defect in glucose-induced RAS-mediated cAMP signalling *EMBO J* **10** 2095-2104

Yasui M Hazama A Kwon TH Nielsen S Guggino WB Agre P 1999 Rapid gating and anion permeability of an intracellular aquaporin *Nature* **402**: 184-187

5.4 Microbial MIP channels

AN INVESTIGATION OF THE POSSIBLE EXISTENCE OF HOMOLOGUES OF FPS1, A GLYCEROL FACILITATOR OF *SACCHAROMYCES CEREVISIAE*, IN THE OSMOTOLERANT YEAST *ZYGOSACCHAROMYCES ROUXII*

[1]Gerald Kayingo, [2]Cornelia Casaleggio, [3]Markus J. Tamás, [4]Stefan Hohmann and [1]Bernard A. Prior
[1]Department of Microbiology, University of Stellenbosch, Private Bag X1, 7602 Matieland, South Africa. [2]Department of Microbiology and Biochemistry, University of the Orange Free State, P.O. Box 339, Bloemfontein 9300, South Africa. [3]Laboratorium voor Moleculaire Celbiologie, Katholieke Universiteit Leuven. [4]Department of Cell & Molecular Biology/Microbiology, Göteborg University, Box 462, SE-405 30 Göteborg, Sweden. E-mail: bap@maties.sun.ac.za

1. INTRODUCTION

Yeast cells accumulate substantial amounts of polyols, such as glycerol and arabitol, as their major osmolytes (compatible solutes) during hyper-osmotic stress (Yancey, *et al.*, 1982). Upon a decrease in external osmolarity, solute levels must be adjusted to balance the turgor pressure difference created as cells are moved from hyper-osmotic to hypo-osmotic conditions. We have recently shown that *Saccharomyces cerevisiae* cells rapidly release their intracellular glycerol upon an osmotic downshock and that the membrane channel protein Fps1p controls this release (Tamás, *et al.*, 1999). *fps1Δ* mutants are sensitive to hypo-osmotic shock indicating that glycerol export is required for recovery (survival) following a sudden drop in external osmolarity. This observation raised the question whether other yeasts also employ a channel mediated glycerol export system similar to that

Molecular Biology and Physiology of Water and Solute Transport
Edited by Hohmann and Nielsen, Kluwer Academic/Plenum Publishers, New York, 2000.

of *S. cerevisiae*, when adapting to hypo-osmotic shock. A previous study indicated that the mechanisms of glycerol accumulation in the osmotolerant yeast *Zygosaccharomyces rouxii* differ considerably from those of *S. cerevisiae* (Table 1). *Z. rouxii* synthesise several compatible solutes and retain high amounts of glycerol due to an osmotically active glycerol uptake system. In this study, we explored the possible occurrence of channel mediated glycerol export and of an *FPS1* homologue.

Table 1. Differences in osmotolerance and glycerol metabolism between *S. cerevisiae* and *Z. rouxii*

Feature	***S. cerevisiae***	***Z. rouxii***
Osmotolerance[1]	0.90 a_w	0.65 a_w
Compatible solute	Glycerol	Glycerol + Arabitol
Glycerol metabolism	G-3-P pathway DHA pathway	G-3-P pathway DHA pathway[2]
	NADH	NADPH[3]
	GPD1 present, induced by osmotic stress	*GPD*1 present
Glycerol dehydrogenase and Dihydroxy acetone kinase activity	Low	High, induced by osmotic stress
Phosphofructokinase activity	Increase during stress Inhibited by ATP	Decrease during stress Significantly inhibited by ATP
HOG pathway	*HOG1* present	*HOG1* present
Glycerol retention	High intra/extracellular glycerol ratio (600X)	Very high intra/extracellular glycerol ratio (10,000X)
Glycerol transport	Passive Active Fps1 channel	Passive Osmotically active transport Unknown¯

- [1] The minimum water activity (a_w) tolerated by a particular yeast strain varies according to the solute used to adjust the a_w. Osmotolerance also depends on the pH of the media. [2] The DHA pathway for glycerol metabolism is significant in osmoregulation in *Z. rouxii*, and [3] the reduction of dihydroxy acetone phosphate and dydroxyacetone is more effective with NADPH than NADH as cofactors whereas *S. cerevisiae* is more dependent on NADH (for review see, Attfield, 1998).

1.1 Release of Osmolytes from Yeast exposed to Hypo-osmotic Shock

To explore the possible involvement of glycerol channel proteins in *Z. rouxii,* the patterns and kinetics of glycerol export after an osmotic downshock were first investigated. Yeast cells were grown in media with 5% NaCl and then resuspended in media without salt. Intracellular and extracellular glycerol concentrations at different time points were monitored. A rapid glycerol efflux was observed in the osmotolerant yeast *Z. rouxii* (Fig 1a). About 75 % of the intracellular glycerol were lost in 5 min. A rapid glycerol efflux could still be observed at 0°C suggesting the involvement of a novel protein-mediated mechanism other than passive diffusion. Another osmotolerant yeast, *Pichia sorbitophila* behaved in a similar way (data not shown). In addition to glycerol, *Z. rouxii* cells also released arabitol during adaptation to hypo-osmotic shock (Fig 1b). Extracellular glycerol and arabitol levels increased drastically indicating that osmolytes were not metabolised but released after osmotic downshock. Although glycerol accumulation in *Z. rouxii* occurs by an osmotically active transport (Van Zyl *et al.*, 1990), the pattern and kinetics of glycerol release upon an osmotic downshock suggest the presence of a channel-mediated transport system similar to that of *S. cerevisiae* Fps1 (Fig 1c). During hyper-osmotic conditions, the *S. cerevisiae* glycerol facilitator Fps1 apparently closes thereby conserving glycerol inside the cell whereas in hypo-osmotic conditions, Fps1p opens and intracellular glycerol release occurs rapidly (Fig 1c and 1d).

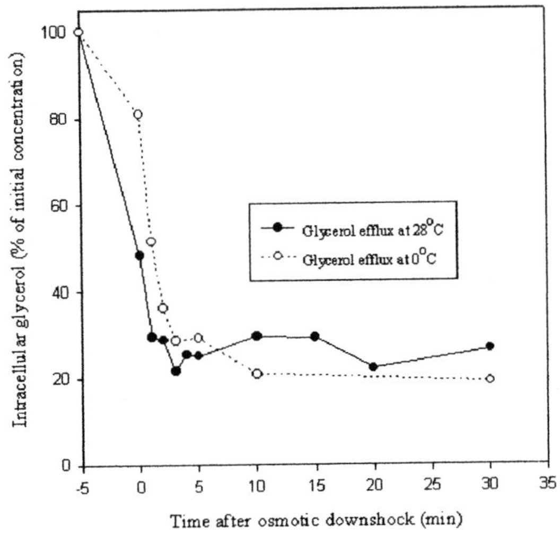

Figure 1a. Glycerol efflux during hypo-osmotic stress (from 5% to 0%) in Z. rouxii NRRL 2547 grown in YNB / glucose.

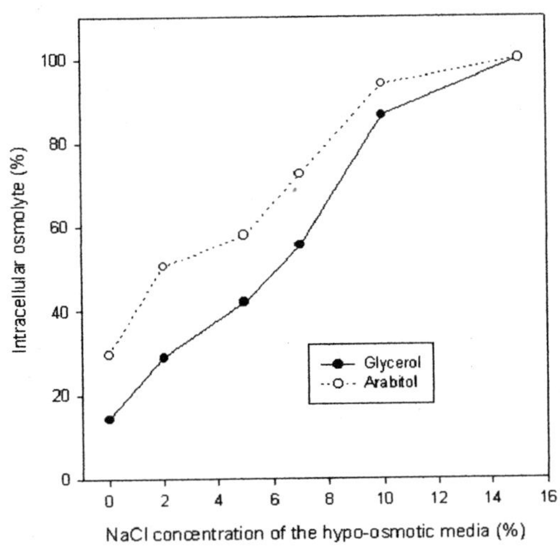

Figure 1b. Release of osmolytes (at 28°C) as a function of shock intensity in Z. rouxii NRRL 2547 grown in YNB / glucose.

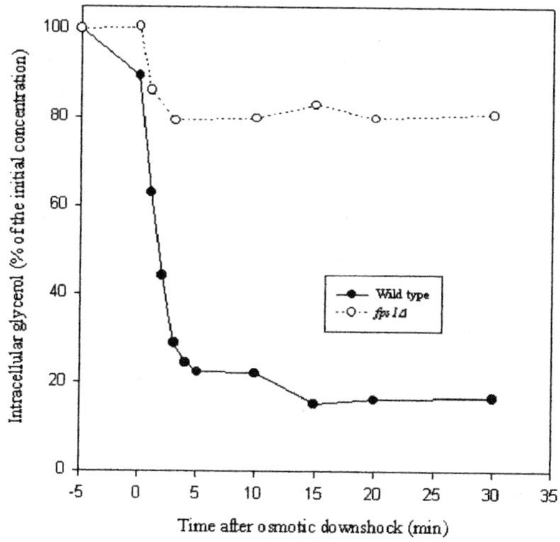

Figure 1c. Glycerol efflux during hypo-osmotic stress in *S. cerevisiae W303 1A* wild-type and *fps1Δ* mutant grown in YNB / glucose at 28°C.

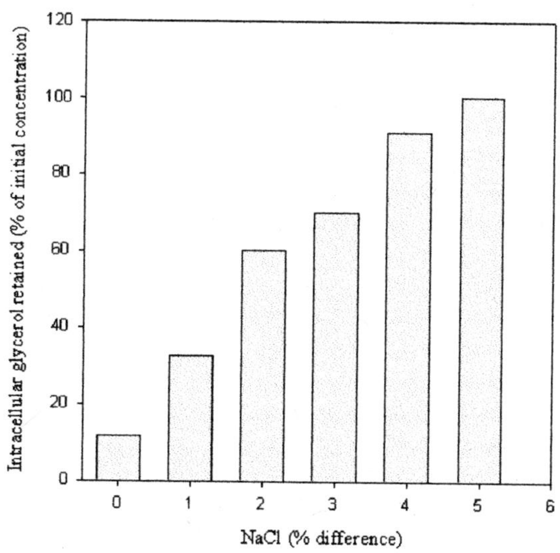

Figure 1d. Release of accumulated glycerol as a function of shock intensity in *S. cerevisiae*. Cells were grown in media containing 5% NaCl and then resuspended in media with lower NaCl concentrations and incubated at 28°C for 5 min. Intracellular glycerol was determined enzymatically as described previously (Luyten *et al.*, 1995).

1.2 Analysis of *FPS1* Homologues by PCR and DNA Probes

We further explored the occurrence of *FPS1* homologues in osmotolerant yeasts using DNA and PCR probes. Southern blots were prepared with *Z. rouxii* and *S. cerevisiae* genomic DNA and the membranes were probed with either the entire *FPS1* reading frame or a 426 bp fragment corresponding to the most conserved region between the NPS and NLA motifs (Park and Saier, 1996; Kayingo *et al.*, 2000b). Under stringent conditions, hybridisation was observed only in *S. cerevisiae*. However, with less stringent conditions, hybridisation also occurred in the three *Z. rouxii* strains investigated (Fig. 2b). Two bands were observed in *BamHI*/*Hind*III digests of *Z. rouxii* strains, the strongest of which was approximately 6.1 kb. More bands could be observed under non-stringent conditions when the entire *FPS1* reading frame was used as a probe. Some degree of DNA to DNA cross-hybridisation has been observed among bacterial MIP family genes (Calamita *et al.*, 1995). Lack of strong hybridisation to *FPS1* probes under stringent conditions suggests a low nucleotide sequence similarity with genes from other yeasts. This is consistent with the observation that MIP family genes generally share low nucleotide sequence similarity and some members display less than 20% identity. The yeast Fps1p appears to be the most divergent MIP family protein so far known (Kayingo *et al.*, 2000b). For example, neither of the family's signature NPA boxes is fully preserved, being NPS and NLA, respectively (Fig. 2a). Fps1p also differs from other MIP family members by having long amino- and carboxy-terminal hydrophilic extensions. The protein has 669 amino acids, compared with 250- 300 amino acids for most of the family members.

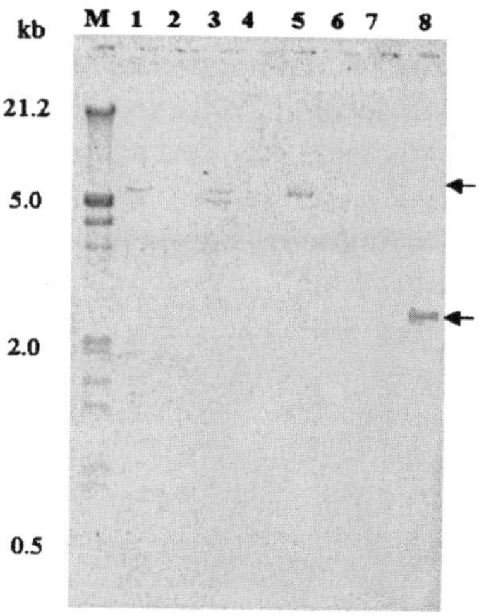

Figure 2a. Southern blot analysis: Th*e FPS1* (426 bp) probe hybridised to restriction digest (*Bam*H1 /*Hind*111) of genomic DNA from (1) *Z. rouxii* NRRL Y998, (3) *Z. rouxii* NRRL Y2547,(5) *Z. rouxii* NRRL Y225, (8) *S. cerevisiae* W303-1A under medium stringent conditions: Hybridisation at 55°C overnight, followed by 2 X 10 min wash in 2 x SSC , 0.1% SDS at RT. The probes were generated by PCR using primers; oligoG5 5' ATCTCAGGTGCTCATTTG 3' and oligoG6 5' TGGGCCCAGATCACGAGC 3' which corresponds to the Fps1p amino acid sequences, ISGAHL (sense) and ARDLGP (antisense) respectively.

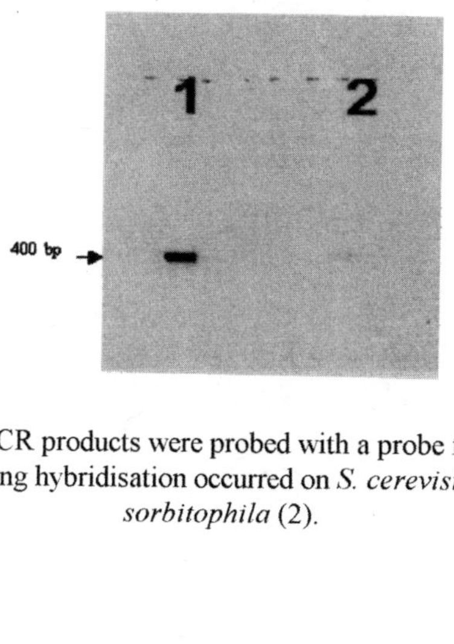

Figure 2b. Degenerate PCR products were probed with a probe from the most conserved fragment of *FPS1*. Strong hybridisation occurred on *S. cerevisiae* (1) as well as on *P. sorbitophila* (2).

Genomic DNA from various yeast strains was also screened by PCR using specific primers directed against the end terminals and conserved portion of *FPS1*. Apart from *S. cerevisiae*, no amplification was observed from other strains. Amplification was however, achieved with degenerate primers corresponding to *FPS1* amino acid sequences LNPSIT (sense) and NLARDL (antisense). Since several fragments ranging from 300 to 600 bp were obtained, Southern blots were prepared and probed with the *FPS1* gene (426 bp fragment) to find out which of these PCR products was homologous to *FPS1*. Under stringent conditions, only a single band of about 400 bp from *S. cerevisiae* hybridised. However, with less stringent conditions, strong hybridization occurred with *S. cerevisiae* and with *P. sorbitophila* (Fig 2c) and all hybridising fragments were between 400-500 bp. Weak signals were also observed for *Z. rouxii* and *Schizosaccharomyces pombe* (data not shown).

As discussed above, physiological and genetic analysis pointed to a possible existence of glycerol facilitator in *Z. rouxii* that might be homologous to MIP channel proteins. Attempts have been made to clone the corresponding gene(s). Degenerate primers from the conserved portions of the MIP family were used in RT-PCR. Amplification products within the expected size range of 350-450 bp have been cloned for further analysis.

1.3 Complementation Analysis

S. cerevisiae mutants lacking Fps1p do not show a growth defect on glycerol as a sole carbon source but show defects (phenotypes) related to glycerol export that could be utilised for complementation studies. *fps1Δ* mutants grow poorly under anaerobic conditions and exhibit slightly diminished osmotic induction of *GPD1* expression (Tamás *et al.*1999). Deletion of *FPS1* also causes cell fusion defects during mating (Philips and Herskowitz, 1997) and leads to prolonged phosphorylation of the Mpk1p kinase in the PKC pathway after a hypo-osmotic stress (M.C. Gustin unpublished data). However, the most clear-cut phenotype so far observed is that *S. cerevisiae* mutants in which *FPS1* has been deleted, survive poorly an osmotic downshock (hypo-osmotic shock) due to inability to dispose of accumulated glycerol. This defect is partially suppressed by over-expression of the *E. coli* glycerol facilitator GlpF (Luyten *et al.*, 1995). To investigate the occurrence of functional homologues or suppressors in *Z. rouxii*, *fps1Δ* mutants were transformed with a genomic library constructed in the autonomously replicating yeast plasmid pKU24 using partially digested DNA from *Z. rouxii* NRRL 2547 (Ushio *et al.*, 1996). Transformed cells were grown for about 5 days until a colony size of about 1-2 mm. About

300,000 transformants were replica-plated on 5% NaCl YNB plates. After 3 days, salt stressed cells were replica-plated on plates without NaCl (osmotic downshock) and monitored for survival following a hypo-osmotic shock. Nine transformants grew almost like the wild-type in 24 hours of hypo-osmotic stress; the rest of the transformants grew much slower. Promising transformants were re-screened using spot assay (Fig 3). Complementing plasmids were isolated and the inserts (*Z. rouxii* genes) sequenced. Sequence analysis of the most active fragments revealed genes that play a role in yeast cytoskeleton, cell cycle or the cell integrity pathway, none of which is a MIP family member.

fps1Δ

fps1Δ with
Z. rouxii gene

Wild type

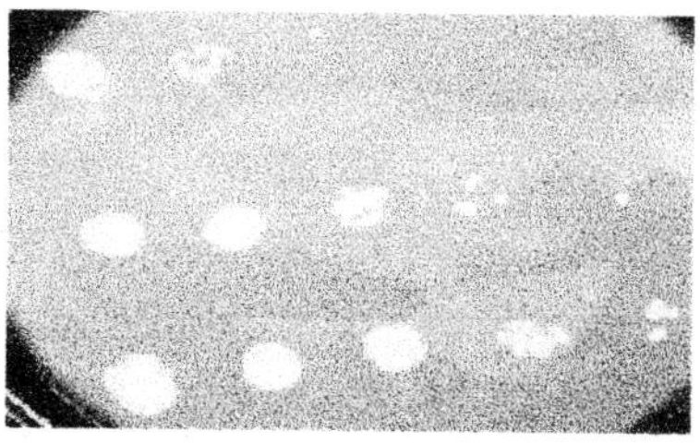

Figure 3. Complementation analysis: *fps1Δ* mutants were transformed with plasmids containing *Z. rouxii* genes (genomic library) and monitored for survival following a hypo-osmotic shock. Promising colonies were further re-screened using a spot assay (Kayingo *et al.* 2000a).

2. CONCLUSION

Yeast cells rapidly release their intracellular osmolytes such as glycerol upon an osmotic downshock. In *S. cerevisiae*, glycerol efflux is regulated and mediated by the MIP family glycerol channel Fps1p. *Z. rouxii* appears to have a similar protein-mediated glycerol release mechanism under hypo-osmotic stress. Whether the protein mediating the rapid glycerol efflux in *Z. rouxii* is homologous to Fpslp in structure and function requires further investigation. However, lack of strong hybridisation signals with *FPS1* probes suggest that the putative glycerol channel encoding gene may be divergent from *S. cerevisiae FPS1*. It is also possible that *Z. rouxii* releases osmolytes via other mechanisms as observed in bacteria where osmolyte export after hypo-osmotic shock is mediated by mechanosensitive channels. The fact that other genes (unrelated to *FPS1*) were readily found that improved *fps1Δ* growth defects, signifies the complexity of yeast osmoadapation and suggest that *FPS1* might be involved in other physiological process in addition to glycerol export. The occurrence of *FPS1* homologues in other yeasts might yield additional information on their physiological roles, which could be of significance in the food and beverage industry.

ACKNOWLEDGEMENTS

This work was supported by grants from National Research Foundation of South Africa to B.A. Prior. We are grateful to the yeast research groups in Bloemfontein (S. Africa) and Göteborg (Sweden) for helpful suggestions.

REFERENCES

Albertyn J Hohmann S Thevelein JM and Prior BA 1994 *GPD1* which encodes glycerol-3 phosphate dehydrogenase is essential for growth under osmotic stress in *Saccharomyces cerevisiae* and its expression is regulated by the high-osmolarity glycerol response pathway *Mol Cell Biol* **14**: 4135-4144

Attfield VP 1998 Physiological and molecular aspects of hyperosmotic stress tolerance in yeasts In *Recent res Devel in Microbiology Vol 2 Part II* (SG Pandala) Research Signpost Trivandrum India pp427-440

Calamita G Bishai WR Preson GM Guggino WB and Agre P 1995 Molecular cloning and characterization of AqpZ a water channel from *Escherichia coli J Biol Chem* **270**: 29063-29066

Kayingo G Potier S Hohmann S and Prior BA 2000a Isolation and characterization of the TIM10 homolgue from the yeast *Pichia sorbitophila*: a putative component of the mitochondrial protein import system *Yeast* **16**: 589-596

Kayingo G Bill R Calamita G Hohmann S and Prior BA 2000b Microbial water and glycerol channels (*in Press*)

Luyten K Albertyn J Skibbe WF Prior BA Ramos J Thevelein JM Hohmann S 1995 Fps1 a yeast member of the MIP family of channel proteins is a facilitator for glycerol uptake and efflux and is inactive under osmotic stress *EMBO J* **14**: 1360-1371

Park JH and Saier MHJr 1996 Phylogenetic characterization of the MIP family of transmembrane channel proteins *J Membrane Biol* **153**: 171-180

Philips J and Herskowitz I 1997 Osmotic balance regulates cell fusion during mating in *Saccharomyces cerevisiae J Cell Biol* **138**: 961-974

Tamás MJ Luyten K Sutherland FCW Hernandez A Albertyn J Valadi H Li H Prior BA Kilian SG Ramos J Gustafsson L Thevelein JM Hohmann S 1999 Fps1p controls the accumulation and release of the compatible solute glycerol in yeast osmoregulation *Mol Microbiol***31**: 1087-1104

Ushio K Otsuka H Hoshikawa S Taguchi G Shimosaka M Mitsui N Okazaki M 1996 Cloning of the *SAT1* gene concerned with salt tolerance of the yeast *Zygosaccharomyces rouxii J Ferment Bioeng* **82**: 16-21

Van Zyl PJ Kilian SG and Prior BA 1990 The role of an active transport mechanism in glycerol accumulation during osmoregulation by *Zygosaccharomyces rouxii Appl Microbiol Biotechnol* **34**: 231-235

Yancey PH Clark ME Hand SC Bowlus RD and Somero GN 1982 Living with water stress: evolution of osmolyte systems *Science* **217**: 1214-1222

5.5 Microbial MIP channels

THE YEAST PHOSPHATE TRANSPORTING SYSTEM

Regulated trafficking of the Pho84 phosphate transporter

[1,2]Jens O. Lagerstedt, [3,4]Arthur L. Kruckeberg, [4]Jan A. Berden and [1,2]Bengt L. Persson
[1] School of Biosciences and Process Technology, Växjö University, S-351 95 Växjö, Sweden; [2] Department of Biochemistry, Stockholm University, S-106 91 Stockholm, Sweden; [3] Department of Molecular Cell Physiology, The Free University of Amsterdam, 1081 HV Amsterdam, The Netherlands; [4] E.C. Slater Institute, University of Amsterdam, 1018 TV Amsterdam, The Netherlands. E-mail: jens.lagerstedt@ibp.vxu.se

1. INTRODUCTION

The life of *Saccharomyces cerevisiae* with about 6000 putative genes and gene products distributed over several intracellular compartments is maintained and regulated by a complicated interplay between its different cellular components. For the cell to respond to different internal and external signals, proteins are synthesised and routed to their functional localisation. Phosphate starvation is such a signal leading to derepression of several genes in the *PHO*-regulon of the yeast cell (Oshima, 1997). A *de novo* production of proteins active in cellular phosphate acquisition under these limiting conditions results in a continued viability of the cells.

1.1 Phosphate acquisition of the yeast cell

During evolution, yeast cells have developed an acquisition system for uptake of external phosphate, which supplies the cell with sufficient amounts of this essential inorganic molecule. This process is regulated by the activity of two uptake systems, i.e., a low-affinity system believed to be constitutively expressed and a derepressible high-affinity system which is

Molecular Biology and Physiology of Water and Solute Transport
Edited by Hohmann and Nielsen, Kluwer Academic/Plenum Publishers, New York, 2000.

mobilised under conditions of environmental phosphate starvation (for a review, see Persson *et al.*, 1999). Derepression of the high-affinity system results in expression of two distinct high-affinity transporters, the proton-coupled Pho84p (Bun-ya *et al.*, 1991) and the sodium-coupled Pho89p (Martinez and Persson, 1998) phosphate transporter. Both of these proteins are derepressibly synthesised when the cells meet limitations in the external phosphate concentration (Martinez and Persson, 1998; Martinez *et al.*, 1998; Petersson *et al.*, 1999). Although both genes are expressed in cells grown in a phosphate-limiting medium (Pattison-Granberg and Persson, unpublished data) the activity of the two transporters are clearly pH dependent with the Pho84p transporter being most active at acidic conditions while the Pho89p transporter is most active at alkaline conditions. The reasons for the cells to maintain expression of the Pho89p transporter, predominantly active at alkaline conditions, is presently not clear. Despite this limited knowledge, yeast cells constitute a suitable model system for deriving information on the evolution of proton-coupled phosphate transporters of prokaryotes to sodium-coupled Pho89p homologues of multicellular organisms (Persson *et al.*, 1999).

1.2 Phosphate uptake associated proteins/components

Based on similar phenotypes of deletion mutants as compared to the *Δpho84* phenotype, i.e. defective in high-affinity phosphate uptake and constitutive expression of the *PHO5* gene, encoding an acidic phosphatase, several cellular components have been suggested to be involved in the high affinity phosphate uptake process. The products of these genes, Pho86p (Yompakdee *et al.*, 1996a), Pho87p (Bun-ya *et al.*, 1996), Pho88p (Yompakdee *et al.*, 1996b) and Gtr1p (Bun-ya *et al.*, 1992) could either be directly involved in a phosphate transporting plasma membrane protein complex together with the Pho84p or they might have a direct or indirect role in gene regulation and/or protein trafficking. It is clear, though, that the Pho84p transporter alone is capable of catalysing a high-affinity phosphate transport as clearly shown in several *in vitro* studies (Berhe *et al.*, 1995; Fristedt *et al.*, 1999).

Recently, the Pho86p was demonstrated to be necessary for the transport of the Pho84p via COPII vesicles from the endoplasmatic reticulum (ER) to the Golgi (Lau *et al.*, 2000). The Gtr1p has been reported to have an intracellular function as well and is suggested to be a negative regulator of the Ran/Gsp1 GTPase cycle (Nakashima *et al.*, 1999). However, more detailed information on the specific role of the Gtr1p in the phosphate acquisition system and also the interactions of other phosphate uptake associated proteins, such as Pho87p and Pho88p, with this system need to be obtained.

2. REGULATION AND TRAFFICKING

Yeast cells growing in environments containing low concentrations of phosphate respond rapidly by derepression of several genes regulated by the *PHO* regulon (Oshima, 1997) with a resultant *de novo* synthesis of e.g. the Pho84p transporter. The mechanism by which the external phosphate concentration signal is transduced across the plasma membrane is to this date not known. Inside the cell, however, the scenario of the signal pathway is somewhat clearer. In situations when cells are grown at phosphate-rich conditions a nucleus-localised complex of the cyclin-dependent protein kinase (CDK), Pho85p, and the cyclin Pho80p hyperphosphorylates the transriptional activator, Pho4p (Fig 1; event 1), by modification of five or more serines (O'Neill *et al.*, 1996). The phosphorylation renders the Pho4p protein accessible for a complex formation with the receptor Msn5p and Gsp1p, a yeast ran homologue, after which the complex is transported to the cytoplasm (event 2) (Kaffman *et al.*, 1998b). For the formation of the complex the Gsp1p has to contain bound GTP (Kaffman *et al.*, 1998b) and hydrolysis to GDP, possibly stimulated by the GTPase-activating (GAP) protein Rna1, dissociates the complex in the cytoplasm (event 3). Interestingly, the Ran/Gsp1 G-protein cycle seems to be negatively regulated through Gtr2p, by Gtr1p-GTP (Nakashima *et al.*, 1999). Thus the phenotype of the *Δgtr1* mutant, defective in high-affinity uptake and provided with a constitutive expression of *PHO5*, might possibly reflect a shift of the existing equilibrium towards the Gsp1p-GTP form, which, in turn, is the necessary nucleotide-bound state of Gsp1p for Pho4p-Msn5p-Gsp1p-GTP complex formation.

At low phosphate concentrations, the hyperphosphorylation of Pho4p is abolished by inactivation of the Pho85p-Pho80p kinase complex by the CDK inhibitor, Pho81p (event 4; Schneider *et al.*, 1994). Pho4p, in its unphosphorylated form, is presumably transported into the nucleus through the nonclassical import pathway that utilises the importin beta family

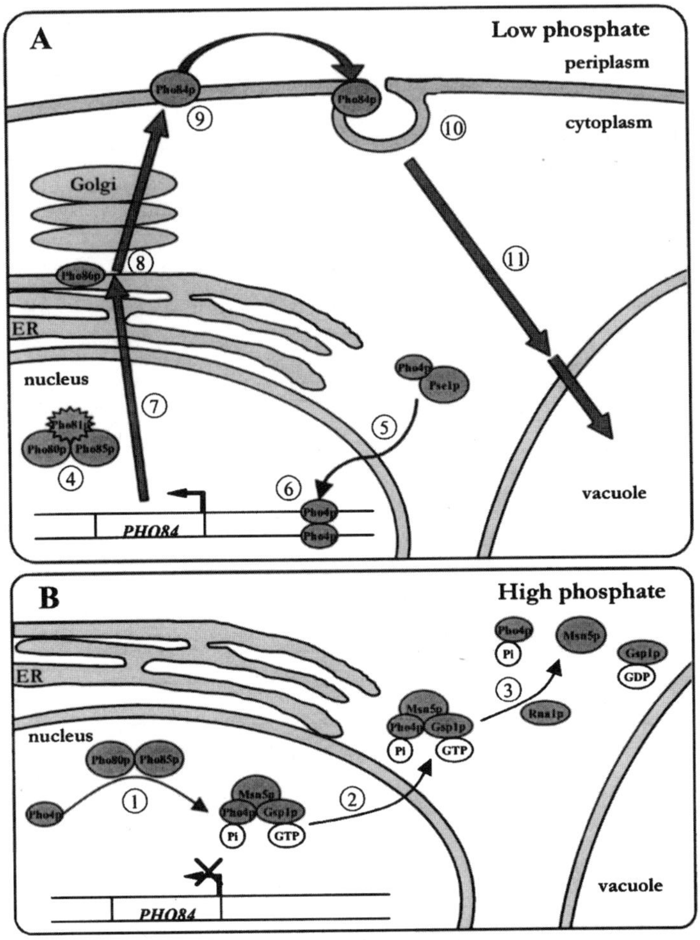

Figure 1. Regulated expression and trafficking of the Pho84 phosphate transporter in *Saccharomyces cerevisiae* at phosphate limiting (panel A) and at phosphate rich (panel B) conditions.

member Pse1p/Kap121p (event 5) (Kaffman *et al.*, 1998a). In contrast, in its phosphorylated state, Pho4p is unable to bind to Pse1p (Kaffman *et al.*, 1998a) thus preventing transport of Pho4p into the nucleus at repressible conditions, resulting in a cellular organisation where the cytoplasmic localisation of Pho4p eliminates its role as a gene activator within the *PHO* regulon.

In the nucleus the Pho4p binds as a homodimer to the sequence CACGTG (event 6) (Shimizu *et al.*, 1997), a sequence present in the promotor region of *PHO5*, *PHO8*, *PHO81* and *PHO84* and which together with the positive regulator Pho2p (=Grf10p) induces the expression of these genes (Oshima, 1997). The *PHO84* gene is thus expressed at these conditions to enable a continued phosphate uptake (event 7) (Martinez *et al.*, 1998). The threshold in terms of external phosphate concentration for this *de novo* synthesis of the Pho84p is approximately 80 μM (Martinez *et al.*, 1998; Petersson *et al.*, 1999). Following synthesis, the Pho84p protein is transported to the plasma membrane (Petersson *et al.*, 1999) via the ER and Golgi compartments. For the passage from ER to Golgi, Pho86p has been shown to be required for packaging of the Pho84p into COPII vesicles (event 8) (Lau *et al.*, 2000). In this process, Pho86p is not itself present in the COPII vesicles. Thus, the Pho86p is classified as an 'outfitter', a protein facilitating the embarkation of a secretory protein into vesicles for transport from ER to Golgi without being co-packaged itself (Herrman *et al.*, 1999). Details on the concomitant steps in the secretory pathway for plasma membrane delivery of the Pho84p transporter still remains to be clarified (event 9). When the external phosphate is fully exhausted the catalytically active Pho84p transporter is removed from the plasma membrane (event 10) (Petersson *et al.*, 1999), an event coinciding with a diminution of the amount of *PHO84* transcripts (Martinez *et al.*, 1998). In similarity to the deactivation pathway of several other yeast plasma membrane proteins, e.g. the Hxt2 (Kruckeberg *et al.*, 1999) and Hxt6 and Hxt7 (Krampe *et al.*, 1998) hexose permeases to which the Pho84p has high homology, the Pho84p is routed to the vacuole for degradation (event 11) (Petersson *et al.*, 1999). Interestingly, when repressible concentrations of phosphate are added to phosphate-starved cells, a rapid removal of the Pho84p from the plasma membrane followed by a vacuolar localisation occurs (Fig 2; Lagerstedt, Kruckeberg, Berden and Persson, unpublished data; Lau *et al.*, 2000). The removal of the high-affinity phosphate transporter upon addition of phosphate might possibly reflect a regulated strategy of the cells to prevent overloading of phosphate. The deactivation under conditions of limiting amounts of external phosphate may hypothetically illustrate the opposite, i.e., a strategy to prevent efflux of essential internalised phosphate.

It is likely that removal of the protein occurs by invagination of the plasma membrane to form endocytic vesicles that are sorted to the vacuole. However, the regulatory events preceding endocytosis of Pho84p have not been described, neither has the signal that discriminates Pho84p destined for endocytosis from Pho84p with a preserved high-affinity phosphate transport across the plasma membrane.

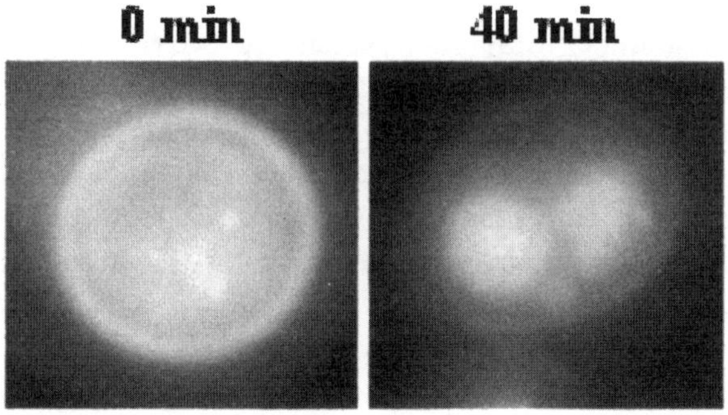

Figure 2. Deactivation of the Pho84-GFPp visualised by fluorescence microscopy. Addition of repressible amounts of phosphate to cells expressing Pho84-GFPp (0 min) results in a rapid plasma membrane removal of the permease followed by a vacuolar localisation (40 min). In a control, where sorbitol was added to the cells to compensate for osmotic effects, the plasma membrane localisation of Pho84-GFPp remained unchanged (not shown)

3. MOLECULAR SIGNALS FOR ENDOCYTOSIS

For a number of yeast plasma membrane proteins covalent attachments of ubiquitin, a 76 amino acid polypeptide, is the molecular signal which initiates the degradative pathway (reviewed in Hicke, 1999). However, these modified integral proteins are not recognised by the 26s proteosome machinery as is the case for ubiquitinated, cytosolic proteins destined for degradation. Instead the binding of ubiquitin is necessary for endocytosis as demonstrated by ligand-induced internalisation of the α-pheromone receptor (Ste2p) in mutants defective in the ubiquitin-conjugating enzymes, Ubc4p and Ubc5p (Hicke and Riezman, 1996). Once internalised in endocytic vesicles, the proteins are targeted to the vacuole where they undergo proteolysis. The factor which discriminates between the different ubiquitin modified proteins of the proteosome and the vacuolar proteolytic system, appears to be the number of ubiquitin chains attached. Whilst the proteosome binds efficiently to proteins being poly-ubiquitinated by at least four subunits (Deveraux *et al.*, 1994), mono- or di-ubiquitin is adequate for ligand-stimulated endocytosis of the α- and the a-pheromone receptors (Terrell *et al.*, 1998; Roth and Davis, 1996, respectively) and also for deactivation of a yeast maltose permease (Lucero *et al.*, 2000). That mono-ubiquitination is sufficient for endocytosis, was further proven in a recent study where a C-terminal ubiquitin fusion to the, otherwise, stable plasma membrane ATPase, Pma1 (Benito *et al.*, 1991), triggered internalisation (Shih *et al.*, 2000).

The targets for ubiquitination in the substrate proteins are lysines to which isopeptide bonds are formed between the C-terminal glycine of the ubiquitin-peptide and the ε amino group of the lysine. The lysine that serves as the acceptor site for ubiquitination of the Ste2p was found to be localised in the sequence SINNDAKSS (Hicke and Rietzman, 1996). This sequence, located in the C-terminal tail, was previously demonstrated to be sufficient for endocytosis (Rohrer *et al.*, 1993). Later on, phosphorylation at the SINNDAKSS-serines was described as a necessary step preceding ubiquitination (Hicke *et al.*, 1998).

For the Ste3p, a 36-residues peptide sequence (residues 414 to 449) located in the C-terminal tail comprises the ubiquitination/endocytosis signal (Roth *et al.*, 1998). Based on the high proportion of acidic and hydroxylated amino acid residues, the sequence is suggested to be a PEST-like sequence. However, the major effect of triple alanine mutations in the 414-449 in a Ste3Δ450–468p is predominantly found in triplets containing either lysine or serine(s). Indeed the lysines are important as subjects for ubiquitination (Roth and Davis, 2000). Lysine to arginine mutations of Lys-424, Lys-432 and Lys-453 results in a severe blockade to both ubiquitination and endocytosis where the Lys-432 plays a predominant role.

Recently, the Zrt1 zinc transporter in *S. cerevisiae* was also demonstrated to depend on ubiquitination of a lysine as an obligatory step in endocytosis (Gitan and Eide, 2000). However, in this case the critical lysine, Lys-195, is located in a cytoplasmic loop region in the middle of the protein.

Of the 25 lysines contained in the Pho84p primary structure, 16 are, according to the existing secondary model, cytoplasmically located (Persson *et al.*, 1999). It remains to be seen whether these residues are functionally important for ubiquitination/endocytosis. Preliminary studies indicate that the residues in the C-terminal loop are at least not essential.

Conclusively, it is clear that, although lysines and serines are important residues for the first step in the endocytic pathway, no general consensus sequence and location of these can be pointed out.

ACKNOWLEDGEMENTS

This work was supported by research grants from the Swedish Natural Science Research Council and Växjö University. J.O.L. was a recipient of short-term travel fellowships from Stiftelsen Bengt Lundqvists Minne and Växjö University.

REFERENCES

Berhe A Fristedt U and Persson BL 1995 Expression and purification of the high-affinity phosphate transporter of *Saccharomyces cerevisiae Eur J Biochem* **227**: 566-572

Benito B Moreno E and Lagunas R 1991 Half-life of the plasma membrane ATPase and its activating system in resting cells *Biochem Biophys Acta* **1063**: 265-268

Bun-ya M Harashima S and Oshima Y 1992 Putative GTP-binding protein Gtr1 associated with the function of the Pho84 inorganic phosphate transporter in *Saccharomyces cerevisiae Mol Cell Biol* **12**: 2958-2966

Bun-ya M Nishimura M Harashima S and Oshima Y 1991 The *PHO84* gene of *Saccharomyces cerevisiae* encodes an inorganic phosphate transporter *Mol Cell Biol* **11** 3229-3238

Bun-ya M Shikata K Nakade S Yompakdee C Harashima S and Oshima Y 1996 Two new genes *PHO86* and *PHO87* involved in inorganic phosphate uptake in *Saccharomyces cerevisiae Curr Genet* **29** 344-351

Deveraux Q Ustrell V Pickart C and Rechsteiner M 1994 A 26 S protease subunit that binds ubiquitin conjugates *J Biol Chem* **269**: 7059-7061

Fristedt U Weinander R Martinsson H–S and Persson BL 1999 Characterization of purified and unidirectionally reconstituted Pho84 phosphate permease of *Saccharomyces cerevisiae FEBS Lett* **458**: 1-5

Gitan RS and Eide DJ 2000 Zinc-regulated ubiquitin conjugation signals endocytosis of the yeast ZRT1 zinc transporter *Biochem J* 346: 329-336

Herrmann JM Malkus P and Schekman R 1999 Out of the ER – outfitters escorts and guides *Trends Cell Biol* **9**: 5-7

Hicke L 1999 Gettin' down with ubiquitin: turning off cell-surface receptors transporters and channels *Trends Cell Biol* **9**: 107-112

Hicke L and Rietzman H 1996 Ubiquitination of a yeast plasma membrane receptor signals its ligand-stimulated endocytosis *Cell* **84**: 277-287

Hicke L Zanolari B and Riezman HJ 1998 Cytoplasmic tail phosphorylation of the alpha-factor receptor is required for its ubiquitination and internalization *J Cell Biol* **141**: 349-358

Kaffman A Rank NM O'Neill EM Huang LS and O'Shea EK 1998b The receptor Msn5 exports the phosphorylated transcription factor Pho4 out of the nucleus *Nature* **396**: 482-486

Kaffman A Rank NM and O'Shea EK 1998a Phosphorylation regulates association of the transcription factor Pho4 with its import receptorPse1/Kap121 *Genes Dev* **12**: 2673-2683

Krampe S Stamm O Hollenberg CP and Boles E 1998 Catabolite inactivation of the high-affinity hexose transporters Hxt6 and Hxt7 of *Saccharomyces cerevisiae* occurs in the vacuole after internalization by endocytosis *FEBS Lett* **441**: 343-347

Kruckeberg AL Ling Y Berden JA and van Dam K 1999 Functional expression quantification and cellular localization of the Hxt2 hexose transporter of *Saccharomyces cerevisiae* tagged with the green fluorescent protein *Biochem J* **339**: 299-307

Lau W-TW Howson RW Malkus P Shekman R and O´Shea EK 2000 Pho86p an endoplasmic reticulum ER resident protein in *Saccharomyces cerevisiae* is required for ER exit of the high-affinity phosphate transporter Pho84p *Proc Natl Acad Sci USA* **97**: 1107-1112

Lucero P Peñalver E Vela L and Lagunas R 2000 Monoubiquitination is sufficient to signal internalization of the maltose transporter in *Saccharomyces cerevisiae J Bacteriol* **182**: 241-243

Martinez P and Persson BL 1998 Identification cloning and characterization of a derepressible Na^+-coupled phosphate transporter in *Saccharomyces cerevisiae Mol Gen Genet* **248**: 628-638

Martinez P Zvyagilskaya R Allard P and Persson BL 1998 Physiological regulation of the derepressible phosphate transporter in *Saccharomyces cerevisiae J Bacteriol* **180**: 2253-2256

Nakashima N Noguchi E and Nishimoto T 1999 Saccharomyces cerevisiae putative G protein Gtr1p which forms complexes with itself and a novel protein designated as Gtr2p negatively regulates the Ran/Gsp1p G protein cycle through Gtr2p *Genetics* **152**: 853-867

O'Neill EM Kaffman A Jolly ER and O'Shea EK 1996 Regulation of PHO4 nuclear localization by the PHO80-PHO85 cyclin-CDK complex *Science* **271**: 209-212

Oshima Y 1997 The phosphatase system in *Saccharomyces cerevisiae Genes Genet Syst* **72**: 323-334

Persson BL Petersson J Fristedt U Weinander R Berhe A and Pattison J 1999 Phosphate permeases of *Saccharomyces cerevisiae*; Structure function and regulation *Biochim Biophys Acta/Biomembranes* **1422**: 255-272

Petersson J Pattison J Kruckeberg AL Berden JA and Persson BL 1999 Intracellular localization of an active green fluorescent protein-tagged Pho84 phosphate permease in Saccharomyces cerevisiae *FEBS Lett* **462**: 37-42

Rohrer J Benedetti H Zanolari B and Rietzman H 1993 Identification of a novel sequence mediating regulated endocytosis of the G protein-coupled alpha-pheromone receptor in yeast *Mol Biol Cell* **4**: 511-521

Roth AF and Davis NG 1996 Ubiquitination of the yeast a-factor receptor *J Cell Biol* **134**: 661-674

Roth AF and Davis NG 2000 Ubiquitination of the PEST-like endocytosis signal of the yeast a-factor receptor *J Biol Chem* **275**: 8143-8153

Roth AF Sullivan DM and Davis NG 1998 A large PEST-like sequence directs the ubiquitination endocytosis and vacuolar degradation of the yeast a-factor receptor *J Cell Biol* **142**: 949-961

Schneider KR Smith RL and O'Shea EK 1994 Phosphate-regulated inactivation of the kinase PHO80-PHO85 by the CDK inhibitor PHO81 *Science* **266**: 122-126

Shih SC Sloper-Mould KE and Hicke L 2000 Monoubiquitin carries a novel internalization signal that is appended to activated receptors *EMBO J* **19**: 187-198

Shimizu T Toumoto A Ihara K Shimizu M Kyogoku Y Ogawa N Oshima Y and Hakoshima T 1997 Crystal structure of PHO4 bHLH domain-DNA complex: flanking base recognition *EMBO J* **16**: 4689-4697

Terrell J Shih S Dunn R and Hicke L 1998 A function for monoubiquitination in the internalization of a G protein-coupled receptor *Mol Cell* **1**: 193-202

Yompakdee C Bun-ya M Shikata K Ogawa N Harashima S and Oshima Y 1996a A putative new membrane protein Pho86p in the inorganic phosphate uptake system of *Saccharomyces cerevisiae Gene* **171**: 41-47

Yompakdee C Ogawa N Harashima S and Oshima Y 1996b A putative membrane protein Pho88p involved in inorganic phosphate transport in *Saccharomyces cerevisiae Mol Gen Genet* **251**: 580-590

5.6 Microbial MIP channels

AQUAPORIN WATER CHANNELS IN *SACCHAROMYCES CEREVISIAE*

Function and expression studies

Vincent Laizé, Marie Céleste De Jesus Ferreira and Stefan Hohmann
Department of Cell and Molecular Biology, University of Göteborg, Göteborg, Sweden. E-mail: hohmann@gmm.gu.se

1. INTRODUCTION

Major Intrinsic Proteins (MIPs) form a large family of integral membrane channel proteins involved in the transport of water and small uncharged solutes such as glycerol and urea. Forming the largest subgroup in the MIP family, aquaporins (AQPs) are water-selective channels facilitating rapid water flow across cellular membranes. Because water movement in cells and tissues is central to life, it is not surprising that AQPs have been identified in mammals, plants, invertebrates, amphibia, prokaryotes and recently in the baker's yeast, *Saccharomyces cerevisiae*. The yeast genome contains 2 open reading frames (ORFs) that are homologous to plant aquaporin genes (André 1995), and 75.7% identical to each other. Those were recently renamed *AQY1* (ORF *YPR192w*) and *AQY2* (overlapping ORFs *YLL052c-053c*).

2. POLYMORPHISM OF YEAST AQUAPORINS

Two alleles have been described so far for *AQY1*: *AQY1-1*, which was found in strain Σ1278b, and *AQY1-2*, which occurs in laboratory strains such as S288C, the strain used to determine the yeast genome sequence (Goffeau *et al.*, 1997). Sequence comparison of *AQY1* alleles revealed 7 nucleotide changes leading to 4 differences in the deduced amino acid sequence (Bonhivers *et al.*, 1998, Laizé *et al.*, 1999, Fig 1). Aqy1-1p has been shown

Molecular Biology and Physiology of Water and Solute Transport
Edited by Hohmann and Nielsen, Kluwer Academic/Plenum Publishers, New York, 2000.

to mediate water transport when expressed in *Xenopus* oocytes (see paragraph 3). Aqy1-2p does not mediate water transport, due to two mutations that prevent its expression in oocytes (Bonhivers *et al.*, 1998).

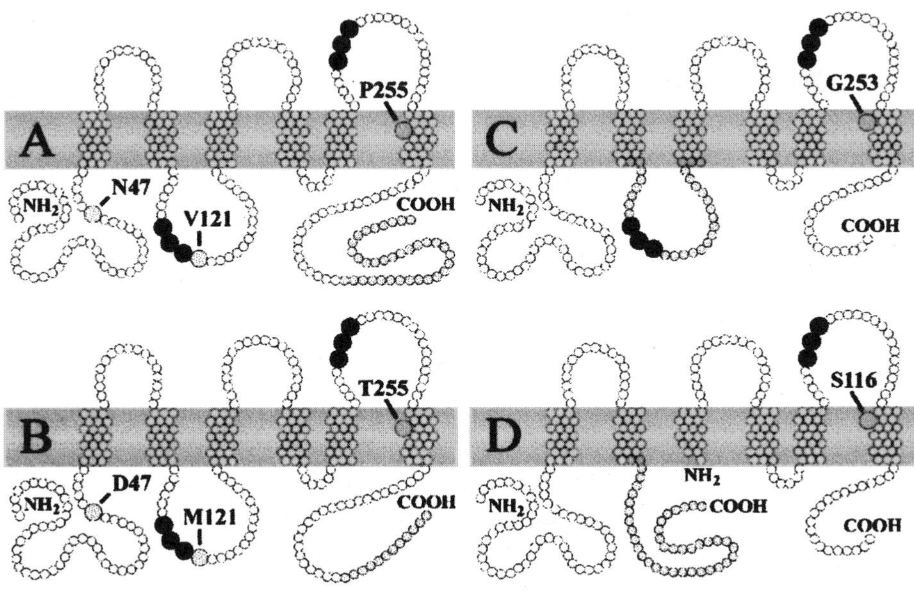

Figure 1. Predicted topology for alleles of yeast aquaporins. A, Aqy1-1p from Σ1278b; B, Aqy1-2p from S288C; C, Aqy2-1p from Σ1278b; D, Aqy2-2p from S288C. Black-filled circles, NPA boxes; grey-filled circles, sequence differences between A & B, and C & D.

AQY2 sequence was also determined in strains Σ1278b and S288C, and the alleles *AQY2-1* and *AQY2-2*, respectively, were identified. The sequence obtained from S288C is identical to that originally deposited in the SGD database and is interrupted by a frameshift mutation. The sequence from Σ1278b revealed 7 differences: 6 substitutions (5 of them silent) and an 11-bp insertion suppressing the frameshift mutation present in *AQY2-2* (Laizé *et al.*, 2000, Fig 1). *AQY2-1* encodes a protein of 289 amino acids (predicted Mw 31.3 kDa) that is 87.5% identical to Aqy1-1p and whose topology is similar to that of known aquaporins. Despite this similarity, we did not observe Aqy2-1p-mediated water transport in *Xenopus* oocytes (see paragraph 3).

A survey of 52 yeast strains (27 laboratory strains, 21 industrial yeasts and 4 natural isolates) revealed that all industrial and wild yeasts carry the allele encoding a functional Aqy1p while none of these strains appear to have a functional Aqy2p. We conclude that natural and industrial conditions provide selective pressure to maintain *AQY1* but apparently not *AQY2*.

3. WATER TRANSPORT MEDIATED BY YEAST AQUAPORINS

The osmotic water permeability coefficient (P_f) of oocytes injected with the RNAs synthesised from *AQY1-1*, *AQY1-1ΔC* (C-terminal truncated form of *AQY1-1*), *AQY1-2*, *AQY2-1* and *AQY2-1-SSED* (C-terminal mutant of *AQY2-1*) was measured at 21°C (Fig 2).

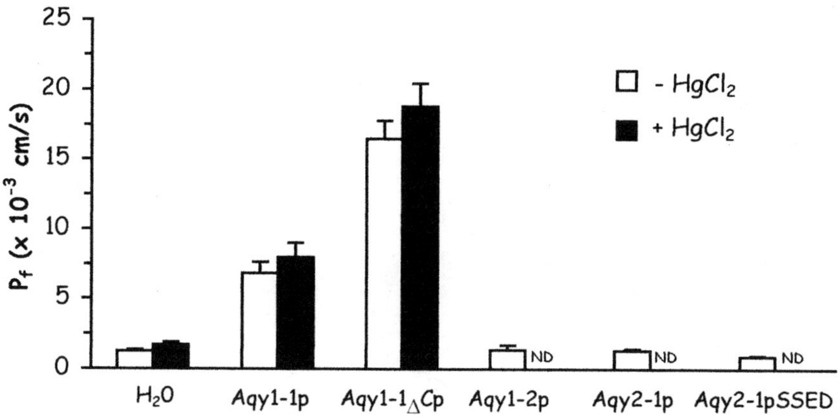

Figure 2. Osmotic water permeability coefficients of *Xenopus* oocytes injected with water or RNA-encoding proteins as indicated. Data are expressed as means ± SE. ND, non-determined.

The P_f of oocytes expressing Aqy1-1p was 5.7 times higher than the P_f of control oocytes. Surprisingly, oocytes expressing the truncated Aqy1-1ΔCp exhibited a 13.8-fold increase in P_f when compared to control, and a 2.7-fold increase when compared to oocytes expressing the full-length protein. This corresponds to a 2.6-fold higher protein density in the oocyte plasma membrane, as shown by freeze-fracture electron microscopy. No inhibition with 1 mM $HgCl_2$ was observed for both forms of Aqy1-1p (Laizé *et al.*, 1999). Aqy1-2p does not mediate water transport in *Xenopus* oocytes due to two mutations that prevent its expression in oocytes. This situation may be similar in yeast cells, since Aqy1-1p, but not Aqy1-2p, is present in yeast membranes (Bonhivers *et al.*, 1998). No expression of Aqy2-1p was observed in oocyte plasma membranes and hence also no increase of the oocyte osmotic water permeability (Laizé *et al.*, 2000). Because of its similarity to a mammalian endoplasmic reticulum retention signal (Jackson *et al.*, 1990), the KKED sequence present at the C-terminus of Aqy2-1p was suspected to be responsible for mislocalisation and subsequent degradation of Aqy2-1p. The *AQY2-1-SSED* mutant was constructed in which the two lysines were replaced each by serine but again the protein could not be detected in oocyte plasma membranes. Hence it remains unclear if *AQY2-1* encodes a truly functional water channel.

P_f measurements in yeast spheroplasts using a stopped-flow apparatus confirmed the presence of functional water channels in Σ1278b (*AQY1-1* and *AQY2-1*) but apparently not in the laboratory strain FL100 (*AQY1-2* and *AQY2-2*, Laizé *et al.*, 1999).

4. EXPRESSION OF YEAST AQUAPORINS

Expression levels were first investigated under normal conditions of growth and aeration. *AQY1* expression remained very low during the exponential growth phase and was approximately 7-fold induced when cells reached the stationary phase (Fig 3). *AQY2* expression was high during the early exponential phase and was approximately 7-fold reduced when cells entered the late exponential phase (Fig 3). As for *AQY1*, expression of *AQY2* was induced during the stationary phase, but to a lesser extent (2-fold). Deletion of *AQY1* or *AQY2* had no remarkable effect on *AQY2* or *AQY1* expression, respectively, indicating the absence of any compensatory effects between yeast aquaporins. Induction of both *AQY1* and *AQY2* expression during the stationary phase may explain the high expression levels observed at t=0 h. At that time, cells from a pre-culture in stationary phase were diluted into fresh medium. It is not known yet why yeast aquaporins are differentially expressed during normal growth conditions and what could be the role of Aqy2p during early exponential phase.

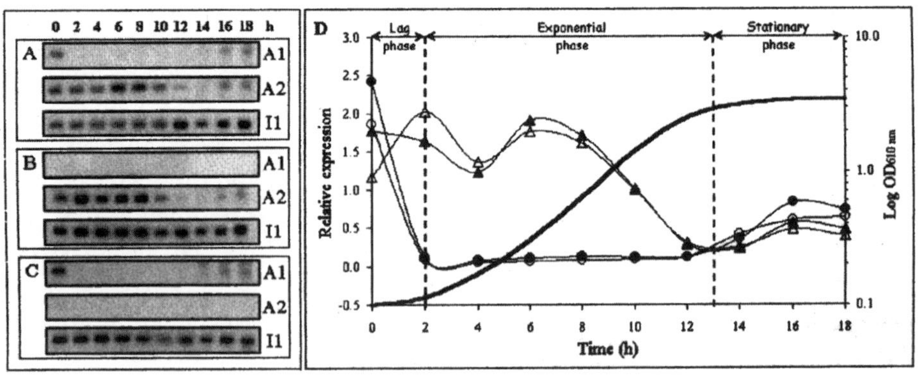

Figure 3. Expression of yeast aquaporins during growth in YPD medium. A, strain Σ1278b; B, *aqy1Δ* mutant; C, *aqy2Δ* mutant; D, time-course of relative expression of *AQY1* (circles) and *AQY2* (triangles) in Σ1278b (filled symbols) or in deletion mutant strains (open symbols). A1=*AQY1*, A2=*AQY2*, I1=*IPP1*.

On the basis of the results published by Chu *et al.,* (1998), the expression of yeast aquaporins was then investigated in diploid cells during meiosis and sporulation (Fig 4). While *AQY2* expression remained very low during exposure of cells to sporulation medium (SPM, potassium acetate 1%), a 4-fold sustained induction of *AQY1* was observed after 13 hours. Surprisingly, *AQY1* expression was also induced in haploid cells, but to a lesser extent (2-fold). Altogether these results suggest a role of Aqy1p in spore formation and germination.

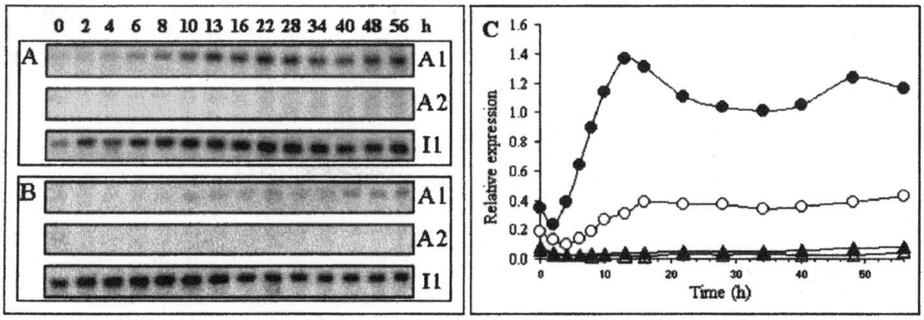

Figure 4. Expression of yeast aquaporins during meiosis and sporulation. A, Σ1278b diploid strain; B, Σ1278b haploid strain; C, time-course of relative expression of *AQY1* (circles) and *AQY2* (triangles) in the diploid (filled symbols) or the haploid (open symbols) cells. A1=*AQY1*, A2=*AQY2*, I1=*IPP1*.

Finally, expression levels of yeast aquaporins were investigated after a hyper-osmotic shock (Fig 5). *AQY1* expression was slightly reduced and remained very low during adaptation of yeast cells to high osmolarity in wild-type and *aqy2Δ* strains. *AQY2* expression was 30-fold induced 1 hour after the osmotic shock and induction was transient as observed for many osmo-induced genes (Rep *et al.,* 1999a, b). *AQY2* expression was diminished to pre-shock levels 3 hours after transfer of cells in the hyper-osmotic medium, when cells were fully adapted to high osmolarity and started to divide again. As already observed in Figure 3 during normal growth, yeast aquaporins did not compensate for each other in deletion mutants during adaptation to a hyper-osmotic shock. Taken together, these results suggest a role of Aqy2p in retrieval of water during the recovery after an osmotic shock.

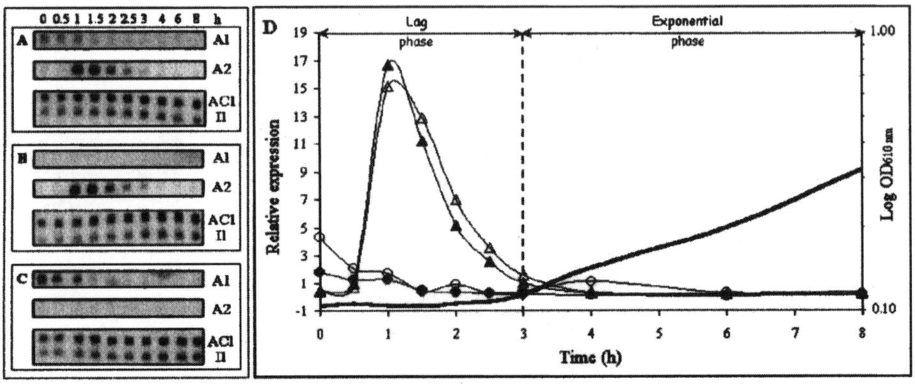

Figure 5. Expression of yeast aquaporins after a hyper-osmotic shock (5% KCl). A, strain Σ1278b; B, *aqy1Δ* mutant; C, *aqy2Δ* mutant; D, time-course of relative expression of *AQY1* (circles) and *AQY2* (triangles) in Σ1278b (filled symbols) or in deletion mutant strains (open symbols). A1=*AQY1*, A2=*AQY2*, I1=*IPP1*, AC1=*ACT1*.

5. CONCLUSIONS

Polymorphic variations at gene level have been reported several times in *Saccharomyces cerevisiae*. Aquaporin genes add an interesting example to this list and highlight the notion that certain processes may operate differently in laboratory strains as compared to wild-type strains.

Aqy1p from the strain Σ1278b has been shown to be a functional water channel. *AQY1* is induced in diploids when shifted to sporulation medium and to a moderate extent in haploids when shifted to nitrogen starvation suggesting a role of aqy1p in spore formation and germination. The function of Aqy2p has not been yet confirmed. *AQY2* is strongly induced by a hyper-osmotic shock in Σ1278b, suggesting a possible role of Aqy2p in retrieval of water during yeast adaptation to high osmolarity.

Expression of the yeast aquaporin has been shown to be regulated, and both transcripts have been detected during important physiological processes. Aqy1p and Aqy2p are involved in different processes and one cannot compensate the defect of the other one.

ACKNOWLEDGEMENTS

This work was supported by the European Commission (contracts BIO4-CT98-0024 and FMRX-CT97-0128) and by the Wenner-Gren Foundation.

REFERENCES

André B 1995 An overview of membrane transport proteins in *Saccharomyces cerevisiae Yeast* **11**: 1575-1611

Bonhivers M Carbrey J M Gould S J and Agre P 1998 Aquaporins in *Saccharomyces J Biol Chem* **273**: 27565-27572

Chu S DeRisi J L Eisen M Mulholland J Botstein D Brown PO and Herskowitz I 1998 The transcriptional program of sporulation in budding yeast *Science* **282**: 699-705

Goffeau A et al 1997 The Yeast Genome Directory *Nature* **387**(suppl): 1-105

Jackson MR Nilsson T and Peterson PA 1990 Identification of a consensus motif for retention of transmembrane proteins in the endoplasmic reticulum *EMBO J* **9**: 3153-3162

Laizé V Gobin R Rousselet G Badier C Hohmann S Ripoche P and Tacnet F 1999 Molecular and functional study of AQY1 from *Saccharomyces cerevisiae*: role of the C-terminal domain *Biochem Biophys Res Commun* **257**: 139-144

Laizé V Tacnet F Ripoche P and Hohmann S 2000 Polymorphism of *Saccharomyces cerevisiae* aquaporins *Yeast* in press

Rep M Albertyn J Thevelein JM Prior BA and Hohmann S 1999a Different signalling pathways contribute to the control of *GPD1* gene expression by osmotic stress in *Saccharomyces cerevisiae Microbiology* **145**: 715-727

Rep M Reiser V Gartner U Thevelein JM Hohmann S Ammerer G and Ruis H 1999b Osmotic stress-induced gene expression in *Saccharomyces cerevisiae* requires Msn1p and the novel nuclear factor Hot1p *Mol Cell Biol* **19**: 5474-5485

5.7 Microbial MIP channels

FUNCTION AND REGULATION OF THE YEAST MIP GLYCEROL EXPORT CHANNEL Fps1p

[1]Markus J. Tamás, [1]Johan M. Thevelein and [2]Stefan Hohmann
[1]*Laboratorium voor Moleculaire Celbiologie, Katholieke Universiteit Leuven, Kardinaal Mercierlaan 92, B-3001 Heverlee, Flanders, Belgium:* [2]*Department of Cell and Molecular Biology / Microbiology, Göteborg University, Box 462, S-405 30 Göteborg, Sweden. E-mail: hohmann@gmm.gu.se*

1. INTRODUCTION

The genome of the unicellular eukaryotic model organism *Saccharomyces cerevisiae* (baker's yeast) contains four genes encoding MIP channels (André 1995, Hohmann *et al.*, 2000): *FPS1* encodes a glycerol channel protein (Luyten *et al.*, 1995, Sutherland *et al.*, 1997, Tamás *et al.*, 1999, Van Aelst *et al.*, 1991), *AQY1* and *AQY2* encode water channels (Bonhivers *et al.*, 1998, Laizé *et al.*, 1999) and *YFL054c* encodes a protein with homology to other MIP glycerol channels (André 1995, Murakami *et al.*, 1995). Here, we describe the physiological role of Fps1p in regulated glycerol export and the molecular basis for osmoregulation of Fps1p.

2. FUNCTION OF FPS1

Early kinetic studies of glycerol transport in *S. cerevisiae* suggested that glycerol passes the plasma membrane only by passive diffusion (Brown 1974, Gancedo *et al.*, 1968). However, *FPS1*, which encodes a protein belonging to the MIP family of channel proteins, clearly affects the movement of glycerol across the membrane (Luyten *et al.*, 1995, Sutherland *et al.*, 1997, Tamás *et al.*, 1999). Studies on glycerol uptake by yeast cells have shown that glycerol influx can be separated into two components:

Molecular Biology and Physiology of Water and Solute Transport
Edited by Hohmann and Nielsen, Kluwer Academic/Plenum Publishers, New York, 2000.

passive diffusion through the lipid bilayer and facilitated diffusion mediated by Fps1p (Sutherland *et al.*, 1997).

The *FPS1* gene was isolated by coincidence, as a suppressor of the growth defect on glucose of a mutant defective in the regulation of glycolysis (Van Aelst *et al.*, 1991).

2.1 Fps1p is a glycerol export channel

Fps1p mediates both the influx and the efflux of glycerol (Luyten *et al.*, 1995, Sutherland *et al.*, 1997, Tamás *et al.*, 1999). Mutants lacking Fps1p do not show any growth defect on medium containing glycerol as a sole carbon source, suggesting that the main function of Fps1p is not the uptake of glycerol for further metabolic utilisation (Tamás *et al.*, 1999). Instead, several observations establish Fps1p as a glycerol exporter. Firstly, Fps1p is required for glycerol export and survival of yeast cells after a hypo-osmotic shock (Luyten *et al.*, 1995, Tamás *et al.*, 1999). When confronted with a sudden decrease in osmolarity of the growth medium, yeast cells reduce their intracellular glycerol content by 50-75% within 3 minutes of the shock. Cells lacking Fps1p require about 1 hour to achieve the same reduction in intracellular glycerol content. In addition, *fps1*Δ mutants survive a hypo-osmotic shock at a 50- to 100-fold lower proportion than the wild type, illustrating the importance of Fps1p-mediated glycerol export after a drop in osmolarity (Luyten *et al.*, 1995, Tamás *et al.*, 1999). Secondly, regulated glycerol export is also necessary during growth under anaerobic conditions, when glycerol is produced for redox-regulation (Ansell *et al.*, 1997, Hohmann 1997, Nevoigt and Stahl 1997). Mutants lacking Fps1p hyperaccumulate glycerol and grow poorly compared to wild type cells under anaerobic conditions, probably due to osmotic problems caused by elevated intracellular glycerol levels (Tamás *et al.*, 1999). Thirdly, *fps1*Δ mutants exhibit a cell fusion defect during mating. This fusion defect results from an inability to regulate the osmotic balance of the cell during mating (Philips and Herskowitz 1997). Fourthly, the yeast osmosensor Sln1p is hyperactivated in an *fps1*Δ mutant, probably due to an increased cellular glycerol content and increased turgor pressure (Tao *et al.*, 1999). Finally, kinetic studies have shown that Fps1p-mediated glycerol export is more efficient than its uptake (Tamás *et al.*, 1999). Taken together, these results confirm the important role of Fps1p in glycerol export and its involvement in yeast osmoregulation.

2.2 Fps1p controls the glycerol transport rate during changes in external osmolarity

Drastic and very rapid changes in external osmolarity, is a commonly encountered stress for a yeast cell in various natural environments. *S. cerevisiae* adapts to such changes by altering the intracellular concentration of the compatible solute glycerol (Blomberg and Adler 1992, Brown 1976, Yancey *et al.*, 1982).

When wild type yeast cells are transferred to a growth medium with high external osmolarity, the glycerol transport rate is rapidly reduced (Tamás *et al.*, 1999). Together with an increase in glycerol biosynthesis (Albertyn *et al.*, 1994, Ansell *et al.*, 1997), downregulation of the glycerol transport rate is required for efficient glycerol accumulation and retention under hyperosmotic conditions (Tamás *et al.*, 1999).

On the other hand, when yeast cells are transferred from high to low osmolarity, glycerol transport is rapidly restored as illustrated by the export of the majority of the glycerol previously accumulated. These rapid changes in glycerol transport rate are fully dependent on the presence of the Fps1p channel (Luyten *et al.*, 1995, Tamás *et al.*, 1999). Hence, Fps1p is an osmoregulated channel that rapidly closes under conditions of high external osmolarity and that opens when the external osmolarity drops.

3. REGULATION OF FPS1

3.1 Fps1p closure requires the unique N-terminal extension

The mechanisms triggering channel closure upon hyper-osmotic shock and its opening upon a hypo-osmotic shock have been investigated. None of the known yeast osmosensing signalling pathways seems to be involved (Luyten *et al.*, 1995, Tamás *et al.*, 1999). Furthermore, by using a variety of inhibitors, we could not find evidence for the involvement of several candidate protein kinases and there is no evidence for the involvement of membrane stretching. Finally, ATP does not appear to be required as an energy source (Tamás *et al.*, 1999).

One of the closest Fps1p homologues is the *Escherichia coli* MIP glycerol facilitator GlpF. Interestingly, expression of *E. coli* GlpF in an *fps1Δ* mutant restores glycerol transport but not its regulation by changes in osmolarity (Tamás *et al.*, 1999). Fps1p differs from GlpF by having long hydrophilic N- and C-terminal extensions of about 250 and 150 amino acids, respectively. Cells expressing an *FPS1* allele lacking the N-terminal extension cannot rapidly reduce the glycerol transport rate when shifted to high osmolarity, i.e. the Fps1p channel appears to remain constitutively open. As a result, these cells permanently lose glycerol and grow slowly in high osmolarity medium, illustrating the importance of channel closure and glycerol retention under hyper-osmotic conditions (Tamás *et al.*, 1999).

While the N-terminal extension of Fps1p is required for channel closure under high osmolarity conditions, it is not known what parts of Fps1p are involved in the opening process when external osmolarity drops. Deletion of the C-terminal extension results in an inactive, non-transporting channel (Tamás *et al.*, 1999). However, since expression and localisation of this allele has not been investigated, it is not known whether the corresponding channel is truly inactive.

3.2 Analysis of the Fps1p regulatory domain

We have performed a detailed mutational analysis of the Fps1p N-terminal extension in order to define the molecular requirements of channel closure (Tamás *et al.*, 2000). A stretch of 18 residues close to the first transmembrane domain (TMD1) is essential for channel closure, whereas most of the remaining sequences can be deleted without affecting channel regulation. This regulatory domain is rich in proline residues, it is hydrophilic and lacks both the smallest and largest amino acids (Tamás *et al.*, 2000).

Not only is the sequence of this regulatory domain important, also its precise position appears to be important for the closing mechanism. Moving the regulatory domain somewhat closer to or further away from TMD1 results in a loss of channel regulation (Tamás *et al.*, 2000).

Only a few MIP channels have extended termini, such as the *Drosophila melanogaster* big brain BIB protein, the yeast *YFL054c* encoding a protein of unknown function and a putative MIP protein from *Schizosaccharomyces pombe* (Park and Saier 1996, Reizer *et al.*, 1993). Interestingly, *YFL054c* contains a sequence within its N-terminal extension with homology to the regulatory domain required for Fps1p channel closure. Hence, regulation of these two yeast MIP channels may share common features.

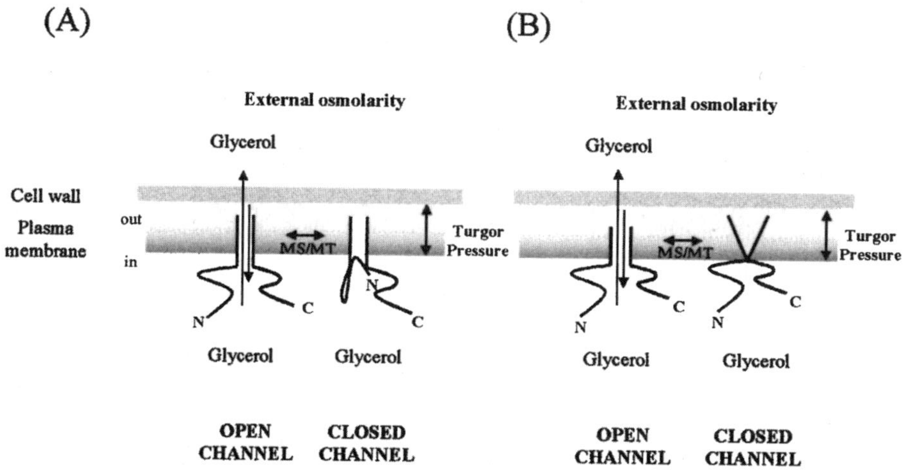

Figure 1. Putative models for the role of the N-terminal regulatory domain and physical parameters that could act as signals for channel closure. (A) The regulatory domain could function as a door that physically interacts with the channel thereby closing it. (B) The regulatory domain may sense high osmolarity and induce a conformational change that closes the channel. Physical parameters that could affect channel closure include external osmolarity, membrane strain and tension (MS and MT respectively) and turgor pressure.

3.3 Possible mechanisms regulating Fps1p closure

The N-terminal regulatory domain could mediate channel closure by several mechanisms (Figure 1). Firstly, channel closure may require an interaction between the regulatory domain and (an) other regulatory protein(s). Secondly, the closing mechanism could involve positioning the N-terminal domain onto the channel-forming unit, thereby physically blocking the channel. Such a mechanism has been described for the mammalian ClC-2 channel (Gründer *et al.,* 1992, Jordt and Jentsch 1997). However, in contrast to that of ClC-2, the precise position of the Fps1p regulatory domain is important for channel closure. Alternatively, Fps1p could be regulated via osmo-induced conformational changes. In this case, the prolines in the regulatory domain may play an important role. Proline residues exert characteristic effects on protein structure and function (MacArthur and Thornton 1991, Williams and Deber 1991). Membraneburied prolines have in some cases been shown to mediate discrete conformational transitions, thereby regulating the opening and closing of transport channels (Williams and Deber 1991). Although the proline-rich regulatory domain of Fps1p is not predicted to be part of a membrane helix, it may still be involved in transducing a conformational change through the polypeptide chain. Channel regulation via conformational rearrangements has been reported for the *Streptomyces lividans* K^+-channel KcsA (Perozo *et al.*, 1999). Eventually, purification and reconstitution into proteoliposomes will be necessary to demonstrate whether Fps1p operates alone or whether it needs other cellular constituents or their activities.

4. CONCLUSIONS AND PERSPECTIVES

The MIP channels affect a wide range of processes in osmoregulation and metabolism at the level of individual cells and whole organisms. Physiological studies have established the yeast MIP channel Fps1p as an osmoregulated glycerol exporter. Furthermore, mutagenesis has identified a regulatory domain involved in osmo-induced closure of the channel. The two-dimensional structure of some MIP channels has recently been resolved. It will be interesting to obtain structural information on Fps1p so that physiological, structural and functional studies together can give a more detailed picture on the molecular mechanisms controlling channel function and regulation of this MIP member.

ACKNOWLEDGEMENTS

This work was supported by the Commission of the European Union via contracts FMRX-CT96-0007 to J.M.T., FMRX-CT97-0128 and BIO4-CT98-0024 to S.H. and a post-doctoral fellowship (PDM) from the Katholieke Universiteit Leuven to M.J.T. We thank S. Wera for critical comments on the manuscript.

REFERENCES

Albertyn J Hohmann S Thevelein J M and Prior B A 1994 *GPD1* which encodes glycerol-3-phosphate dehydrogenase is essential for growth under osmotic stress in *Saccharomyces cerevisiae* and its expression is regulated by the high-osmolarity glycerol response pathway *Mol Cell Biol* **14**: 4135-4144

André B 1995 An overview of membrane transport proteins in *Saccharomyces cerevisiae Yeast* **11**: 1575-1611

Ansell R Granath K Hohmann S Thevelein JM and Adler L 1997 The two isoenzymes for yeast NAD^+-dependent glycerol 3-phosphate dehydrogenase encoded by *GPD1* and *GPD2* have distinct roles in osmoadaptation and redox regulation *EMBO J* **16** 2179-87

Blomberg A and Adler L 1992 Physiology of osmotolerance in fungi *Adv Microb Physiol* **33**: 145-212

Bonhivers M Carbrey JM Gould SJ and Agre P 1998 Aquaporins in *Saccharomyces* Genetic and functional distinctions between laboratory and wild-type strains *J Biol Chem* **273**: 27565-27572

Brown AD 1974 Microbial water relations: features of the intracellular composition of sugar-tolerant yeasts *J Bacteriol* **118**: 769-777

Brown AD 1976 Microbial water stress *Bacteriol Rev* **40**: 803-846

Gancedo C Gancedo JM and Sols A 1968 Glycerol metabolism in yeasts Pathways of utilization and production *Eur J Biochem* **5**: 165-172

Gründer S Thiemann A Pusch M and Jentsch T J 1992 Regions involved in the opening of ClC-2 chloride channel by voltage and cell volume *Nature* **360**: 759-762

Hohmann S 1997 Shaping up: The response of yeast to osmotic stress In *Yeast Stress Responses* (Hohmann S and Mager W H eds) Landes Company pp 101-145

Hohmann S Bill RM Kayingo G and Prior BA 2000 Microbial MIP channels *Trends Microbiol* **8**: 33-38

Jordt SE and Jentsch TJ 1997 Molecular dissection of gating in the ClC-2 chloride channel *EMBO J* **16**: 1582-1592

Laizé V Gobin R Rousselet G Badier C Hohmann S Ripoche P and Tacnet F 1999 Molecular and functional study of *AQY1* from *Saccharomyces cerevisiae*: role of the C-terminal domain *Biochem Biophys Res Commun* **257**: 139-144

Luyten K Albertyn J Skibbe W F Prior BA Ramos J Thevelein JM and Hohmann S 1995 Fps1 a yeast member of the MIP family of channel proteins is a facilitator for glycerol uptake and efflux and is inactive under osmotic stress *EMBO J* **14**: 1360-1371

MacArthur MW and Thornton JM 1991 Influence of proline residues on protein conformation *J Mol Biol* **218**: 397-412

Murakami Y Naitou M Hagiwara H Shibata T Ozawa M Sasanuma S Sasanuma M Tsuchiya Y Soeda E Yokoyama K Yamazai M Tashiro H and Eki T 1995 Analysis of the nucleotide sequence of chromosome VI from *Saccharomyces cerevisiae Nat Genet* **10**: 261-268

Nevoigt E and Stahl U 1997 Osmoregulation and glycerol metabolism in the yeast *Saccharomyces cerevisiae FEMS Microbiol Rev* **21**: 231-241

Park JH and Saier MH 1996 Phylogenetic characterization of the MIP family of transmembrane channel proteins *J Membrane Biol* **153**: 171-180

Perozo E Cortes DM and Cuello LG 1999 Structural rearrangements underlying K^+-channel activation gating *Science* **285**: 73-78

Philips J and Herskowitz I 1997 Osmotic balance regulates cell fusion during mating in *Saccharomyces cerevisiae J Cell Biol* **138**: 961-974

Reizer J Reizer A and Saier MH Jr 1993 The MIP family of integral membrane channel proteins: sequence comparisons evolutionary relationships reconstructed pathway of evolution and proposed functional differentiation of the two repeated halves of the proteins *Crit Rev Biochem Mol Biol* **28**: 235-257

Sutherland FCW Lages F Lucas C Luyten K Albertyn J Hohmann S Prior BA and Kilian SG 1997 Characteristics of Fps1-dependent and -independent glycerol transport in *Saccharomyces cerevisiae J Bacteriol* **179**: 7790-7795

Tamás MJ Allegri L Karlgren S Thevelein JM and Hohmann S 2000 Analysis of the regulatory domain of the yeast Fps1p glycerol export channel *Manuscript in preparation*

Tamás MJ Luyten K Sutherland FC Hernandez A Albertyn J Valadi H Li H Prior BA Kilian SG Ramos J Gustafsson L Thevelein JM and Hohmann S 1999 Fps1p controls the accumulation and release of the compatible solute glycerol in yeast osmoregulation *Mol Microbiol* **31**: 1087-1104

Tao W Deschenes RJ and Fassler JS 1999 Intracellular glycerol levels modulate the activity of Sln1p a *Saccharomyces cerevisiae* two-component regulator *J Biol Chem* **274**: 360-367

Van Aelst L Hohmann S Zimmermann FK Jans AW and Thevelein JM 1991 A yeast homologue of the bovine lens fibre MIP gene family complements the growth defect of a *Saccharomyces cerevisiae* mutant on fermentable sugars but not its defect in glucose-induced RAS-mediated cAMP signalling *EMBO J* **10**: 2095-2104

Williams KA and Deber CM 1991 Proline residues in transmembrane helices: structural or dynamic role? *Biochemistry* **30**: 8919-8923

Yancey PH Clark ME Hand SC Bowlus RD and Somero GN 1982 Living with water stress: evolution of osmolyte systems *Science* **217**: 1214-1222

6

AQUAPORINS IN FISH

6.1 Aquaporins in fish

WATER TRANSPORT AND AQUAPORIN EXPRESSION IN FISH

Christopher P. Cutler and Gordon Cramb
School of Biology, University of St Andrews, St Andrews, Scotland, U.K. E-mail: cpc@st-andrews.ac.uk

1. INTRODUCTION

One of the major differences between terrestrial and aquatic vertebrates is that the latter group have a potentially greater difficulty in maintaining osmotic homeostasis because they are surrounded by an external environment which is almost always in osmotic dis-equilibrium with their body fluids. In comparison to terrestrial animal species, aquatic organisms such as teleost fish may well possess a distinct set of mechanisms in order to adapt and survive the osmotic challenges posed by the seawater (SW: hyper-osmotic) or freshwater (FW: hypo-osmotic) environments. However, to date, little is known about how these mechanisms manifest themselves at the molecular level. In order to begin an evaluation of which proteins may be involved in osmoregulation in both the freshwater and marine environments, investigations have been initiated using the euryhaline teleost the European eel (anguilla anguilla) which naturally inhabits both environments.

2. WATER TRANSPORT

2.1 Water Transport Across The Gills Of Teleost Fish

The gills are the major site of water exchange between the internal milieu and the external aquatic environment inhabited by teleost fish, with for

Molecular Biology and Physiology of Water and Solute Transport
Edited by Hohmann and Nielsen, Kluwer Academic/Plenum Publishers, New York, 2000.

example in FW, the branchial epithelium (gill) responsible for over 90% of the total body water influx (Motais *et al.*, 1969; Haywood *et al.*, 1977). In FW the diffusional component of water fluxes across the gill occurs primarily across the secondary lamellae, through so-called 'respiratory cells', due in part to the large percentage of the exterior surface area of the gills these cells occupy (Isaia 1982). Consequently, the diffusional permeability of the gill is not radically different, although it is slightly higher, in FW compared to SW adapted euryhaline teleosts (Isaia 1984).

However, the situation regarding osmotic water permeability of the gills is some what different. Measurements in euryhaline teleosts demonstrate that osmotic water permeability is generally higher in FW than in SW adapted fish, and in particular in the European eel permeability is 6 times greater (Isaia 1984).

It has been suggested that water fluxes through so-called gill 'chloride cells' (mitochondrial-rich ion transporting cells) represent the principal osmotic water flux pathway across the gill. This process has been associated with a 'bulk flow' water and ion transporting mechanism through a specialised membrane structure (the baso-lateral tubular network) of chloride cells, resulting in water efflux through "leaky" tight junctions into the external environment (Isaia 1984). While this may explain a portion of gill water permeability in marine fish, fluxes through the tubular network of chloride cells in FW teleosts are likely to be lower, as there are fewer chloride cells (which also have more extensive tight junctions) in FW adapted fish (Potts 1984). This suggests that the bulk of osmotic water fluxes (particularly in FW fish) must occur through or between other surface epithelial cells such as secondary lamellar respiratory cells (Rankin and Bolis 1984). While it is not clear why osmotic water permeability should be so much greater in FW compared to SW acclimated eels, it does suggest that a regulated pathway for water movement in the gill may exist, the components of which would be expected to be more abundant in FW than SW fish.

2.2 Hormonal Control of Water Transport Across the Gills

If a regulated pathway for water transport did exist within gill surface epithelial cells it is likely that it would be hormonally regulated. Evidence for the action of several hormones on gill water fluxes does exist and the most convincing evidence comes from experiments investigating the effect of catecholamines and their associated agonist and antagonist drugs. Adrenaline significantly increases the level of cAMP in the branchial epithelium and also increases the diffusion component of the water flux

(Isaia 1979) and this effect is 100% greater in FW- rather than SW-acclimated fish. The increase in water flux was also shown to result from a direct effect of adrenaline on membrane permeability rather than on changes in perfusion caused by changes in vascular resistance (Haywood *et al.*, 1977). The b-adrenergic agonist, isoproterenol, has also been demonstrated to acutely increase water movement across isolated Killifish opercular membranes, which are often used as analogues of the gill epithelium (Zadunaisky 1984).

A number of other hormones may regulate water fluxes across the branchial epithelium although the evidence is somewhat less clear cut. Both prolactin and cortisol have been implicated in the regulation of water fluxes/permeability in the gill (Ogawa, 1975 Ogasawara and Hirano 1983 Rankin and Bolis 1984).

2.3 Water Transport Across The Gut

Marine teleosts have a major osmoregulatory problem with their aquatic environment in that they are constantly losing water across permeable body surfaces such as the gills. In order to prevent dehydration, fish have a regulated drinking response. Imbibed SW is first partially desalinated in the oesophagus before passing through the stomach and entering the intestine. The absorption of water across the intestinal epithelium takes place with the concomitant uptake of salts.

When eels are transferred to the marine environment they immediately drink the SW which enters the oesophagus at a slow but steady rate (Parmelee and Renfro 1983). Here salts are thought to be absorbed by both secondary active and passive processes and, as water permeability of the oesophagus is extremely low, only small net effluxes of water occur due to osmosis, leading to an overall dilution of the oesophageal fluid concentration to around one half to one third of the initial osmolality (Hirano and Mayer-Gostan 1976 Parmelee and Renfro 1983, Nagashima and Ando 1993). Very little further ion uptake or water secretion occurs as the luminal fluid passes through the stomach (Hirano and Mayer-Gostan 1976). When the luminal fluid reaches the intestine most of the remainder of the salts and water are absorbed (Skadhauge 1969, Hirano and Mayer-Gostan 1976). The oesophageal desalination appears to be an essential process as, if the intestinal lumen is artificially exposed to strongly hyper-osmotic fluids such as SW, water is secreted into the lumen rather than being absorbed (Skadhauge 1969). Salts are actively absorbed in the intestine and as the osmolality of the luminal fluids is reduced to a point where no net water flux occurs (known as the turning point osmolality: TP osm). The osmolality at which this occurs is above that of the eel's plasma. Subsequently, as further salts are absorbed, net water is absorbed against the opposing osmotic gradient. Once the osmolality of the luminal contents are reduced to that of the plasma, fluid uptake continues until the Na and Cl concentrations are reduced to around 30-90 meq/l, (Skadhauge 1969, Parmelee % Renfro 1983). As the net uptake of divalent ions is very low and other solutes are actively secreted, the final osmolality of the luminal fluid still remains iso-osmotic to that of the body fluids (Parmelee and Renfro 1983). The model proposed to explain these phenomena suggests that water transport occurs by two processes, 1) secondary-active water transport (which can transport water against the osmotic gradient) thought to occur through a trans-/para-cellular shunt pathway and 2) passive water fluxes which follow the osmotic gradients and which are entirely transcellularly based (Skadhauge 1969). The secondary-active water transport component has been investigated in a number of studies principally using iso-osmotic conditions (to eradicate the normal passive osmotic water fluxes) and is tightly coupled to the transport of chloride and sodium and/or potassium ions (Skadhauge 1969, 1974, Ando 1975, 1980, 1981, 1983, 1985). Hence it has been described as solute-coupled water flux. The osmolality at which the TP osm occurs (a measure of the solute-linked secondary-active salt/water transport) is directly proportional to the salinity of the external environment in which the eels are acclimated. This suggests that the capacity of the solute-linked water flux increases with acclimation to increasing salinities (Skadhauge 1969, 1974) with this component increasing 3-3.4 fold following FW/SW acclimation (Utida *et al.*, 1972, Ando 1975). Furthermore, the passive (or general) osmotic water permeability (assessed using osmotic gradients produced without additional NaCl) also increases by 2-6 fold following SW acclimation (Skadhauge 1969, Ando 1975). The magnitude of water fluxes across various parts of the intestine have also been determined, with highest levels occurring in the mid region followed in descending order by the posterior, anterior intestine and rectum (Ando and Kobayashi 1978, Ando 1980). This evidence taken together also suggests that there may be a role for a regulated cellular pathway for water movement in the intestine, the components of which would be expected to be more abundant in SW than FW eels.

2.4 Hormonal Regulation Of Intestinal Water Transport

Both cortisol and ACTH have been shown to significantly increase the level of water flux across the intestine in FW-acclimated eels. Increases in water flux were observed after a latent period of about ten hours and lasted for at least a week (Utida *et al.*, 1972). The hormone prolactin has also been reported to act antagonistically to cortisol and has been demonstrated to significantly decrease water fluxes across the intestine in SW acclimated eels (Utida *et al.*, 1972).

Various other hormones have also been implicated in the acute regulation of intestinal water fluxes, the most potent of these being atrial natriuretic peptide (ANP) which inhibited net absorption by 50-60% after approximately one hour exposure to the hormone (Ando *et al.*, 1992, Ando and Kondo 1993). Serotonin (via a 5-HT3-like receptor), and the acetylcholine muscarinic agonist methacholine, both inhibited water fluxes by around one third, and these actions were antagonised by noradrenaline (Mori and Ando 1991, Ando and Kondo 1993) However noradrenaline failed to antagonise the ANP induced inhibition of water flux (Ando and Kondo 1993).

2.5 Water Transport In The Kidney

In mammalian kidney the physiological and molecular mechanisms of water transport have been widely studied and well documented, however, much less information is available for teleost fish. Freshwater teleost fish excrete relatively large volumes of dilute urine (around 35 mOsmol/l) due to a high glomerular filtration rate (GFR) and an almost complete reabsorption of NaCl. Ions remaining in the tubular are further absorbed by the urinary bladder. Marine teleosts generally only produce low volumes of urine due to a low GFR, the kidney of the marine teleost is unable to concentrate salts in the urine and the fluid extracted is still hypo-osmotic compared to plasma and contains relatively low concentrations of NaCl. These processes suggest that water transport mechanisms might be expected to be of less importance to teleost kidney function in comparison to the significance of those found in mammalian kidney.

3. EXPRESSION OF AQUAPORINS IN TELEOST FISH

The purpose of the molecular investigations undertaken in this study has been to examine the role that fish aquaporin water channel homologues play in the physiological processes mentioned above. As there was no published evidence for the presence of aquaporins in fish, initial studies focused on the cloning and identification of aquaporin homologues using degenerate RT-PCR.

Experiments concentrated on identifying aquaporins in target tissues such as the gill, intestine and kidney. The first aquaporin homologue identified from the gill, shared highest levels of nucleotide and amino acid homology with mammalian AQP3 (Ma *et al.*, 1994) and subsequent Northern blotting showed that it is expressed as a 2.4 kb mRNA in the eye, oesophagus, intestine and gill with a minor amount of a 7 kb mRNA species also present. Quantitative studies revealed that the major site of expression was in the gill of FW eels and that levels decreased to 3% of FW values, three weeks after transfer of fish to SW. Further studies revealed that the down-regulation of mRNA coding for this AQP3 homologue in SW occurred rapidly with a half-time of around 10 hours. The levels of mRNA expression in the intestine were relatively low in both FW or SW fish. This data suggests that the AQP3 homologue may be, at least in part, responsible for the increased osmotic water permeability documented for the gills of FW eels (Isaia, 1984). As suggested above, the most likely cellular location of functional expression of the AQP3 homologue within the gill is therefore probably within the basolateral membranes of surface epithelial cells, where it would serve to release water entering apically from the hypo-osmotic FW environment. This process would prevent cell swelling and eventual bursting and is presumably present because cell volume regulation, in the face of continuous water uptake, could not be achieved by the secretion of ions or other osmolytes.

Although the AQP3 homologue was shown to be present in the intestine, the low levels of expression suggested the presence of other AQP homologues in this tissue. Consequently, the existence of further intestinal aquaporins was investigated and a homologue of AQP1 was identified (Preston and Agre 1991). This homologue had a wider tissue distribution than that of the AQP3 homologue and expression of a 1.4 kb mRNA was found in brain, eye, heart, pancreas, oesophagus, stomach, and intestine, with much lower levels in skeletal muscle, gill and kidney. A minor 3.1 kb mRNA component was also present in some tissues. Quantitative studies revealed that the level of mRNA expression in the intestine increased 10-25x following the transfer of FW eels to SW. Messenger RNA abundance was also significantly decreased to 28% of FW values in the kidneys of SW adapted fish, although overall levels in the kidney were much lower than the intestine. The results in the intestine strongly suggest that the AQP1 homologue may have a role to play in the mechanisms associated with the absorption of water in this tissue, however the exact nature of this role remains unclear. It seems unlikely that an AQP1 homologue could play a direct role in the secondary-active water transport taking place in the intestine, however, it could be involved indirectly, possibly by providing a basolateral water efflux pathway for luminal epithelial cells, and hence the expression of an AQP1 homologue may at least be part of the explanation for the increases in water permeability also found.

As only limited expression of the AQP1 homologue was found in eel kidney, this tissue is a major site of mammalian aquaporin expression and therefore a further search was recently undertaken to identify other aquaporin homologues in this tissue. These studies resulted in the discovery of a third aquaporin which, initial indications suggest, is a duplicate isoform of the AQP1 homologue, with which, the cDNA fragment isolated, shares 69.7% derived amino acid homology. Initial studies have indicated that mRNA for this isoform is only has detectable in the oesophagus and kidney, and that the high level of mRNA expression in the kidney is dramatically down-regulated when eels are transferred from FW to SW, in a similar fashion to the expression of AQP1 homologue in kidney. The physiological role of these changes in expression are difficult to reconcile with the current models of kidney function in teleosts, especially when high levels of aquaporin expression are present during physiological conditions which induce the production of dilute urine associated with a minimal level of water re-absorption. The presence of high levels of expression of the AQP1 homologue and its duplicate in the oesophagus is also puzzling when this tissue is thought to represent a particularly tight epithelium with very low water permeability. It suggests that the aquaporins expressed in these tissues are concerned primarily with cell volume control rather than vectoral water transport.

4. CONCLUSION

The considerable changes in aquaporin homologue expression that occur in various tissues, following salinity acclimation in teleost fish, suggest that there are a significant number of interesting avenues for further research to elucidate the role of water channels in aquatic species.

ACKNOWLEDGEMENTS

The work from our laboratory was supported by grants from the N.E.R.C. and The Wellcome Trust.

REFERENCES

Ando M 1975 Intestinal water transport and chloride pump in relation to sea-water adaptation of the eel Anguilla japonica *Comp Physiol Biochem* **52A**: 229-233

Ando M and Kobayashi M 1978 Effects of stripping of the outer layers of the eel intestine on salt and water transport *Comp Physiol Biochem* **61A**: 497-501

Ando M 1980 Chloride-dependent sodium and water transport in the seawater eel intestine *J Comp Physiol B* **138**: 87-91

Ando M 1981 Effects of ouabain on chloride movements across the seawater eel intestine *J Comp Physiol B* **145**: 73-79

Ando M 1983 Potassium-dependent chloride and water transport across the seawater eel intestine *J Membrane Biol* **73**: 125-130

Ando M 1985 Relationship between coupled Na+-K+-Cl- transport and water absorption across the seawater eel intestine *J Comp Physiol B* **155**: 311-317

Ando M Kondo K and Takei Y 1992 Effects of eel atrial natriuretic peptide on NaCl and water transport across the intestine of the seawater eel *J Comp Physiol B* **162**:436-439

Ando M and Kondo K 1993 Noradrenalin antagonizes effects of serotonin and acetylcholine in the seawater eel intestine *J Comp Physiol B* **163**: 59-63

Haywood G P Isaia J and Maetz J 1977 Epinepherine effects on branchial water and urea flux in rainbow trout *Am J Physiol* **232**: R110-R268

Hirano T and Mayer-Gostan N 1976 Eel esophagus as an osmoregulatory organ *Proc Natl Acad Sci USA* **73**: 1348-1350

Isaia J 1979 Non-electrolyte permeability of trout gills Effects of temperature and adrenaline *J Physiol* **286**: 361-373

Isaia J 1982 effects of environmental salinity on branchial permeability of rainbow trout Salmo gairdneri *J Physiol* **326**: 297-307

Isaia J 1984 Water and Nonelectrolyte permeation In "Fish Physiology Volume X Gills Part B Ion and water transfer" (W S Hoar and D J Randall eds) Academic press London pp 1-38

Ma T Frigeri A Hasegawa H and Verkman A S 1994 Cloning of a water channel homolog expressed in brain meningeal cells and kidney collecting duct that functions as a stilbene-sensitive glycerol transporter *J Biol Chem* **269**: 21845-21849

Mori Y and Ando M 1991 Regulation of ion and water transport across the eel intestine: effects of acetylcholine and serotonin *J Comp Physiol B* **161**: 387-392

Motais R Isaia J Rankin J C and Maetz J 1969 Adaptive changes of the waterpermeability of teleostean gill epithelium in relation to external salinity *J Exp Biol* **51**: 529-546

Nagashima K and Ando M 1993 Characterization of esopageal desalination in the seawater eel Anguilla japonica *J Comp Physiol B* **164**: 47-54

Ogasawara T and Hirano T 1984 Effects of prolactin and environmental calcium on osmotic water permeability of the gills in the eel Anguilla japonica *Gen Comp Endocrinol* **53**: 315-324

Ogawa M 1975 The effects of prolactin cortisol and calcium-free environment on water influx in isolated gills of the Japanese eel Anguilla japonica *Comp Biochem Physiol A* **52A**: 539-543

Parmelee J T and Renfro J L 1983 Esophageal desalination of seawater in flounder: role of active sodium transport *Am J Physiol* **245**: R888-R893

Potts W T W 1984 Transepithelial potentials in fish gills In "Fish Physiology Volume X Gills Part B Ion and water transfer" (W S Hoar and D J Randall eds) Academic press London pp 105-128

Preston G M and Agre P 1991 Isolation of the cDNA for the erythrocyte integral membrane protein of 28 kilodaltons: Member of an ancient channel family *Proc Natl Acad Sci USA* **88**: 11110-11114

Rankin J C and Bolis L 1984 Hormonal control of water movement across the gills In "Fish Physiology Volume X Gills Part B Ion and water transfer" (W S Hoar and D J Randall eds) Academic press London pp 177-201

Skadhauge E 1969 The mechanism of salt and water absorption in the intestine of the eel (Anguilla anguilla) adapted to waters of various salinities *J Physiol* **204**: 135-158

Skadhauge E 1974 Coupling of transmural flows of NaCl and water in the intestine of the eel (Anguilla anguilla) *J Exp Biol* **60**: 535-546

Utida S Hirano T Oide H Ando M Johnson DW and Bern H A 1972 Hormonal control of the intestine and urinary bladder in teleost osmoregulation *Gen Comp Endo* **S3**: 317-327

Zadunaisky J A 1984 The chloride cell: The active transport of chloride and the paracellular pathway In "Fish Physiology Volume X Gills Part B Ion and water transfer" (W S Hoar and D J Randall eds) Academic press London pp 129-176

INDEX

A

B

C

G

H

I

K

L

M

N

O

P